COCAINE

COCAINE

A DRUG
AND ITS SOCIAL
EVOLUTION

Lester Grinspoon
&
James B. Bakalar

Basic Books, Inc., Publishers

NEW YORK

The authors gratefully acknowledge permission to reprint excerpts from the following sources:

"Casey Jones," words by Robert Hunter, © 1970 Ice Nine Publishing Company.

"Earth Mother," © 1971 Steel Wind Publishing Company.

"Let It Bleed," © 1970 ABKCO Music, Inc.

Peru: History of Coca, W. Golden Mortimer, © And/Or Press.

"Sister Morphine," Mick Jagger and Keith Richard, ©1969 ABKCO Music, Inc. All Rights Reserved. International Copyright Secured.

"Snow Blind Friend," words and music by Hoyt Axton, © 1968, Lady Jane Music.

"Take a Whiff on Me," words and music by Huddie Ledbetter, collected and adapted by John A. Lomax and Alan Lomax. TRO—© copyright 1936 and renewed 1964 Folkways Music Publishers, Inc., New York, New York.

Library of Congress Cataloging in Publication Data

Grinspoon, Lester, 1928–
    Cocaine: a drug and its social evolution.

    Bibliography: p. 287
    Includes index.
    1. Cocaine. I. Bakalar, James B., 1943– joint author. II. Title.
HV5810.G73   362.2′93   76-7675
ISBN: 0–465–01189–6 (cloth)
ISBN: 0–465–09732–4 (paper)

To our families

CONTENTS

PART III

PREFACE

THIS BOOK is addressed mainly to the nonspecialist but includes a few sections, especially the material on neurophysiology and pharmacology in Chapter 4, that are necessarily somewhat technical. These passages are designed to be comprehensible to those without extensive specialized training, and they are sometimes helpful for an understanding of other sections of the book. Nevertheless, to readers who may find them tedious we offer our apologies and the suggestion that they skim; we believe that they will find the book useful even without them.

There is no objective language for psychoactive drug experiences, and the choice of terminology often implies a moral or social attitude. In the case of cocaine, recent reports in clinical or experimental language are scarce, so it is especially important to take note of the perceptions of the drug users themselves as expressed in their own words. We conducted interviews with 17 people who had at one time used cocaine regularly or were using it at the time of the interview. We obtained subjects by asking several people whom we knew to be familiar with cocaine to refer us to acquaintances who also used it, and then repeated the request with these new subjects. Most of the cocaine users shared their knowledge and opinions with us freely; a few were paid. The subjects ranged in age from 22 to 38; 12 were men and 5 women; 14 were white and 3 black; occupations included student, housewife, receptionist, drug rehabilitation counselor, consultant on urban affairs, radio news reporter, and shop proprietor; some had used cocaine for many years, others had used it intensively for a few months, and still others used it intermittently, when they were able to obtain it; several had given it up for financial or other reasons. Most of the subjects had only sniffed cocaine, but some, including a former heroin addict, had taken it by intravenous injection. The interviews, although lengthy and open-ended, were planned to touch on certain listed topics. We asked about amount and frequency of use, price, quality, the nature of the high, long-term effects, adverse physiological and psychological reactions, sexual effects, tolerance, dependence, withdrawal reactions, comparisons with other drugs, and drug preference. We also tried to find out something about cocaine's effect on the lives of its users by asking them whether most of their friends also used it, whether they took it alone or in company, and whether they thought it had changed their lives for better or for worse. Quotations from these interviews are interspersed throughout the book. We are not suggesting that our subjects are a representative sample or that their remarks cover the full range of experiences associated with cocaine; instead we mean to convey some of the flavor, atmosphere, and language of cocaine use in America today.

For similar reasons we have occasionally used passages from novels and short stories to suggest or confirm the existence of one or another cocaine effect; these quotations are introduced when it seems clear that the author is either describing his own experience in fictional form or has otherwise had a first-hand acquaintance with the drug or its users. Passages of this kind from fiction are not less valuable to a student of cocaine than memoirs or remarks made in interviews; if the writer has some literary talent, they may be more valuable.

We are indebted to a number of people for help in our work. William von Eggers Doering and Norman E. Zinberg read the manuscript and offered critical suggestions. Richard E. Schultes and Andrew Weil gave us the benefit of their experience and knowledge of coca; we are particularly indebted to Dr. Weil for several observations quoted in the book. Timothy Plowman and Charles Sheviak of the Harvard Botanical Museum were kind enough to offer some suggestions for the section on botany and cultivation and provide information on the taxonomy of the coca plant. We also owe thanks to Elizabeth Weiss, Anne C. Bauer, Susan Wolf, and Ogie Strogatz, and a special debt of gratitude to Betsy Grinspoon and Hazel E. Cherney for their invaluable help in preparing the manuscript.

COCAINE

INTRODUCTION

THE STIMULANT and euphoriant extracted from the leaves of the coca plant is becoming one of the most prized, if not most often used, of pleasure-giving drugs. Reports of the arrest of drug smugglers more and more often mention cocaine as their cargo. It is being sniffed or injected more and more openly and frequently by those who can afford it as well as those who have to beg, borrow, steal, or deal to obtain it. Movies and popular songs have celebrated it or condemned it or done both at once. Articles in national magazines have described its pleasures and warned about its dangers, producing the usual combination of attitudes toward newly popular drugs in which interest predominates over fear. Cocaine has begun to reach the college campuses, where, notoriously, the merely fashionable is admitted into the company of the *Zeitgeist*.

In spite of the growing interest in the drug and its long history in both South America and the United States, ignorance and misapprehensions about it are substantial on both the popular and the scientific levels. Some people confuse the coca leaf with the cacao bean and assume vaguely that cocaine is related to chocolate; others confuse the coca plant with the coconut. A more serious misconception identifies cocaine with opiates; this confusion has been enshrined in laws that are only now first being challenged in the courts. There are also those who still believe that cocaine is a variant of heroin or similar to heroin in its effects, although in actuality the characteristic pleasures and dangers of cocaine and those of morphine or heroin are quite different. On the scientific level information is also surprisingly limited, and very little of it is recent. The most extensive work on the botany, ethnography, and medicinal uses of the coca leaf was published in 1901; the most important medical and psychiatric studies of cocaine abuse were made in the 1920s; most of the work on the effects of coca-leaf chewing on mood and behavior was done in the 1940s. The clinical literature on the effects of cocaine in man remains comparatively sparse, and most of it is over 50 years old. Controlled experimental studies on human beings are almost entirely lacking. The National Institute on Drug Abuse is funding several research projects, and in a few years considerably more will be known. Meanwhile we have to piece together information and conclusions from old case histories, animal experimentation, interviews, literary descriptions, and analogies with the effects of other drugs, especially the much more extensively studied amphetamines.

There are many reasons for taking an interest in cocaine, and no book can satisfy all the different kinds of concerns equally. One is a trivial but harmless curiosity on the level of gossip: a desire to know about the secret practices of the rich, or entertainers, or racial minorities, or adolescents. This can be disguised by or converted to a sociological interest in who takes the drug, when, why, how, and how much, and its connections with other social practices and conditions. Another kind of curiosity is the desire to find out what a drug might do for or to oneself. A false impersonality or objectivity, a pretense of addressing only students who are interested in a problem that is located at a respectable distance from their own lives, may conceal the fact that readers are using a book in this way. On the other hand, many studies openly choose potential drug users or those who are trying to cope with the consequences of drug abuse as their main audience and aim largely to promote or prevent the use of a given drug. Most of the works in the literature on cocaine are of this kind—perhaps unfortunately, since certain facts are selectively overemphasized for polemical purposes. All these matters are secondary to us, although we have included information that is relevant to them.

This book is divided into three parts. The first is a historical and sociological sketch of the use of the coca leaf and cocaine; the second is a description of what is known of the source, pharmacology, and physiological and psychological effects of the drug, including its medical uses; the third is a more general discussion of the problem of drug abuse and the roots of contemporary confusion about psychoactive drugs, with specific reference to the lessons that can be learned from the history of coca and cocaine. We are interested in cocaine primarily as a case study in the cultural definitions of psychoactive drug use. Since cocaine becomes more a subject of open controversy as it becomes more popular, the implications of this case for public policy are important. In its long history as a medicine, stimulant, and intoxicant, cocaine has been classified in many different ways by different societies and in different eras. This history provides suggestions about the meaning of drug use in several different social contexts: primitive and archaic ritual and medicine, the labor of the poor in colonial empires and underdeveloped countries, and industrial society in a period when attitudes toward drugs and drug technology were undergoing a revolutionary change. The nineteenth-century developments that fixed our present established attitudes toward cocaine are especially interesting. By examining them we can suggest some reasons why nonmedical use of psychoactive drugs generates such strong passions and controversial public policies in our society. Present attitudes toward cocaine as well as other drugs that affect the mind may change as the historical conditions that created them disappear.

Our interest in the historical aspect of public policy on cocaine is inseparable from a concern about what that policy should be in the future, and more generally about the intellectual and moral basis of all public controls on psychoactive drugs. Therefore we have restated the general problem of drug dependence and drug abuse as applied to the particular case of cocaine. Cocaine is sometimes referred to as a drug problem or even as part of "*the* drug problem." Although the concern these phrases express is admirable, they are misleading. They tend to make us overemphasize the question of what we should do about "it" (the drug) or, worse, about "them" (the drug users), with unpleasant overtones of condescension, envy, or scapegoating. Instead we should be asking what to do about ourselves and our society. Although that question may seem too general to be meaningful, it saves us from the kind of premature definition and classification of problems that has been so disastrous in our treatment of the use of psychoactive substances. A redefinition of our ideas about these drugs has begun in recent years, first among the general public and now among physicians; it has produced more careful differentiations and at the same time more flexibility and tolerance of ambiguity. We hope this study of cocaine will contribute to that process. Although the use of drugs in general and cocaine in particular may not be in itself as important a social issue as the attention devoted to it sometimes suggests, the amount and quality of the interest in this subject is significant as a symptom of present social conditions and an indication for the future. By lifting some of the fog surrounding the cocaine issue, and especially by clarifying its historical aspects and its relation to general problems of drug dependence and drug abuse, we hope to promote the kind of rational public decision that has been rare where psychoactive drugs are concerned. The emotional obstacles to considering these substances calmly and in a large enough context are great, but the effort is worth making. For the questions discussed (almost always inadequately and with the wrong focus) under the rubric of drug problems involve, at their broadest extent, the kind of society and the kind of humanity we want and are capable of creating.

PART I

1

THE COCA LEAF

THE PLANT from which cocaine is extracted has been cultivated in South America for thousands of years, and a large part of the population of Bolivia and Peru, smaller numbers in Colombia, and a few people in Argentina and Brazil now chew its leaf every day. One student of the coca leaf has gone so far as to write of the Peruvian Indians, "Never in the life of a people has a drug had such importance." [1] A sixteenth-century Spanish administrator of the conquered Inca empire said, "If there were no coca, there would be no Peru." [2] The magical leaf that, in the words of the early chronicler Garcilaso de la Vega (himself of Inca and Spanish descent and the heir to a coca plantation), "satisfies the hungry, gives new strength to the weary and exhausted and makes the unhappy forget their sorrows" became perhaps even more important when the Spanish conquest made its consumption more widespread. Since then it has been alternately extolled and condemned, and it remains a topic of controversy today.

No one knows where *Erythroxylum coca* first grew wild or was first used as a drug: the present wild varieties are the descendants of once-cultivated plants in abandoned fields, and coca use may have begun in one region or in several different areas at different times. It has been suggested that coca chewing originated among the Arhuacos of the Rio Negro area or in the central Amazon. The Aymara Indians in Bolivia apparently used it before the Inca conquest (tenth century A.D.), and the word *coca* itself is thought to be of Aymara origin; it means simply "plant" or "tree," which suggests that coca was regarded as *the* plant, the plant of plants. Statues found in graves on the coast of Peru and Ecuador dated at about 300 B.C. sometimes have the puffed-out cheeks that indicate a wad of coca in the mouth. The drug was once used in northern South America from the Venezuela coast to the region of Arica in Chile, and also on the Pacific coast of Nicaragua and the Caribbean coast of Panama. Its Caribbean name was *hayo*. Although coca use has declined or disappeared over much of its former range, many tribes in the Amazon basin and in the Colombian mountains still use it. But at least since the time of the Incas the center of coca cultivation has been the warm valleys on the eastern slopes of the Andes, and the center of coca chewing has been the highlands of Bolivia and Peru.[3]

Under the Incas coca had sacred status. It was used in divination and scattered by priests before religious rites to propitiate the gods. The mouth of a corpse might be filled with coca and bags of the leaves deposited in the tomb to ease the journey to the next world. It was incorporated in amulets used to attain riches or success in love and played an important part in weddings, funerals, and the *huaraca,* or initiation rite for young Inca nobles. Each useful plant was thought to have a divine essence, its "mother." There were "mothers" of maize, quinine, and coca— Mama Coca, sometimes pictured as a beautiful woman, with associated myths. One of the Inca rulers gave his queen the title Mama Coca. Precisely because of its high status, the actual use of coca among the populace was very restricted. The early chroniclers almost all affirm that it was rarely available to the common people before the Spanish conquest. The Inca elite controlled most cultivation in state-owned *cocales* or coca plantations and permitted its use only in religious rites and as a special royal gift. Although the masses used coca ceremonially and as a medicine, it was not, as it is now, the daily drug of millions.[4]

The first impulses of the conquerors were hostile to coca. In some regions, like the Pacific coast of Peru, *cocales* were in fact abandoned, and the use of the drug declined. At ecclesiastical councils in 1551 and 1567 the bishops of the Catholic church formally denounced its use as idolatry, and a royal proclamation declared its effects a demoniacal illusion. But by that time Spaniards were already putting it to use in the silver and gold mines and on the plantations, and growing rich from the trade in it. The church eventually found it possible to condone the cultivation of coca while outlawing its religious use: taxes on coca actually provided the sixteenth-century bishops and canons of Cuzco with much of their revenue. A royal order under the Viceroy Francisco de Toledo in 1573 in effect removed official obstacles to the cultivation of the coca plant. The inclusion of its leaf after the Wars of Independence in the coat-of-arms of Peru as a symbol of endurance constitutes both a gesture toward its former status and a recognition of its present importance.[5]

Why did the use of coca become so much more widespread in the highlands after the Spanish conquest? One apparent cause is the break-down of the religious restrictions imposed by the Incas. Another possible reason is the havoc wrought by the conquerors on the food economy—much fruit and grain cultivation was abandoned and the livestock in-dustry was destroyed.[6] Coca is often chewed most in regions where the diet is poorest: the drug that "satisfies the hungry" serves as a substitute for food. But no historical analysis has yet explained why coca survived where it did and not elsewhere; for example, the use of the coca leaf has been almost extinct since the eighteenth century in the highlands of northern Peru and Ecuador, areas that are climatically, culturally, and historically similar to some of the regions where it is chewed daily.

The liberation of South America from Spanish colonial rule did not change the conditions of Indian life or the practice of coca chewing. In 1850, for example, 8 percent of the revenue of the government of Bolivia came from the coca trade. In the three centuries before its introduction into western Europe, most Peruvians and Bolivians were complacent about coca and few regarded it as a drug problem or as any kind of problem. In 1787 the Jesuit priest Antonio de Julián suggested that it be permanently substituted for tea and tobacco; the most famous Peruvian physician of the eighteenth century, Hipólito Unánue, praised its stimu-lant effect and other medicinal virtues. The discovery of the surgical uses of cocaine in 1884 and its vogue in late nineteenth-century Europe and the United States even induced a kind of patriotic pride.[7]

In the twentieth century, cultivation of coca has continued at the same level, while controversy about it has become sharper. Newly conscious of the poverty and misery of the Indians, and observing the consequences of cocaine abuse in Europe and North America, Peruvian and Bolivian phy-sicians have begun to reexamine the question of coca chewing. At least since 1929, when Dr. Carlos A. Ricketts presented a plan for the reduc-tion of coca cultivation to the Peruvian parliament, opposition to the drug has been growing. But public opinion has not followed physicians in their nearly unanimous rejection of coca. In 1947 Drs. Carlos Gutiérrez-Noriega and Vicente Zapata Ortiz could still write, "Cocaism, in a word, is not recognized as a public health problem." [8] Official publicity about real or imagined dangers of the drug and efforts at restriction of its use over the last 25 years attest that this is no longer true. But the rural population of Bolivia and Peru has not acquiesced in the official definition of coca use as a health problem, which is enforced halfheartedly by law and not at all by custom.

The issue of coca's possible harmfulness and what ought to be done about it is complicated by the fact that important and legally respected economic and political interests are at stake, as in the case of alcohol and tobacco in this country. Gutiérrez-Noriega, who attacked planters and businessmen for making their fortunes from the misery of the Indians, was dismissed from his post at the University of Lima in 1948 and went into exile; the Institute of Pharmacology he founded was dissolved. The Tax Collection Department of the Coca Monopoly in Peru reported in 1962 that eradication of the coca habit would put 200,000 people out of work. Coca plants may be cultivated in preference to other crops because they are hardier and therefore economically safer. Besides, the coca trade provides regional economic integration and participation in a cash economy. One study of coca chewing in southern Peru indicates that upper-class plantation owners, *mestizo* merchants, and small Indian farmers are united in their opposition to any attempt to eradicate the use of the drug. It is hardly surprising that to many people in Peru and Bolivia dislike of coca has some of the crankish aspect of prohibitionism.[9]

The difficulty of changing such attitudes—whether they need to be changed is another matter—is suggested by the struggle of the UN and its international drug control machinery to make the governments of Peru and Bolivia do something about coca. A United Nations Commission of Enquiry visited those countries in 1950 and concluded that coca chewing "leads to genuinely harmful, closely related economic and social effects in both Peru and Bolivia." It recommended a 15-year program of gradual reduction in cultivation and use. A Peruvian Commission for the Study of the Coca Problem was then formed and challenged the methods and conclusions of the UN study. Dr. Carlos Monge, a student of high-altitude biology and former Surgeon General of the Peruvian army, defended the position of this commission, which was also the position of the Peruvian government, at a United Nations meeting in December 1950, and the UN commission responded.[10]

Since then the governments of Bolivia and Peru have formally relinquished their dissenting opinion, ratifying the measures proposed by the UN without doing much about putting them into effect. The Single Convention on Narcotic Drugs of 1961 was signed by Peru and approved by Bolivia: it required the abolition of coca-leaf chewing and the destruction of most coca bushes within 25 years. It is unlikely that this will happen. Commissions and consultative groups and regional meetings under the auspices of the UN continue to advocate abolition or at least that the

conditions for abolition be created by providing substitute crops and occupations and other sources of tax revenue. Government information
campaigns and various restrictions and registration requirements have
been instituted. But official figures on coca production are probably as
unreliable as official figures on whiskey production in Kentucky, and the
recent increase in the number of illicit cocaine factories servicing the international trade will make them even more unreliable. The state institution known as the Coca Monopoly in Peru probably has less control over
the coca market than General Motors has over the automobile market in
the United States. The International Narcotics Control Board of the UN
admits that very little has changed since it began its campaign against
the coca leaf in 1950.[11]

Various estimates of the total production and consumption of coca are
available. The United Nations Commission of Enquiry asserted in 1950
that about half the rural adult population of Peru and Bolivia—a quarter
of the total adult population—chewed the coca leaf. It was said to be used
mostly by men; only 20 percent of the women chewed it.[12] A 1965 study
indicated that 13 percent of the people of Peru, more than half the working rural population, used coca. Fifty-seven percent of the *coqueros,* or
coca chewers, were men and 43 percent were women. By extrapolation it
has been estimated that in 1970, 1,165,000 Peruvians out of 11,000,000
were using the drug. It is, overwhelmingly, a lower-class habit. Sixty percent of Peruvian *coqueros,* according to the 1965 study, are illiterate, as
opposed to 18.5 percent of nonusers; only 0.2 percent of users have any
secondary education, as opposed to 22 percent of nonusers. Twenty-nine
percent of coca chewers are farm workers, 25 percent housewives, 8.5
percent domestic servants, and 8.4 percent artisans. Official Peruvian agricultural statistics for 1971 show a production of 14,351,000 kg, only a
small proportion of which is legally converted into cocaine. Production
has been rising, and although there is some question whether it is keeping pace with population increase, there has apparently not been any
serious decline in coca use.[13]

Although there are remnants of its former cultural status in folk religion and medicine, coca in the Andean highlands is largely an everyday
drug, used mainly at work. The practice varies from place to place and individual to individual and is not exactly the same in mines or factories as
on farms, but coca use is characterized by daily regularity. The agricultural work day, according to the UN commission, lasts from 7:00 A.M. to
5:00 P.M., with three breaks totaling three to four hours for rest and coca
chewing. Emilio Ciuffardi, who made a study of coca use in Peru in 1949,
states that the men in his sample who chewed only on days when they

worked took breaks for this purpose from 9:00 A.M. to 11:00 A.M., from 2:00 P.M. to 4:00 P.M., and from 7:00 P.M. to 9:00 P.M. Those who chewed daily whether they were working or not might take as many as four or five breaks. The Peruvian Coca Monopoly reports that there is a short break at 10:00 A.M. and a longer one at 3:00 P.M. A few men apparently chew continuously during their waking hours.[14] One writer has estimated that the Peruvian Indian may devote 25 percent of his income to buying coca; employers since the time of the conquest have sometimes paid wages in coca leaves rather than money. It has been estimated that in the region of Nuñoa, in southern Peru, a day's work for the average Indian provides enough money for a little over a week's supply of coca.[15] The period for which a wad of coca is kept in the mouth, 45 minutes, has become a standard unit of time known as the *cocada*. The *cocada* also serves as a distance measure, since it corresponds to three kilometers of walking with a pack on level ground, or two kilometers uphill.

Coca is used as follows. The leaves, carried in a sack (called in some regions a *chuspa*), are moistened with saliva and then wedged in the back of the mouth between the cheek and the gum, like chewing tobacco, on one or both sides. An alkaline substance known as *tocra* or *llipta*, which usually consists of plant ashes of some kind (the most common is *quinoa*) but may also be powdered seashells or quicklime, is added with a stick from a little flask made out of a gourd and called the *iscupuru* or *poporo*. (It is thought that the purpose of the alkali is to release the alkaloids from the leaf and speed their absorption into the bloodstream.) The alkaloids apparently reach the bloodstream partly through the stomach and small intestine as they trickle into the throat dissolved in saliva, and partly through the mucous membrane of the mouth. The verb "to chew" (*masticar*), although sometimes used in Spanish and always in English, is inaccurate except as applied to the preliminary moistening of the leaves. The process of ingestion, which actually resembles sucking, is usually called *acullicar* to the south, *cacchar* farther north, and *mambear* in Colombia. The only significant variation on this method is the Amazonian practice of toasting and pulverizing the dried leaves and mixing cecropia leaf ashes with the powder before placing it between the cheek and gum. One small tribe adds the resin of a plant of the myrrh family to the coca-ash mixture to give it a balsamic savor.[16]

Whether the coca leaf is chewed in this way or taken in powdered form or as a tea, it fills the role of an all-purpose healing herb, like opium in some regions of southeast Asia or cannabis in Jamaica. It might be said to combine the functions of coffee, tobacco, aspirin, and bicarbonate of soda in our society. The South American Indians anticipated most of the uses for cocaine that became popular in Europe and the United States in the late nineteenth century. Folk-healing practices are not just anthropological curiosities; we discuss Indian uses of coca further in Chapter 7, on the place of cocaine in medicine.

In spite of its everyday character, coca is not yet so completely a profane, secularized drug as coffee or aspirin is for us. The Inca and pre-Inca legends associating it with Mama Quilla (the moon) and other goddesses have not been entirely forgotten, and newer legends ascribe its origin to Jesus and the Virgin Mary. Many social and domestic activities are solemnized by the use of coca. Travelers exchange handfuls of leaves as a greeting. It is chewed at religious festivals (where alcohol is also used) and offered to Pacha Mama, the earth, against bad harvests. Indians used to pray to Mama Coca before starting on a journey, and there was a custom, called *jacho lajay,* of sticking a chewed wad of coca (*jacho*) on a stone to propitiate the gods at dangerous places on mountain paths. Miners used to stick the *jacho* to a hard vein of ore to "soften" it as they chewed the leaf to lighten their own labors.[17] The Indians also have songs about coca, some connected with marriage, with bawdy double meanings; others about the Holy Family and their gift of coca; and still others that are solemn and philosophical songs of praise:

> *Pretty little round-leafed coca*
> *the only one who understands my life. . . .*
> *Curative herb of the highlands*
> *Medicinal herb of the puñas*
> *you who understand man's life*
> *you are the only one who knows my fate.*[18]

The religious beliefs associated with coca have lost much of their power, but this heritage of the Incas is not yet entirely exhausted.

Coca is also used by Indians in certain areas of the Colombian Andes and the Colombian and Brazilian Amazon. For example, the Páez of the western highlands of Colombia, with 30,000 members the largest tribe in that country, have cultivated coca for four centuries. Their way of using it is similar to that of the Peruvians and Bolivians. Most of the people, including many women and children, chew it. When a child has learned how to chew, at about the age of eight, he is considered ready to work. The Páez also employ coca as medicine, in divination, and as a medium of exchange.[19] Smaller and more isolated tribes may have very different social definitions of coca use. A striking example is the Kogi of the Sierra Nevada de Santa Marta, also in Colombia. Among these people the religious role of coca is pervasive. Its use is forbidden to women, who harvest the crop but are not permitted to cultivate it. The initiation ceremonies for young men include a marriage to the coca leaf. The men use coca in religious ceremonies for mental lucidity and physical vigor. It enables them to stay awake, to fast, to abstain sexually, and to "speak of the ancestors"—sing, recite, and dance for long periods of time. Kogi priests induce trance by means of a mixture of coca and tobacco (which also has many religious uses among the Colombian Indians). The men say that it stimulates sexual desire in the early years and later produces impotence. Kogi men are said not to value sex highly and not to be much troubled by impotence. The women regard coca as a rival power and if they are childless may try to get their husbands to stop using it.[20]

Andrew Weil describes the everyday use of coca in still another region, the Amazon basin:

> In 1974 I lived among a group of Cubeo Indians on the Río Cuduyari in eastern Colombia near the Brazilian border. These people grew their own coca and prepared it in the traditional Amazonian way: they toasted and pulverized the leaves and mixed them with *Cecropia* leaf ashes to a fine gray-green powder that is placed in the mouth and allowed to dissolve slowly. This is an especially powerful preparation of coca, and every family in the village I stayed in had its own can of it, available for use at all times. Yet I never saw it used except on two kinds of occasions. One was to facilitate physical work. For instance, men who went out to chop trees down to prepare a new area for planting crops would fortify themselves with large helpings of coca as they marched out of the village. The other occasion was a fiesta, where everyone drank freshly made beer (*chicha*), played music, and danced. Coca would be passed around at these ritualized gatherings, and those who wished could partake. I never saw children use it, and adolescents and women generally declined it.[21]

The past and present social functions of coca in South America suggest the wide variety of possible cultural definitions of drug use. Under the Incas coca was a semisacred substance employed in rituals, and the Kogi still use it as a medium of transcendence and communication with spirits. In Peru and Bolivia today it is mainly a stimulant that eases labor, although it also remains important in folk medicine. Its derivative, cocaine, a product of modern Western science and industry, has had a different fate; at first it too was considered a medicine and general stimulant, but now its use is almost exclusively recreational. With the introduction of coca to the United States and Europe, a double transformation took place that made the drug a new one, pharmacologically and socially: cocaine replaced the coca leaf as the center of interest, and the use of cocaine ceased to have anything to do with medicine or religion and came to be defined as an illicit pleasure.

2

EARLY HISTORY
OF COCAINE

EUROPEANS began to learn about the coca leaf soon after the discovery of America, but for hundreds of years they had practically no firsthand experience of its effects, and their knowledge had the vagueness and dubiety of legend. Why this is so is not clear. The contrast with tobacco, which immediately spread to the Old World and became important in international trade, is striking. Most writers on this subject believe that coca leaves were not properly packed or cared for and so became inert on the long ocean voyage. Another possibility is that the Spanish were so obsessed with the gold and silver of Peru that, in spite of the importance of coca in the internal economy of their colony, they never seriously considered it as another potential commodity in international trade. It is also true that many of the *conquistadores* considered coca chewing, as well as most other native customs, a vice, and did not want to encourage it among Europeans. In any case, for many years the reports about coca that reached Europe had some of the same air of the fabulous that characterized other tales of the Inca Empire.

The first-generation chroniclers of the Spanish conquest refer to coca with more scorn than interest. For example, Pedro Cieza de León, a traveler and soldier in South America from 1532 to 1550, one of the first writers to describe coca chewing for a European audience, remarked, "When I asked some of these Indians why they carried these leaves in their mouths . . . they replied that it prevents them from feeling hungry, and gives them great vigor and strength. I believe that it has some such effect, although perhaps it is a custom only suitable for people like these Indians." [1] Later comments, however, tended to be enthusiastic. None are more so than the verses of Abraham Cowley, a British physician and poet who celebrated the legendary virtues of coca in 1662 in his *Books of Plants*. He presents a South American goddess speaking to Bacchus and Venus:

> *Behold how thick with leaves it is beset;*
> *Each leaf is fruit, and such substantial fare,*
> *No fruit beside to rival it will dare.*
> *Mov'd with his country's coming fate, whose soil*
> *Must for her treasurers be exposed to spoil,*
> *Our Varicocha first this coca sent,*
> *Endowed with leaves of wondrous nourishment,*
> *Whose juice sucked in, and to the stomach taken*
> *Long hunger and long labor can sustain:*
> *From which our faint and weary bodies find*
> *More succor, more they cheer the drooping mind,*
> *Than can your Bacchus and your Ceres joined.* [2]

Other favorable references to coca appear in the writings of the Peruvian physician Hipólito Unánue, who tells of its use in the siege of La Paz (1774) to help the inhabitants bear the hunger, fatigue, and cold, and in the works of the German scientist and explorer Alexander von Humboldt.[3] The first botanical description of the coca plant appeared in a book by Nicolas Monardes, a Spanish physician, published in 1580 and translated into English in 1582 by John Frampton under the title *Joyfulle News Out of the Newe Founde Worlde, wherein is declared the Virtues of Herbs, Treez, Oyales, Plantes, and Stones*.[4] In 1750 the botanist Joseph de Jussieu sent the first specimens to Europe for examination, and the plant eventually received the classification *Erythroxylon coca* Lamarck.[5]

In the nineteenth century, travelers' reports continued to be overwhelmingly favorable. Observers like Johann Jakob von Tschudi (1846), Clements Markham (1856 and 1862), and H. A. Weddell (1853) were impressed by its power of physical invigoration and its effectiveness in respiratory troubles at high altitudes. Tschudi wrote, "I am clearly of the opinion that moderate use of coca is not merely innocuous, but that it may even be very conducive to health." [6] The praise reached a height in an essay by the neurologist Paolo Mantegazza (1859), which was very influential in the late nineteenth century and had a particularly strong impact on Sigmund Freud.[7] One of the few dissenters was Eduard Poeppig, a German physician who published his travel journals in 1836 and was often quoted by later critics of coca. He thought that the inveterate *coquero* was "the slave of his passion even more than the drunk" and "incapable of pursuing any serious goals in life," and contended that coca chewing caused anemia and various digestive troubles.[8] About his writings William Prescott, the great historian of the conquest of Peru, comments in a footnote: "A traveler [Poeppig] expatiates on the malignant effects of the habitual use of the coca as very similar to those produced on the chewer of opium. Strange that such baneful properties should not be the subject of more frequent comment by other writers! I do not remember to have seen them even adverted to." [9]

At about the time Europe began to know the properties of coca as something more than legend, rumor, and questionable travelers' tales, the leaf's main active principle was isolated. In 1855 Ernst von Bibra published *Die Narkotischen Genussmittel und der Mensch,* which gave accounts of 17 drugs, including coffee, tea, hashish, opium, and coca; in the same year Friedrich Gaedcke (correct spelling—the name is variously spelled in the sources) produced from a distillate of the dry residue of an extract of coca a crystalline sublimate he called "Erythroxylin," which was probably a mixture of alkaloids including cocaine. After further attempts by various chemists, Albert Niemann of Göttingen finally isolated the principal alkaloid in 1860 from Peruvian leaves brought to Europe by a Dr. Scherzer. Wilhelm Lossen ascertained the chemical formula of cocaine in 1862, and later in the nineteenth century researchers completed the isolation and description of the other coca alkaloids. Throughout the late nineteenth century both coca itself (that is, an extract from the leaf including all its alkaloids) and the pure chemical cocaine were used as medicines and for pleasure—the distinction was not always made—in an enormous variety of ways.

We have referred to Mantegazza as one of the most important early sources of the new European interest in coca. He adopted an exaggeratedly lyrical tone that may have aroused some skepticism: "I flew about in the spaces of 77,438 worlds, one more splendid than another. I prefer a life of ten years with coca to one of a hundred thousand without it. It seemed to me that I was separated from the whole world, and I beheld the strangest images, most beautiful in color and in form, that can be imagined." [10] Very few of the physicians who later tried coca on themselves or their patients reported any experience like this. But Mantegazza gave enough detail to convince even so sane and astute a clinician as Freud that coca was a useful medicine.[11] Freud and others were un-

doubtedly reassured of Mantegazza's reliability by some cautionary qualifications in which he ascribed digestive complaints, emaciation, and
"moral depravity" to overuse of the drug. If coca had in fact often produced the extravagant "psychedelic" effects described by Mantegazza,
physicians would have been reminded of opium dreams or alcoholic delirium and the drug would not have become nearly so popular. It made its
reputation as a tonic and analgesic.

At first the popularity of coca and cocaine grew steadily but slowly.
Mantegazza recommended coca in 1859 for toothache, digestive disor-
ders, neurasthenia, and other illnesses. In 1863 Angelo Mariani, a Cor-
sican chemist, patented a preparation of coca extract and wine, "Vin
Mariani," that eventually made his fortune, becoming one of the most
popular prescription medicines of the age. As early as 1865 Dr. Charles
Fauvel of Paris was prescribing Mariani preparations for various complaints, including soreness of the larynx and pharynx. According to
Freud, he called coca "la tenseur par excellence des chordes vocales." [12]
In 1868 Tomas Moreno y Maíz, Surgeon General of the Peruvian army,
experimented with cocaine and said that it gave him "some of the most
blessed moments of my life." Charles Gazeau in 1870 took 20 to 30 grams
of coca leaf a day for two days and found that it suppressed his appetite
completely; he thought the appetite-suppressing and invigorating powers
might have military applications. In 1876 there was a flurry of interest in
the British medical press when several men, including the 78-year-old
Dr. Robert Christison of Edinburgh, reported that coca had enabled them
to walk long distances without food or sleep and with no serious afteref-
fects. The *British Medical Journal* prophesied in an editorial that coca
would prove to be "a new stimulant and a new narcotic: two forms of novelty in excitement which our modern civilization is highly likely to es-
teem." Three weeks later the *Journal* referred to the use of coca in
France as an elixir and wine and by the week after that was apparently
receiving inquiries which prompted the comment that some ladies hoped
it would give them "strength and beauty forever." The literature was
fairly extensive by the time Bordier reviewed it in the *Dictionnaire encyclopédique des sciences médicales* in 1876 and recommended the use of
coca by armies and in industry. [13]

Through the late 1870s and early 1880s the literature on coca and
cocaine continued to grow. An advertising pamphlet published by
Mariani in 1880 could cite numerous favorable references to coca from
the medical press. By 1878 coca was being recommended in advertisements in the United States for "young persons afflicted with timidity in
society" and as "a powerful nervous excitant." In the same year, fatefully,

W. H. Bentley began to promote it as a cure for morphine addiction.* In
1880 Bentley's article in the Detroit *Therapeutic Gazette,* "Erythroxylon
Coca in the Opium and Alcohol Habits," cited several "cures," including
the case of a rich woman, age 72, who had been using opium for 35 years
and was now alternating two weeks of coca with two weeks of opium (a
combination that, in the stronger form of cocaine and heroin, still has its
attractions). In another *Therapeutic Gazette* article that same year, Bentley called coca "the desideratum . . . in health and disease." He claimed
to have cured a "great rake" of impotence with it and to have used it himself since 1870. The *Gazette,* in an editorial published earlier that year,
breezily quoted a breezy editorial in the *Louisville Medical News:* " 'One
feels like trying coca, with or without the opium-habit. A harmless remedy for the blues is imperial.' And so say we." [14] Medical journals today
rarely use this style or openly promote the identification of pleasure with
cure in this way.

Between 1880 and 1884 the *Therapeutic Gazette* published 16 reports
of cures of the opium habit by coca. By 1883 Parke Davis, the American
manufacturer of cocaine, was advertising it in medical journals for morphinism and alcoholism. (The sharp distinction between coca and cocaine that later became a favorite point with advocates of coca was rarely
made at that time).[15] The fluid extract of coca was admitted to the U.S.
Pharmacopoeia in 1882. In 1883 Aschenbrandt clandestinely put cocaine
into the water of Bavarian soldiers before they went on maneuvers and
obtained the expected results. There were still some skeptics, like George
Ward, a physician who had spent three years in Cerro-del-Pasco, Peru, at
an altitude of 14,000 feet; he doubted that coca had any more effect than
whiskey or tea.[16] But coca and cocaine were about to realize their potential popularity.

The *annus mirabilis* was 1884. Until then, possibly because the diver-
sity of claims about its effects was confusing and possibly because preparations were of unreliable quality, cocaine had not attained the renown

* Since we intend to use the word *addiction* in connection with morphine and heroin (diacetylmorphine), we must insist that we do not intend it to carry the misleading emotional connotations it has
acquired because of public hysteria and legal persecution. As we use the term in this book, it is simply a
convenient way of referring to the fact that these drugs sometimes produce severe physiological abstinence symptoms. The same is true of barbiturates and alcohol. Many alcoholics could just as well be
called alcohol addicts. And just as people can drink liquor without becoming alcoholics, they can smoke,
sniff, or inject opiates without becoming addicts. Addiction to opiates occurs earlier in the course of habitual use than addiction to alcohol, but in almost all other ways alcohol addiction is much worse. The abstinence symptoms in alcohol (and barbiturate) addiction are much more severe than in opiate addiction;
unlike opiate abstinence symptoms, they can be fatal without proper medical care. More important, the
potential acute and chronic toxic effects of the *use* of alcohol and barbiturates, as opposed to withdrawal,
are much greater than those of opiate use. We discuss drug addiction and dependence in detail in
Chapter 8.

some thought it deserved. The decisive events were the publication of Freud's paper "On Coca" in July and Karl Koller's rediscovery in September of the anesthetic power of cocaine, which meant the advent of local anesthesia in surgery. Freud's still useful article, written with what Ernest Jones calls "a remarkable combination of objectivity and personal warmth, as though he were in love with the object itself," created a sensation.[17] He reviewed the historical and medical literature and contended that coca (he makes no distinction between coca and cocaine) should be regarded as a stimulant like caffeine and not a narcotic like opium and cannabis. He blamed the drug's past failures on bad quality preparations. On the authority of various physicians and from his own experience he recommended coca or cocaine for a variety of illnesses, and especially for the congeries of symptoms including fatigue, nervousness, and small physical complaints then known as neurasthenia, which he was later to associate with sexual repression. He described the effect of cocaine on himself (apparently taken orally at this time) in doses of 50 to 100 mg as "exhilaration and lasting euphoria, which does not differ in any way from the euphoria of a normal person." He also felt an increase in self-control and vigor without the "characteristic urge for immediate activity which alcohol produces" or the "heightening of the mental powers which alcohol, tea, or coffee induce," and he had the capacity but not the need to sleep or eat. Probably the most significant passage in this paper was his praise of cocaine as a cure for morphine addiction and alcoholism, based on the American reports and on his own observation. He believed that coca directly antagonized the effect of morphine and insisted that treatment with coca did not turn the morphine addict into a *coquero*. (By "coca" here he meant cocaine).[18] It was in reference to the writings of Freud and his friend Ernst von Fleischl-Marxow that the Parke Davis company's pamphlet *Coca Erythroxylon and Its Derivatives* declared: "If these claims are substantiated . . . [cocaine] will indeed be the most important therapeutic discovery of the age, the benefit of which to humanity will be incalculable." [19]

Freud's article was influential, but the discovery of the use of cocaine in surgery by the friend and colleague whose interest in the drug he had aroused was of more permanent importance. It had been known for a long time that cocaine was a local anesthetic. Folk medicine in South America makes use of this property, and skulls have been found on archaeological sites in the Andes with holes indicating that trepanation, possibly with the help of coca's pain-deadening effect, was performed. Samuel Percy read a paper on anesthesia by means of coca to the New York Medical Society in 1857. In 1862 Schroff noted the numbing effect of cocaine on the tongue. By 1865 Fauvel was using coca to soothe sore throats. Moreno y Maíz (1868) and Alexander Bennett (1874) demonstrated its anesthetic effect on the mucous membranes. Von Anrep (1878) also proposed its use as an anesthetic. In 1880 Coupart and Bordereau described anesthesia of the cornea of animals' eyes with cocaine.[20]

But somehow no one thought of the seemingly obvious application to surgery until Koller introduced it as a topical anesthetic in eye operations. He filled a desperate need. General anesthesia was unsatisfactory in ophthalmology because the conscious cooperation of the patient was often necessary and because ether, as then administered, could cause retching and vomiting, which, if they occurred during or soon after surgery, might damage the eye. Delicate, sensitive, lengthy operations like cataract removal, carried on without anesthesia, were a torture for both doctor and patient. Koller's discovery put an end to that. It immediately became common knowledge in the medical community, and there was sudden great interest in all aspects of cocaine. For example, the December 6, 1884, issue of the *British Medical Journal* had seven articles on cocaine; after a Dr. Squibb published a note about the drug in the *Philadelphia Medical Record,* he received 300 letters of inquiry from physicians. The use of cocaine anesthesia was quickly extended to other areas of surgery: rhinology, laryngology, gynecology, urology, dentistry. Before the end of the year 1884 H. Knapp was able to review a wide range of surgical uses.[21]

New forms of anesthesia that went beyond the topical application of cocaine solutions to body surfaces were almost immediately introduced. In 1884 William Halsted of Johns Hopkins invented nerve block, or conduction anesthesia, by injecting cocaine into nerve trunks. In 1885 J. Leonard Corning introduced regional anesthesia. Later Carl Ludwig Schleich produced infiltration anesthesia by subcutaneous injection (1892), and August Bier originated spinal anesthesia (1898). Although its potential toxicity was recognized early, until the end of the nineteenth century cocaine was the only available local anesthetic. It was not until 1899 that Einhorn synthesized procaine (Novocain), a substitute without cocaine's unpredictable and dangerous central nervous system effects.

Meanwhile, especially in the early years, cocaine was triumphing as what would now be disparagingly called a panacea. The same Corning who invented regional anesthesia declared, "Of all the tonic preparations ever introduced to the notice of the professions, this [coca] is undoubtedly the most potent for good in the treatment of exhaustive and irritative conditions of the central nervous system." [22] Considering the importance of the central nervous system in body functioning, this almost amounted to saying that coca is good for what ails you. Similar claims have been and are made for alcohol, opium, cannabis, LSD, and other drugs affecting the mind. But cocaine was the "drug of the eighties." Its promotion as a cure for morphine addiction and alcoholism was an example of a typical stage in the career of psychoactive drugs: their use to overcome the consequences of abuse of other psychoactive drugs. But cocaine was also used as a cure for many other ailments. From July to December of 1885, for example, there were 27 articles, notes, and letters on cocaine in the *New York Medical Journal,* recommending it for seasickness and trigeminal neuralgia, among other conditions. The drug house Parke Davis brought out its 101-page pamphlet, *Coca Erythroxylon and Its Derivatives,* in that year. Cocaine was adopted as the remedy of choice by the Hay Fever Association and recommended for head colds and "catarrh." William A. Hammond, a former Surgeon General of the United States Army, suggested coca wine for stomach irritability, "cerebral hyperemia due to excessive mental exertion," "the mental depression that accompanies hysteria in the female," and other morbid central nervous system conditions.[23] He thought pure cocaine even better than coca for most purposes, and recommended it for inflammations of the mucous membranes and to prevent masturbation in women by anesthetizing the clitoris.[24] Coca products were in use for such varied purposes for a generation or more.

The book *Peru: History of Coca,* published in 1901 by the American physician W. Golden Mortimer, sums up the favorable side of medical opinion about coca. Mortimer's recommendations for coca wine or coca extract and occasionally for cocaine are even more varied than Mantegazza's or Freud's. He approvingly mentions its use by French bicyclists and by a championship lacrosse team. (Apparently he saw no ethical problems in the use of drugs by athletes.) Contradicting Mantegazza's view that coca acts as a stimulant on the heart, he contends that it is primarily a regulator, calming an overexcited heart and strengthening a weak one. He believes that Mantegazza noticed only the central nervous system effect produced by cocaine and failed to observe the direct tonic effect of other coca alkaloids on the heart muscle. Mortimer was recommending coca for alcohol and opium addiction long after the use of cocaine in such cases was considered a disastrous mistake. A whole chapter on the history of music and voice production (the book tends to ramble) was apparently inspired by the popularity of coca preparations among singers. Mortimer also listed the results of a mail survey of prominent physicians; only a small minority of them used coca, but those who did considered it helpful for a large variety of illnesses.[25]

The use of a single drug for various and often vaguely defined conditions, as Mortimer's book and his correspondence show, was characteristic of that era. Cocaine's two distinct effects, as a local anesthetic and central nervous system stimulant, made it seem especially attractive. Mortimer devotes considerable space to a discussion of neurasthenia, a diagnosis that is no longer popular today. He describes its symptoms as headache of a special kind (a feeling of constriction over the back of the head), digestive troubles, incapacity for work, loss of sexual desire, muscular weakness and stiffness, back pain, insomnia, and "hypochondriacal views of life." Mortimer admits that this is a "combination of symptoms of very different nature" which requires a variety of different measures; he recommends coca as a useful adjunct in treatment.[26] Of course, doctors still commonly prescribe psychoactive drugs for very general functional disturbances: amphetamines and tranquilizers have been used in the same ways as coca (or opium, or alcohol). But it was even easier to use drugs this way at a time when pharmacological theories were more speculative and confused than they are now. It was a time of trial, groping, and uncertainty in medicine. Where drugs acting on the central nervous system are concerned, we may not have emerged from that stage of tentativeness as much as we sometimes think we have.

Mortimer dedicated his book to Angelo Mariani, the chemist and entrepreneur whose coca preparations were one of the most popular medicines of the era, calling him "a recognized exponent of the 'Divine Plant' and the first to render coca available to the world." Mariani was not only a promotional genius but a student of the history and folklore of coca; he cultivated coca plants in his own conservatories in Paris and distributed them to botanical gardens all over the world. He provided Mortimer with information and botanical specimens; Mortimer considered his coca preparations superior to all others in flavor and effectiveness. Mariani wrote several articles and monographs on coca which combine historical, botanical, and medical information with the promotion of his company's product. The most important of these was *Coca and Its Therapeutic Applications* (1890), which went through several editions. He sold coca extract not only as Vin Mariani but also as Elixir Mariani (stronger and with greater alcohol content than the wine), Pâte Mariani (a throat lozenge), Pastilles Mariani (Pâte Mariani with a little cocaine added), and Thé Mariani (concentrated coca leaf extract without wine).[27]

Mariani was not one of those manufacturers of proprietary medicines who advertised their products mainly to the general public. He was proud of the numerous and enthusiastic testimonials his coca preparations received from physicians, and his company approved of the moves made after the turn of the century toward restricting direct consumer advertising of drugs. He was able to cite prominent physicians like J. Leonard Corning, who declared his coca wine "the remedy par excellence against worry." He carefully pointed out that his drug was *"introduced solely through physicians"* (his italics) and could provide a list of about 3,000 physicians who recommended it. It won prizes and medals at various expositions, including one from England that called it "wine for athletes," and it received what Mariani says was an unsolicited recommendation from the Academy of Medicine of France. In the later years, when coca was beginning to lose its status as a respectable medicine, his company's American publication offered a reward for identification of defamers of Vin Mariani, declaring that a slur against it was a slur against the intelligence of many prominent physicians.[28]

Mariani's products were as popular with the public as with physicians. From the testimonials of eminent people he put together a "cyclopedia of contemporary biography" with a biographical sketch and portrait of each famous man or woman who testified to the virtues of Vin Mariani.[29] Thomas Edison (famous for his insomniac habits) was among them, and so was Pope Leo XIII, who presented Mariani with a gold medal and habitually carried a flask of the wine at his belt. The Czar of Russia, Jules Verne, Emile Zola, Henrik Ibsen, and the Prince of Wales also endorsed Mariani's wine. Mariani reports that doctors gave General Ulysses Grant, the former president of the United States, Thé Mariani, one teaspoon to a cup of milk per day, for five months during his last illness, in 1885. In their opinion, it prolonged his life and enabled him to complete his famous *Memoirs*.[30] Mariani's advertising made good use of the drug's popularity among actors and actresses, singers, and musicians. (No doubt he would have solicited athletes' testimonials more often if they had had the respectability and celebrity status they enjoy today.) Sarah Bernhardt, the composers Gounod and Massenet, and the *prima donna* Adelina Patti, among many others, praised Mariani's preparations. Entertainers never entirely stopped using cocaine, and its popularity in the worlds of theater, film, and popular music has probably never been greater than it is today. Little has been heard about coca extract itself in those or any other European and North American social groups since 1910, although a half-bottle of Vin Mariani, vintage 1880, was sold at a London wine auction in 1970 for four pounds.[31]

Cocaine was so popular with writers and intellectuals that an article entitled "The Influence of Cocaine on Contemporary Literary Style" appeared in the American journal *Current Literature* in 1910. It declared that cocaine was responsible for the "smooth and flowing sentences now so characteristic of the magazine writing of this period" and warned that cocaine addiction was increasing among the intellectual class.[32] This kind of (to put it mildly) speculative stylistic analysis constitutes most of the evidence we have of influence by cocaine on literary production. The drug is more likely to have simply given people the strength and will to write than to have provided content or style. We have mentioned Grant's *Memoirs*. Another literary work that may have been written with the help of cocaine is Robert Louis Stevenson's "The Strange Case of Dr. Jekyll and Mr. Hyde."

The facts are these. Stevenson wrote his famous story in three days and three nights, then burned the manuscript and rewrote it in another three days and nights. It was October 1885, and the British medical journals, eagerly read by Stevenson's wife, who hoped to find something to help her invalid husband, were full of articles on cocaine. Stevenson had been taking morphine, a drug not conducive to the production of 60,000 words in six days. His stepson later said, "The mere physical feat was tremendous; and instead of harming him, it roused and cheered him inexpressibly." Stevenson had fantasies and dreams about little creatures he called Brownies who gave him the inspiration for "Dr. Jekyll and Mr. Hyde." The story, of course, describes the transformation of an upright physician and scientist into a monster of immorality by a drug. None of Stevenson's letters or biographers mention cocaine. It is certainly conceivable that cocaine gave this sick man the energy that enabled him to write so much so fast. It is possible, though unlikely, that the Brownies were Lilliputian hallucinations of the kind sometimes induced by stimulant drugs like cocaine. And it might even be, although this is less likely still, that the plot of "Dr. Jekyll" owed something to the reports of cocaine abuse, especially among physicians, that were just beginning to appear.[33]

Mariani's coca preparations were the most famous among many on the market in the late nineteenth century. Parke Davis, for example, sold cocaine in cigarettes, in an alcoholic drink called Coca Cordial, and in sprays, ointments, tablets, and injections.[34] One of the most popular drinks containing coca extract was Coca-Cola, first concocted by John Styth Pemberton, a Georgia pharmacist, in 1886. Since cocaine is no longer an acceptable or legal stimulant, the Coca-Cola Company does not like to be reminded of this aspect of its early history. The author of a generally well-informed long article on Coca-Cola in *The New Yorker* implies that the presence of cocaine was accidental: "In its formative days, the drink did contain a minute quantity of cocaine, since this drug was not removed from the coca leaves that constituted a tiny fraction of its makeup." [35] This is totally misleading. Pemberton was a pharmacist and he sold Coca-Cola as a medicine: a headache remedy and stimulant that contained the "wonder drug" of those years, coca, as its main active principle. Pemberton had registered a trademark for a brew he called "French Wine of Coca, Ideal Tonic," possibly in imitation of Mariani, in 1885. In 1886 he removed the alcohol, added kola nut extract (which contains caffeine) and some citrus oils for flavor, renamed the product, and began to advertise it as "the intellectual beverage and temperance drink." In 1888 he replaced ordinary water with soda water, which was already associated with mineral springs and health.

Asa Griggs Candler, another pharmacist, bought all the rights to Coca-Cola in 1891 and founded the Coca-Cola Company in 1892. According to the author of a book on the history of pharmacy in Georgia, the soda fountain became an essential part of the retail drugstore in the United States mainly because of Coca-Cola, "the first generally advertised product that directed people to a drugstore." [36] Candler "believed in Coca-Cola with an almost mystical faith," according to his son, and he advertised it throughout the 1890s as a "sovereign remedy" as well as an enjoyable drink.[37] As late as 1903 a Colonel J. W. Watson of Georgia was quoted in the *New York Tribune* as urging legal action against "a soda fountain drink manufactured in Atlanta and known as Coca-Cola" because of its cocaine content.[38] But the men who ran the Coca-Cola Company were commercially astute enough to sense the change in public opinion that was to make cocaine a social outcast, and by the time the Pure Food and Drug Act was passed in 1906 they had taken it out of their drink and replaced it with caffeine. It is interesting to note that even in 1909 there were 69 imitations of Coca-Cola that contained cocaine.[39] People did not immediately forget the original basis of Coca-Cola's popularity: 40 or 50 years ago it was still possible to order a bottle by asking for a "shot in the arm."

The soda fountain in the drugstore where Coca-Cola was dispensed, a poor relative of the spa or watering place patronized by the upper classes in the nineteenth century, represents a kind of fusion of the health giving and the pleasure giving that is not necessarily unreasonable but may lead to disaster. That is what happened with cocaine almost as soon as doctors began to use it. We leave aside here the cases of acute poisoning and occasional death from its use as an anesthetic. They are important as part of the history of surgical anesthesia and for what they indicate about the physiological effects of cocaine, but they are not the reason it became an outlaw drug. Anesthesia was recognized to be a dangerous procedure in those days. General anesthetics like ether and chloroform killed more often than cocaine. All these drugs were used because there was no choice. When substitutes for cocaine became available, surgeons stopped using it except topically. Before that, they could only try to learn what precautions had to be taken about technique, dosage, and site of application. Cocaine became a "drug menace" to the public not because it sometimes killed people in surgery but because what had been regarded as the very sign of its curative power, the pleasure it gave, became a source of what we now call drug dependence and drug abuse.

Before 1885 there had been reports of acute cocaine intoxication, but physicians doubted the possibility of chronic abuse because many of the effects of large doses seemed frightening rather than alluring. (They were apparently not considering what alcoholics, for example, are willing to put up with for the sake of drink.) But they soon began to conclude that cocaine abuse was "a habit that develops more easily and destroys the body and soul faster than morphine." By 1890, at least 400 cases had been reported in the medical literature of acute and chronic physical and psychological disturbances caused by the drug.[40]

The earliest serious cases of abuse involved morphine addicts who took the cocaine cure recommended by Bentley and Freud. As early as May 1885 Ludwig Lewin, the famous student of psychotropic drugs and later the author of Phantastica: Narcotic and Stimulant Drugs, was expressing skepticism about this cure. He recognized that cocaine provided temporary symptomatic relief in morphine withdrawal but rejected Freud's view that it could serve as a substitute for the opiate. He also suspected that chronic use of cocaine in large doses could produce toxic effects. In the same year J. B. Mattison, in an article in the New York Medical Journal entitled "Cocaine in the Treatment of Opiate Addiction," agreed that its effect was transient and that there was genuine danger of producing a cocaine habit. By 1886 cases of cocaine psychosis with tactile hallucinations ("coke bugs") were appearing. In May of that year the New York Medical Record commented editorially: "No medical technique with such a short history has claimed so many victims as cocaine."[41]

Albrecht Erlenmeyer was probably the most important of the physicians who early observed the symptoms of acute and chronic cocaine intoxication and warned against the use of cocaine in morphine addiction. He issued his first warning in 1885 and returned to the subject in another article written in 1886 and an 1887 monograph on "Morphine Addiction and Its Treatment." Erlenmeyer melodramatically denounced cocaine as "the third scourge of mankind," after alcohol and morphine. He found that morphine addicts tended to add cocaine to morphine rather than replace the original drug, and he considered the combined morphine-cocaine habit worse than straight morphine. He described patients who "cured" themselves of morphine addiction with cocaine and then had to take morphine again to counteract the sleeplessness and mental confusion brought on by cocaine. Some of the symptoms he mentioned—insomnia, graphomania, paranoia (he states that three of his patients eventually had to be put in insane asylums)—were obviously, as we shall see, produced by cocaine if by any drug. Erlenmeyer pointed out that some observers had been confusing cocaine psychoses with morphine abstinence symptoms. He admitted that cocaine itself produced few abstinence symptoms except depression, but he concluded that of all drugs only alcohol could be as devastating.[42]

Erlenmeyer noted that all his cases involved a combination of morphine and cocaine abuse rather than cocaine alone, and most of the cocaine abusers in the next generation whose condition was serious enough to come to the attention of physicians or the law were also using opiates or alcohol. We will discuss the significance of this when we come to talk about the effects of cocaine in more detail. (What it does *not* mean is that cocaine is harmless unless combined with opiates, or that only opiate addicts and alcoholics are likely to abuse cocaine.) Historically, the fact that cocaine was not usually used alone helped to create a confusion between cocaine and opiates that made them both seem more fearful: the physical addictiveness of the opiates was ascribed to cocaine, and the psychological and physical effects of cocaine abuse were attributed to the opiates. The two quite different kinds of drug began to suffer similar legal restrictions and a similar decline in public and professional estimation.

At first physicians and the public made little distinction between cocaine and coca. Therefore the growing fear of cocaine changed attitudes toward coca, just as fear of morphine and heroin made opium smoking seem more dangerous. Advocates of coca then began to fight a rearguard action in its defense. They insisted (correctly) that coca never caused the kinds of disturbances that were ruining the reputation of the pure chemical cocaine, and (with less obvious justification) that it was not the cocaine in coca but the peculiar mixture of alkaloids that produced its characteristic effects. These themes are prominent in Mariani's and Mortimer's works. Mariani described his coca extract as consisting of "the soluble parts of the Peruvian plant" and stated that it could not produce what was then called cocainism because it did not contain the pure alkaloid cocaine. By 1906 the journal published in the United States by Mariani's company was insisting that Vin Mariani was "not a cocaine

preparation" and was made from "sweet" leaves with only "an infinites-
imal trace of the cocaine base." Mortimer extravagantly contended that
cocaine no more fully represents the effect of coca than the prussic acid
in peach pits represents the effect of peaches.[43] More reasonably, he
claimed that some of coca's beneficial qualities, especially the cardiac,
muscular, and digestive effects, should be attributed not to the action of
cocaine on the central nervous system but to alkaloids affecting other
organs directly. We will consider these matters—the evidence about
them is sparse—when we discuss the effects of coca in more detail. In
practice, the idea that coca is fundamentally different from cocaine never
gained much influence in the medical community or among the public.
And when coca extract went into decline as a medicine it did not become
a recreational drug (although a sweet liqueur called Élixir de Coca is still
sold in Peru). The kinds of people who had been drinking Mariani's wine
took to sniffing cocaine in powder form; it seems that cocaine simply gave
them more of whatever they thought they were getting from coca.

Physicians were the first to recognize the powers of cocaine and had
the easiest access to it; they probably used and abused it, as they used
and abused morphine, more than any other occupational group. In 1901,
for example, it was estimated that 30 percent of the cocainists in the
United States were doctors or dentists.[44] Probably the most famous case
of cocaine abuse by a physician is that of the great surgeon William
Halsted of Johns Hopkins (1852–1922), who invented nerve block anesthesia. At the time of his first anesthesia experiments in 1884, according
to his student Wilder Penfield, "cocaine hunger fastened its dreadful hold
on him." There followed a "confused and unworthy period of medical
practice," then a year's stay in a hospital and a curative sailing cruise.
When he returned, "the brilliant and gay extravert seemed brilliant and
gay no longer." [45] After a further hospital stay, he apparently recovered
and stopped using cocaine. What no one knew at the time except Sir
William Osler, who revealed it in a manuscript entitled "Inner History of
Johns Hopkins Hospital" and first made public in 1969, long after his
death, was that Halsted was taking three grains (200 mg) of morphine a
day. He eventually reduced the dose to one and a half grains a day, but
the "struggle against the dreadful discomfort of drug hunger" continued
to the end of his life.[46]

There seems to be some confusion about exactly what Halsted's problem was, since Osler also refers to a "cocaine and morphia habit." A
recent article in the *Journal of the American Dental Association* trans-
lates the three grains of morphine into three grains of cocaine. It looks as
though the epidemic confusion between the two drugs is at work here.

Cocaine might have made Halsted a brilliant and gay extravert; the sedative morphine certainly did not. It appears that Halsted cured himself of
the craving for cocaine, which was ruining his career, by means of morphine, which allowed him to function normally (as opiates often do), and
paid the price of physical addiction.

Halsted remarks in a 1918 letter to Osler that three of his assistants became victims of the cocaine habit and died without recovering from it. In
a later letter he states: "Yes, I published three or four little papers in 1884
and 1885 in the *New York Medical Journal* on the subject of cocaine anesthesia. They were not creditable papers for I was not in good form at the
time." [47] One of these papers had the title "Practical Comments on the
Use and Abuse of Cocaine, Suggested by Its Invariably Successful Employment in More than a Thousand Minor Surgical Operations." Its first
sentence provides what might be one of the few genuine examples of
"the influence of cocaine on contemporary literary style":

> Neither indifferent as to which of how many possibilities may best explain, nor
> yet at a loss to comprehend, why surgeons have, and that so many, quite
> without discredit, could have exhibited scarcely any interest in what, as a local
> anaesthetic, had been supposed, if not declared, by most so very sure to prove,
> especially to them, attractive, still I do not think that this circumstance, or
> some sense of obligation to rescue fragmentary reputation for surgeons rather
> than the belief that an opportunity existed for assisting others to an appreciable
> extent, induced me, several months ago, to write on the subject in hand the
> greater part of a somewhat comprehensive paper, which poor health disin-
> clined me to complete. [48]

This syntax undoubtedly represents some kind of achievement. The
reader can judge whether it is a successful employment of cocaine.

The greatest mind among the medical men who underwent the influence of cocaine belonged to Sigmund Freud. He first learned of it
through Aschenbrandt's experiments on the Bavarian soldiers and the
American reports of its use as a cure for morphine addiction. He refers to
the drug in a letter to his fiancée, Martha Bernays, on April 21, 1884, as
"a project and a hope" for reducing fatigue and counteracting morphine
withdrawal symptoms. He started using it himself and called it a "magical drug," sending some to Martha "to make you strong and give your
cheeks a red color." In May he began administering cocaine to his friend
Ernst von Fleischl-Marxow as a substitute for the morphine he had been
using to deaden phantom pain from an amputated thumb. At the time
Freud wrote the paper "On Coca" (June 1884) he found the results encouraging. For several years he continued to use cocaine and experiment
with it. He teased Martha about it: "You shall see who is stronger, a

gentle little girl who doesn't eat enough or a great wild man who has cocaine in his body." He tested its effect on strength with a dynamometer—the only experimental study of his career—and found that it did not affect muscular capacity directly but produced a general sense of well-being, improving performance most when he was tired or depressed. In general, small doses of cocaine had little effect on him when he was in generally good health but made him feel normal for four or five hours when he was below par. In a letter to Martha of February 1886, possibly written under the influence of cocaine, he referred to the relief it provided for his neurasthenia; he also told how it reduced his shyness in social situations and even suggested that it made him feel as though he had the strength to sacrifice his life, like his ancestors defending the Temple. He was so well known as an authority that Parke Davis solicited his opinion on their cocaine; he pronounced it as good as the product of the German drug house Merck.[49]

As far as its effects on himself were concerned, Freud never had any reason to criticize cocaine. But his friend Fleischl took larger and larger amounts of the drug and began to suffer severe toxic symptoms. By the spring of 1885 he was taking one gram a day by subcutaneous injection and, in the words of Carl Koller, another friend, had become "a cocainist instead of a morphinist, probably the first of these unfortunates in Europe. And many a night have I spent with him watching him dig imaginary insects out of his skin in his sensory hallucinations." Freud too spent "the most frightful night of his life" with Fleischl in a state of cocaine delirium in June of 1885. So when Erlenmeyer denounced cocaine as the "third scourge of mankind" and criticized Freud for encouraging its use, feelings of sorrow and guilt as well as concern for his professional reputation were involved in the response.[50]

This last of Freud's cocaine papers, "Craving for and Fear of Cocaine," published in July 1887, is an interesting professional and personal document. It retreats from some of his former positions and defends others. He admits that cocaine should not be used in morphine addiction because the habit that may ensue is "a far more dangerous enemy to health than morphine," producing quick physical and mental deterioration, paranoia, and hallucinations. But he insists that *all reports of addiction to cocaine and deterioration resulting from it refer to morphine addicts. . . . Cocaine has claimed no other, no victim on its own"* (Freud's italics). (Of course, he did not know about Halsted.) He thought only morphine addicts were so weak in will power as to be susceptible to chronic cocaine abuse; he himself had felt no desire for continued use but on the contrary, "more often than I should have liked," an aversion to the drug.

He discussed acute poisoning and concluded that "the reason for the irregularity of the cocaine effect lies in the individual variations in excitability and in the variation of the conditon of the vasomotor nerves on which cocaine acts." He recommended abandoning subcutaneous injection except as an anesthetic. Appended to the paper was a summary of a report to the New York Neurological Society by William Hammond, which recommended a cocaine-wine drink (not Mariani's coca wine) as a tonic and stimulant and for dyspepsia and injections of cocaine for "female melancholia with mutism." [51] This paper was Freud's last professional publication on cocaine, but in a letter to Ferenczi on June 1, 1916, he commented in connection with a prospective patient who used the drug that it could produce paranoid symptoms if taken to excess. He also remarked that drug abusers were bad risks in analysis because they found it too easy to cling to the security of their drug.[52]

Freud always regretted that he had not made the discovery of the use of cocaine in surgical anesthesia that brought fame to Koller. In An Autobiographical Study he attributes this failure, in what Ernest Jones regards as a disingenuous excuse, to the interruption of his work by a journey to visit his fiancée. In other contexts he admitted that he had simply been too lazy to pursue the matter. As Jones points out, Freud in any case thought of cocaine as a stimulant rather than an anesthetic and was more interested in its internal use than in any local application. Many years later Freud referred to his cocaine studies as a distracting hobby that took him far from his serious work in neuropathology. Jones attributes his intense interest and later guilt and need to inculpate his fiancée Martha to his having taken a surreptitious shortcut to dispel his depression and so "achieve virility and enjoy the bliss of union with the beloved." He had hoped to achieve fame and fortune by means of some application of the drug, so that he would be able to marry sooner. Instead, the fate of Fleischl gave him reason for self-reproach, and the discrediting of cocaine as a cure for morphine addiction and other ills damaged his reputation and eventually made the acceptance of psychoanalysis even more difficult. Freud seldom referred to the cocaine episode later in his life.[53]

Nevertheless, Jürgen vom Scheidt, in an interesting recent article in a German psychoanalytic journal,[54] suggests that the psychopharmacological properties of cocaine may actually have aided Freud's self-analysis and contributed to the development of his ideas. The cocaine episode was one of the most exciting times of Freud's life: the first patient who came to him on his own and not through the recommendation of a colleague had been attracted by the writings on cocaine; and the American ophthalmologist Knapp, the author of a book on the surgical applications of the drug, had recognized and greeted him at the Salpêtrière in Paris as the author of "On Coca." It was during the time when he worked with Charcot in Paris, 1885 to 1886, that Freud began the transition from mainly physiological to mainly psychiatric interests. At that time he was using cocaine regularly and writing several papers about it. As a psycho-pharmacological agent, it may have mediated the change in Freud's interests. The letters to Martha from Paris show that cocaine also loosened Freud's censorship over the expression of his feelings, and even, in this period just before his marriage, relieved his neurasthenia, like the presence of his fiancée; vom Scheidt notes that Freud more than once referred to drugs as a substitute for sex. In his view the intoxicant impaired or changed Freud's ego-functions and released sexual and aggressive drives, producing a mild regression that made the inner world of dream and fantasy come forth more strongly; the unconscious wish for a deeper regression found an outlet in Freud's enthusiastic words about cocaine.

Vom Scheidt also provides new interpretations of the well-known fact that some of the most important dreams analyzed in *The Interpretation of Dreams* contain references to cocaine. Freud continued to prescribe cocaine until at least 1895, the year of his self-analysis, for topical application to the nasal mucous membranes, and he used it himself for sinusitis. He suffered not only from nasal infections but from migraine and, after an attack of influenza in 1889, heart arrhythmia (irregular heart action). Fliess induced him to give up smoking in 1894, and a short while afterward he suffered a severe cardiac condition with racing and irregular heart, tension, hot pain in the left arm, and respiratory difficulties. Fliess, who had previously diagnosed Freud's heart troubles as being of nasal origin, now attributed them to nicotine poisoning. Although Freud doubted the diagnosis, he managed to stop smoking for 14 months, until he could no longer tolerate abstinence. By this time Fliess had again decided that the heart condition was of nasal origin, and this conclusion was apparently supported by the improvement that followed an operation and the use of cocaine nose drops. Jones believes that the heart troubles, the migraine, and the nasal infections were all neurotic, although slightly aggravated by the effects of nicotine.[55] Vom Scheidt remarks that Freud spoke of his need for "something warm between the lips" and suggests that abstinence from tobacco aroused his oral drives and that he needed the kind of mild regression provided by cigars, and earlier by cocaine, to free his creativity.

The crucially important dream of "Irma's Injection" took place six weeks after Freud started smoking again. In his own analysis of this dream, Freud describes his use of cocaine for nasal congestion and mentions the case of a woman in whom it had caused necrosis of the nasal tissues. He also refers to the Fleischl affair and the reproaches he had incurred, stating that he had never intended Fleischl to inject the drug. Acccording to Freud, the thought of Fleischl showed defensive pride in his own conscientious handling of chemical substances. But vom Scheidt suggests that the unconscious conflict and guilt might have been related to Freud's own use of cocaine rather than Fleischl's. He notes that in his analysis Freud incorrectly states that he began to use the drug in 1885; actually, he began to *inject* it in 1885. Freud recalls, not here but in his analysis of the dream of the "Botanical Monograph" (itself a reference to his original paper on coca), that an eye operation was performed on his father for glaucoma in 1885 with the aid of cocaine. Vom Scheidt offers this interpretation of "Irma's Injection": if cocaine had affected his father the way it affected Fleischl, Freud would in effect have committed patricide; the injection of Irma is equivalent to impregnating Martha (she was pregnant at the time of the dream), which is equivalent to incest with his mother; so the use of cocaine is a symbolic expression of oedipal desires. Vom Scheidt finds cocaine references in seven other dreams analyzed by Freud, and remarks that this is not surprising, since cocaine was his most substantial intellectual project before psychoanalysis.

On this interpretation, cocaine euphoria helped to show Freud the way to his new conception of the mind. As primitive cultures use drugs to bring the believer into contact with divinity, Freud used cocaine to make contact with the realm of the unconscious. The drug was more than the distraction Freud later considered it to have been; it turned him off the common academic path and toward research of revolutionary originality. So vom Scheidt raises the possibility that cocaine left its greatest mark on the world by way of the mind of Freud. But even if that were so, it is doubtful how much should be attributed to the properties of the drug itself; alcohol, opium, or other drugs might have had the same effect in other historical circumstances. The psychopharmacological peculiarities of cocaine could not have been so important as the conjunction of intellect, personality, situation, and environment.[56]

The most famous, after Freud, of all the Victorian intellectuals who used cocaine was the fictional detective Sherlock Holmes. Sir Arthur Conan Doyle, his creator, was a physician who practiced for a while as an ophthalmologist. He must have been intimately acquainted with the properties of the drug and may have used it himself as a stimulant. Dr. Watson, Doyle's narrator, first mentions cocaine in *The Sign of the Four*, published in 1890. At that time Holmes was injecting a 7 percent solution intravenously three times a day—apparently a rather large dose. Since

Watson reports asking, when he saw Holmes with the needle, whether it was morphine or cocaine, Holmes seems to have had more than one drug habit. But we hear no more of morphine from Watson. In the spirit of mock scholarship with which Sherlock Holmes studies are conducted, we might guess that Holmes was one of those addicts who used cocaine to withdraw from morphine and simply replaced one drug with another. Holmes admitted that cocaine was bad for him physically but found it "transcendentally stimulating and clarifying to the mind." However, he did not use it when working on a case, but only to dispel boredom when he had nothing to do. In connection with a later case, "The Yellow Face," Watson again mentions the occasional use of cocaine as Holmes' only vice. After a while he began to see it as more than a casual indulgence. In "The Adventure of the Missing Three-Quarter," which ostensibly took place in about 1897, he refers to a "drug mania" that had threatened Holmes' career. Watson claims to have cured him of it, but says, "The fiend was not dead but sleeping." In later life, Holmes' only drug habit, like Freud's, was tobacco.[57]

Watson's fragmentary references to Holmes' cocaine habit, which most Holmes scholars regard as disappointingly inadequate considering that he was a medical man, have naturally excited considerable speculation. Some "authorities" contend that Professor Moriarty, the "Napoleon of crime" Holmes claimed to have destroyed, was a paranoid delusion brought on by the drug. In a recently published novel, *The Seven-Per-Cent Solution*, Nicholas Meyer builds an elaborate fiction on the idea that Watson sent Holmes to Freud in Vienna for a cure, which took place during the three years when Holmes had previously been supposed to be in hiding on the Continent after his alleged encounter with Moriarty and Moriarty's death. In spite of the considerable literature on Holmes, the experts have hardly begun to examine this obscure part of his fictional life.[58]

Cocaine's potentialities as a recreational drug soon became obvious; along with those older euphoriants and panaceas, opium and alcohol, it now became a drug with a dubious social reputation. In the 1890s, although people continued to inject cocaine and take it in drinks, sniffing or snorting it in powder form was discovered to be a particularly easy and efficient method of administration. Its use spread downward as well as upward in the class structure, in both Europe and the United States, and became especially common in the regions where the fringes of high society overlap with the fringes of bohemia and the lower middle class. Its users were described as "bohemians, gamblers, high- and low-class prostitutes, night porters, bellboys, burglars, racketeers, pimps, and casual laborers." In the United States bartenders put it into whiskey on request and peddlers sold it door to door. There were circles in which everyone carried cocaine and treated it as a luxury like cognac.[59]

There are several glimpses of patterns of cocaine use in the medical and lay press of the early twentieth century. An article on "The Increase of the Use of Cocaine among the Laity in Pittsburgh" in the *Philadelphia Medical Journal* for 1903 mentioned a druggist who sold it to railroad engineers for overtime work. This article also asserted that black convicts favored the drug. Negroes in Pittsburgh called one thoroughfare "Cocaine Street." Catarrh cures sold with glass tubes for sniffing were one source of the habit. Cocaine was supposed to be sold only on prescription but in fact was rather freely dispensed. W. B. Meister, in "Cocainism in the Army," published in *Military Surgeon* in 1914, mentions such means of obtaining cocaine as buying it from a prostitute who got it from a wholesale drug company by posing as a local druggist, and diverting it via the *ad libitum* refillable prescription of a laundry-truck driver. Meister suggested that a soldier who seemed too talkative or "egotistical and morose" might be using the drug.[60]

Various articles in the American press also give some idea of the kind of censorious attitude that was becoming more common. A 1908 article in the *New York Times*, "The Growing Menace of Cocaine," declared that cocaine "wrecks its victim more swiftly and surely than opium." It was easily available in patent medicines and popular among Negroes in the South, where "Jew peddlers" sold it to them. The lower classes were said to indulge in "sniff parties." A Father Curry was quoted as saying that because of cocaine and opium, drugstores were a greater menace than saloons. J. Leonard Corning, the anesthetics pioneer, now called cocaine "one of the most useful and at the same time one of the most dangerous agents" in the pharmacopoeia and warned against its use as a stimulant, especially by "neurotic persons," who were especially susceptible to the habit. Another physician asserted that no one was more degraded than the cocaine habitué: nothing could be done for the "coke fiend" and he was best left to die. In an article printed in 1911 the *Times* stated that cocaine was used to corrupt young girls and caused criminal acts and resistance to arrest. By 1914 the Atlanta police chief was blaming 70 percent of the crimes on cocaine, and the District of Columbia police chief considered it the greatest drug menace. When a drug has become so convenient as an explanation for crime, society is ready for prohibition.[61]

Pharmacists concerned for their reputations began to worry about cocaine early. In September 1901, at a meeting of the American Pharmacological Association, Vice-President S. F. Payne brought up the issue of "Negro cocainists." The association set up a Committee on the Acquirement of the Drug Habit, which reported to it in 1902 and again in 1903. The 1903 report combines straightforward racial prejudice with the kind of reasonable concern about the overprescription of psychoactive drugs that some physicians have been expressing recently in connection with amphetamines, barbiturates, and tranquilizers. Here are some quotations: "Georgia reports almost every colored prostitute is addicted to cocaine"; "Maryland reports sale of cocaine by disreputable physicians"; "Indiana reports that a good many negroes and a few white women are addicted to cocaine"; "The negroes, the lower and criminal classes, are naturally most readily influenced." Nothing, the report says, is more baneful than cocaine. It turns upright men into thieves and liars. Drugstores fill prescriptions for it too freely and manufacturers supply it to retail stores even when they know it is being diverted to recreational use. The report blames patent-medicine manufacturers but also a society that wants quick solutions to its problems and thinks any medicine that is taken at a soda fountain and makes one "feel so bright" must be harmless.[62] All this, racial prejudice aside, will sound familiar to anyone who knows the career of amphetamines in 1940–1970: the more things change, the more they remain the same, although amphetamines may never attain the status of an outlawed drug menace.

It is obvious by now that the race issue, exposed sometimes directly and sometimes in the guise of a fear of crime, appears prominently in the condemnations of cocaine. Just as opium was associated with the Chinese in the drive to outlaw it, so cocaine was associated with blacks. For a while employers made it available to black workers, as the Spanish in Peru had given the Indians coca. According to an article in the *British Medical Journal*, for example, stevedores and cotton pickers in Louisiana were supplied with cocaine. But soon whites came to see the drug as more dangerous than useful. They thought it increased the cunning and strength of blacks and enhanced their tendency toward violence— especially, of course, sexual violence against white women. Colonel J. W. Watson of Georgia, the man who had issued the stern warning about Coca-Cola, now said that cocaine sniffing "threatens to depopulate the Southern States of their colored population." [63] Anyone concerned with the more sensational and ludicrous aspects of the racial situation at that time may be interested in knowing that some whites believed cocaine made blacks invulnerable to bullets—a concern that seems to be the opposite of Colonel Watson's. In any case, the cocaine-crazed black dope fiend played an important role in the campaign to prohibit the drug.

Naturally, then, many people think that racial prejudice inspired the hostility toward cocaine. This idea has been revived in recent court cases involving the drug, because claims of racial discrimination have become one of the most effective ways of challenging established legal classifications. But we suspect that racial prejudice was ancillary. When respectable people decided that cocaine was a dangerous drug, they were inclined to concentrate on what used to be called the dangerous classes—the poor, especially blacks. After all, the same moves toward prohibiting the use of cocaine were made in Europe, where there was no racial issue. In fact, blacks probably used cocaine, like other prescription drugs, less than whites, simply because they had less money and less access to physicians. A report in 1914 on 2,100 consecutive black admissions to a Georgia insane asylum, for example, indicated that only two were cocaine users, and even for them cocaine had nothing to do with the reasons for admission.[64] Of course, the actual incidence of cocaine use among blacks must have been higher than that; but the special association of cocaine with blacks, unlike the association of opium with Orientals, was probably baseless. (Later it became more plausible.) Prohibitionists argued that keeping whiskey out of the hands of blacks would be beneficial for the white population, but that hardly meant the movement to abolish the use of alcohol was *inspired* by racism. The case of cocaine was similar. With cocaine, as with alcohol and opium, the dangerous classes also included white women, who were said to be subject to temptation, seduction, and corruption. In actual fact, the best-documented (if not the most severe) cases of cocaine abuse were white professional men, especially physicians.

It was some time before the patchwork of laws instituted to restrict the sale of cocaine actually reduced its availability much. In the United States, where cocaine sniffing was apparently more common than anywhere else, Oregon passed the first restrictive law in 1887. Before World War I most states made cocaine a prescription medicine and required that records of the prescriptions be retained for inspection. By 1912, 14 states had ordered "drug education" in the schools to warn about cocaine and opiates. Cocaine was actually considered more dangerous than opiates. The Proprietary Association of America, bidding for respectability and trying to fight disclosure laws, refused to permit the company that manufactured Dr. Tucker's Asthma Specific, a cocaine nostrum, to join. By 1914, 46 states had laws controlling the sale of cocaine and only 29 had laws controlling the sale of opiates; often the penalties for cocaine were harsher. In 1907 New York, under the leadership of Assemblyman Al Smith, passed a harsh cocaine law that made it almost impossible for physicians or patent medicine manufacturers to dispense the drug legally. This law expressed the attitude of total condemnation that was about to become dominant.[65]

The federal government made its first move with the Pure Food and Drug Act of 1906, which forbade interstate shipment of food and soda water containing cocaine or opium and required that medicines containing these drugs be properly labeled. It also put the first restrictions on imports of coca leaves. But the sanctions were not severe, and the loophole for patent medicines was large. In 1914 Congress passed the Harrison Narcotic Act, and it became the cornerstone of federal policy on cocaine for 50 years. It stipulated that anyone producing or distributing opiates or cocaine must register with the federal government and keep records of all transactions. Anyone handling the drugs was required to pay a special tax; thus the law could be enforced by a federal agency, the Bureau of Internal Revenue. Possession of opiates or cocaine by an unregistered person was not in itself a crime but was evidence of violation of the regulatory and tax provisions of the law. Unregistered persons could buy the drugs only on prescription from a physician for legitimate medical use. (This provision became important in connection with the opiates when later Supreme Court decisions declared "maintaining" an addict to forestall withdrawal symptoms not a legitimate medical purpose, but it never made much difference to official policy on cocaine.) The Harrison Act at first exempted some opiate preparations but none that contained cocaine. European countries, partly under prodding by the United States government, eventually imposed similar sanctions.

Ever since then, our legal machinery has been pressing down on cocaine and the opiates alike. If anything, cocaine has been more heavily penalized. It was included with opiates in the order that set up the maintenance clinics that operated in 1919–1923, but in practice the clinics rarely dispensed cocaine because people who used only cocaine did not feel the physical need that drove opiate addicts to them. In New York, for example, the clinic stopped dispensing it after the first day. An exception was Albany, where 113 of 120 addicts in May 1920 were being given two grains (130 mg) of cocaine a day as well as an average of seven and a half grains of morphine.[66] In 1922 Congress prohibited most importation of cocaine and coca leaves and officially defined cocaine as a narcotic for the first time.* By 1931 every state had placed severe restrictions on the sale

---

* Federal law still classifies cocaine as a narcotic. Etymologically, this seems wrong, since the term is derived from the Greek word for "benumb" or "stupefy" and implies sedation or analgesia rather than stimulation. But for a long time it has been applied loosely to a great variety of drugs, possibly because numbing is thought of as something like "enabling one to forget one's troubles" and possibly because the most venerable psychoactive drugs—alcohol, opium, and cannabis—in fact often have a sedative effect.

COCAINE

of cocaine and 36 states made unauthorized possession a crime. The Uniform Narcotic Drug Act, proposed in 1932, was eventually adopted by all states except California and Pennsylvania and dominated official policy on cocaine, along with the Harrison Act, until 1970. Besides the usual criminal penalties for possession it required licensing for the manufacture and distribution of cocaine and detailed records of sales and prescriptions.

The federal government, meanwhile, continued to pursue the policy of suppression it had instituted with the Harrison Act. There were no substantial changes in this law from 1922 to 1951; then an amendment made prison sentences for possession mandatory and imposed the same penalty for failure to register as for importation of large amounts. In 1956 penalties were increased again. The Narcotics Manufacturing Act of 1960 required manufacturers of cocaine to register with the secretary of the treasury, who was empowered to license them and to set quotas on production. Finally, in 1970 the old federal drug laws were replaced by the Comprehensive Drug Abuse Prevention and Control Act, which in effect reenacted the existing control schedules under a new terminology. Coca and cocaine are classified, along with a number of opiates, barbiturates, and amphetamines, as Schedule II: high abuse potential with restricted medical use. Since cocaine is treated as a narcotic, the penalties are the same as those for morphine and other medical opiates and higher than those for non-opiates in the same Schedule: for illegal manufacture, distribution, or possession with intent to sell, up to 15 years in prison and a fine of $25,000; for possession with intent to use, up to one year and

The title of von Bibra's book, *Die Narkotischen Genussmittel und der Mensch*, published in 1855, may be translated as *The Narcotic Drugs and Man*, although "Genussmittel" literally means "means of enjoyment" or "luxury." Among the drugs it discusses are the stimulants tobacco, coffee, tea, coca, and mescaline. We have already mentioned the reference of a British medical journal in the 1870s to coca as "a new narcotic and a new stimulant." Freud, more etymologically scrupulous, contrasts stimulants like cocaine with narcotics, but he classes cannabis with opium as a narcotic. We would not do this today, although it is far more plausible than using the term for cocaine, which numbs only the tongue and palate. It would still be more plausible to call alcohol, barbiturates, and tranquilizers narcotics, although, unlike opiates, they are less effective against pain than against anxiety. No one does so, of course, since the term *narcotic* no longer has the vague but morally neutral connotations it had in the nineteenth century. It is used in public discourse today mainly to denounce a drug as a menace to society that must be suppressed. (Something similar has happened to the word *addiction*, which originally meant simply a habitual inclination.)

Jerome H. Jaffe, in his chapter on "Narcotic Analgesics" in Louis Goodman and Alfred Gilman's *The Pharmacological Basis of Therapeutics*, 4th ed. (New York: Macmillan, 1970), explicitly defines the term as interchangeable with "opioids" (p. 237). This clarifies matters, but it might be still better to drop the term "narcotic" entirely and simply speak of opiates, barbiturates, alcohol, or whatever depressant drug one has in mind. In any case, applying it to cocaine only promotes the confusion between cocaine and the opiates and an irrational, excessive social prejudice against both drugs. Another prejudicial term to be avoided, incidentally, is "hard drug," which has no meaning in medicine. This phrase sometimes seems to mean "anything illegal except cannabis" and takes no account of the very "hard" time many people have with legal drugs.

$1,000. There is no mandatory minimum sentence for possession. A Uniform Controlled Substances Act, designed to complement the Comprehensive Drug Abuse Prevention and Control Act, has been adopted by many states, with some variations in penalties; for example, simple possession of cocaine is only a misdemeanor in Idaho, but in Missouri sale to a minor may carry the death penalty. In 1973 New York State passed one of the harshest laws, making a life sentence mandatory for possession of more than two ounces of cocaine, with parole provisions reducing potential time served to not less than 15 years.[67]

International developments have been parallel. A series of International Conferences on Opium took place, largely at the urging of the United States government, at Shanghai and The Hague in the years before passage of the Harrison Act. In 1914, 44 nations attending the Third Hague Conference signed The Hague Opium Convention providing for strict restraints on the production, manufacture, and distribution of opiates and cocaine. The United States and a few other countries ratified this convention in 1915, and the rest of the world ratified it in 1919, when it became part of the Versailles Treaty. Since then the international legal machinery has been in principle as dedicated to the suppression of cocaine as the United States government. A series of treaties culminating in the United Nations Single Convention on Narcotic Drugs of 1961 has amplified and elaborated the control system. The Single Convention regulates the opium poppy, the coca bush, the cannabis plant, and their derivatives, replacing all predecessor treaties. It authorizes use of these drugs only for medical and scientific purposes and permits their cultivation, manufacture, possession, sale, distribution, import, and export only by government agencies or under government licenses, with strict supervision and record-keeping requirements. Although the Peruvian and Bolivian governments have approved this treaty and created formal monopolies to regulate the coca trade, it has been nearly impossible to enforce the restrictions. A 1972 amendment, not yet in effect, strengthens the International Narcotics Control Board and increases its authority to insure compliance with the treaty by national governments.

Between 1930 and the late 1960s the use of cocaine and medical and general interest in the drug seem to have declined greatly. Gerald T. McLaughlin writes, "Before 1930, cocaine, rather than heroin or opium, was viewed as the primary drug menace in the United States." A French author estimates that in 1924 there were 80,000 victims of *cocaïnomanie*—a term that suggests chronic cocaine intoxication but may mean any use of cocaine at all—in Paris alone, and only 20,000 drug abusers (*drogués*) of all kinds in all of France in 1971. (Obviously he excludes alcohol.) A 1926 article in the *New York Times* entitled "Cocaine Used Most by Drug Addicts" announced that 60 percent of the criminals in a Welfare Island workhouse were "drug addicts," most of them taking cocaine.[68] Most of the books that deal with cocaine as a social issue or a clinical problem were published in the 1920s; most of the novels, stories, and memoirs that show familiarity with its recreational use are from the same period. Even in the first decade of the new drug culture's flowering, cocaine remained relatively unpopular—expensive, rarely available, bracketed with heroin as a drug to be avoided. But around 1970 a revival of interest and respect began, comparable to the events of those decades at the end of the last century when cocaine was first introduced to the West.

3

COCAINE IN
AMERICA TODAY

Preliminaries

Before discussing the present and future position of cocaine in our society, we must offer some apologies. It is a subject very difficult to write about without guesswork, distortion, and false emphasis. Use of cocaine, like use of most other drugs, covers a good deal of social territory; what could be stated accurately for one corner of it would be misleading about the rest. Because cocaine is illegal, everything about its social status remains half-concealed; because the large-scale interest in it is so recent, there has been little time for a new descriptive literature to accumulate. It is impossible to find out the true extent of the illicit traffic; no substantial public surveys on cocaine use and users are available; even arrest statistics are unreliable, because the Federal Bureau of Investigation continues to lump cocaine with opiates (as a "narcotic") in its Uniform Crime Reports. Most important of all, the situation is changing fast; no stable pattern has yet emerged either in the use of the drug or in attitudes toward it. What we write today may seem naive or outdated in a year or two. In this kind of situation two poses are common: the spurious knowingness of the insider who imagines his own experience to be more extensive or more representative than it is; and the journalistic sensationalism of the outsider cashing in on the shock value of illicit drug use.

These attitudes reinforce each other, and we would like to avoid both. We describe the transient and obscure surface features of the present social situation not so much for their own sake as to place our later description of the effects of cocaine in a cultural context.

One topic usually considered to be an integral part of the sociology of contemporary drug use is clearly of more permanent interest than the latest fashion or scandal. This is the problem defined as "the motivation of the user." Unfortunately, in studies of illicit drug use the answer to an apparently sensible question is often pursued in a confusing and misleading way. In this case, it can degenerate into attempts to find ways to impute psychological abnormalities or moral deficiencies to those who use a socially disapproved drug. Consider two relatively harmless substances, marihuana and coffee. Only corporate marketing research divisions have shown much interest in why people drink coffee; but sociologists, psychiatrists, and—more ominously—police agencies are interested in the motivation of the user of marihuana. It is false to imply that something must be wrong with anyone who wants to ingest a chemical that has been declared legally out of bounds.

The question of why people want to use a drug, formulated in a general way, is both too simple and too complex to be very useful. The simple answer is to describe the effects and point out that they are desirable. So the mildly derisive "explanation" some marihuana smokers give to sociologists who want to study their motivations is that the stuff makes them feel good, or some elaboration of that idea. On the other hand, what the researcher means when he speaks of motivation may have nothing to do with what people think or say they want. He may be concerned with a causal chain based on some idiosyncratic concatenation of individual psychology, cultural norms (including the way in which the use of the drug is defined in a particular society—its folklore and mystique), and the availability of the drug. In that case the answer to the question of motivation becomes part of the biography of the user and the history of the culture. In this sense, as opposed to the derisively simple sense we mentioned before, there is no single motivation or easily classifiable set of motivations for sniffing cocaine or smoking marihuana any more than for drinking alcohol or coffee.

In other words, motivation tends to be either obvious or irrelevant, and the term is too often used for the dubious purpose of setting apart a peculiar class of drug users from respectable society as outcasts and scapegoats. The search for motivation is properly redirected or transformed into three other projects: description of the drug's psychological and physiological effects; analysis of the difficult problem of so-called dependence, habituation, and addiction; and observation of the conditions and circumstances in which the drug is available at a given time in a given society. We began this last project in the first two chapters and continue it in this one.

Background to the Contemporary Scene

Between 1930 and the late 1960s, recreational use of cocaine (except in South America) was largely out of sight and out of mind. There are several plausible reasons for the decline of interest. New restrictions on importation, manufacture, and distribution were introduced; substitutes were found for many of the surgical and prescription uses of cocaine; amphetamines appeared on the market in 1932 and provided a stimulant that was cheaper, more accessible, and longer-lasting, if less attractive to connoisseurs; the Depression made luxuries like cocaine less available. Whether as a cause or a consequence of the decline of interest, the sources of supply seem to have dried up. From about 1910 to 1930, a large proportion of the coca used to manufacture cocaine for medicinal purposes was cultivated in Java. The amount grown there reached a high of 1,676,000 kg in 1920 and began to fall, slowly through the 1920s and then precipitously; by 1938 Java was producing only 41,000 kg of coca, and after World War II cultivation was not resumed.[1] (Coca production in Peru fell from 1930 to 1935, but then began to rise again, until by 1938 it had surpassed the 1929 high of 5,500,000 kg; it has continued to rise until the present.)[2] It appears that until 1930 most of the cocaine in recreational use was diverted or supplied illegally and quasi-legally from pharmaceutical houses, drugstores, hospitals, and doctors' offices. When it was no longer manufactured in large quantities as a medicine, lovers of the drug had nowhere to go until the demand became great enough to generate a new refining and marketing network entirely outside of medical channels.

The background of this demand and the social situation it has created is a combination of old elites and the new drug culture. As a prescription drug, cocaine was most used by those who had access to the source and could pay the fees—physicians, lawyers, writers, preachers, actors, musicians, and other professional people, as well as the idle rich. When it was removed from soda drinks and patent medicines and legally restricted in other ways, the price rose and its use became even more confined to the rich or well connected. To the extent that it was available at all, cocaine became a plaything of the more adventurous and less respectable among the wealthy. Marcel Proust in *A la recherche du temps perdu* refers to its use in Parisian high society before 1914 and in a homosexual brothel during World War I. (He may have used it himself to relieve his asthma.) The drug was a familiar part of Cole Porter's café-society world of the 1930s, when he is said to have written a version of "I Get a Kick Out of You," his famous song of put-on weary decadence (for a musical comedy set in the 1920s), in which he included the line "Some get their kicks from cocaine." In Hollywood and around Broadway cocaine remained available if not abundant. Richard Ashley estimates, on the basis of correspondence with old-timers, that it was used in these circles in the 1920s about as much as marihuana among the middle class in general in 1970—it was not so familiar as to be taken for granted.[3] Cocaine plays a role, usually humorous, in a few early Hollywood films. In a 1916 parody of the Sherlock Holmes stories, starring Douglas Fairbanks, a manic detective named Coke Ennyday who is continually sniffing an unnamed white powder foils a ring of opium smugglers. Charlie Chaplin's *Modern Times* (1936) includes a scene in a jail cell where the tramp sniffs something identified as "nose powder" and becomes a comic superman. A low-budget 1939 film called *The Cocaine Fiends,* ostensibly a dire warning against the drug menace in the melodramatic vein of the better-known *Reefer Madness,* is now believed by some cognoscenti to have been a deliberate parody or joke; it has been revived recently for the sophisticated pleasure of the new cocaine users.

When cocaine went underground its status changed among the poor as well as the rich. It was even less available to the solid working-class citizen than to the solid middle-class citizen; but a few socially marginal members of the lower class, like the socially marginal upper-class groups we have just discussed, found ways of obtaining it. This is always the way with illicit drugs, and it is often misinterpreted as evidence that they are a cause of crime. In fact, even when cocaine was relatively freely available as a prescription drug, the poor could not use it very much if they were strictly law-abiding. When it became almost impossible to obtain legally, its associations became criminal; those who had least general respect for the law and conventional society were most likely to be persuaded of its virtues and to know where it could be found. Evidence that criminals use more of cocaine and other illicit drugs than the rest of the population must always be considered in this light.

The worlds of the unconventional rich and the unconventional poor who used cocaine were of course connected—not only by the quasi-conspiratorial nature of the drug traffic, but also by more general social associations. Gamblers, prostitutes, actors, musicians, journalists, and artists might belong to either world and pass freely between them. In the United States, blacks in particular may have been an important part of this nexus. They were in a permanent condition of social marginality by virtue of skin color and for obvious reasons had little respect for white-imposed laws. If recent surveys are accurate, blacks still use cocaine proportionately more often than whites [4] (although the cocaine they use may be weaker); the situation has probably been that way since 1930 and possibly since the Harrison Act, although evidence is even scarcer here than in other areas of the history of cocaine. In any case, black jazz musicians and other entertainers began to exert cultural influence in the world of white popular music and show business in general during the 1920s; the interest extended to the drugs they used, whether marihuana, heroin, or cocaine. During the period from 1930 to the late 1960s, all these cultural associations of cocaine—with the rich, with show business, with blacks—remained in memory and helped to shape the new cocaine market created by an unprecedented set of social circumstances.

These circumstances involved a profound alteration in the public attitude toward illicit drugs, at least among a large and influential minority. To describe fully how this came about would be impossible without analyzing the general upheaval that began in American society in the early 1960s. We can only gesture toward the subject by referring to the loss of respect for established institutions and moral attitudes among important parts of the middle class, the bohemianization of middle-class youth (the hippie rebellion and its camp followers), the introduction of psychedelics and the spread of marihuana use, with attendant drug ideologies and religions, the growing respect for black culture, and in general the process by which the attitudes and practices of avant-garde and fringe groups were taken over by a large section of the public and became familiar to an even larger section. It was inevitable that in the search for new pleasure drugs that ensued, cocaine would be rediscovered and rise to high social visibility. In a sense, all the old social scenes where cocaine had been used were merged on an enlarged public stage. Only a few other developments were necessary before it became fashionable again. Disillusionment with amphetamines, combined with legal restrictions that made them less freely available, was probably important; and a smuggling and marketing network had to be created with the help of Cuban exiles and other Latin Americans who had connections in Peru and Bolivia.

The Cocaine Traffic Today

Although coca was once cultivated in many parts of the tropics, today all the legal and illegal cocaine that reaches North America and Europe comes from its homeland in South America. Illegal cocaine is also refined almost exclusively in Latin America. The only firm licensed to import coca leaves into the United States is a chemical corporation that removes the alkaloids and sells them to pharmaceutical houses, then passes on the residue to the Coca-Cola Company for use as a flavoring agent. In response to our written request for information, this firm, the Stepan Chemical Company of Maywood, New Jersey, stated in a letter: "It is the policy of our firm not to reveal any information concerning our procurement, manufacturing, and selling operations for any of the products that we produce." According to the International Narcotics Control Board, world legal manufacture of cocaine amounted to 889 kg in 1973, all of it in the United States; legal consumption (as a local anesthetic) in the United States was 279 kg.[5] This is only a small fraction of the amount consumed illegally, and could not possibly be a significant source of illicit traffic.

The cocaine used for recreation in North America and Europe is grown in Peru, Bolivia, and Columbia and refined in clandestine laboratories in those countries and in Chile, Ecuador, and Argentina. (This may be changing. An illicit processing laboratory in California was seized by the federal government's Drug Enforcement Agency in 1975.)[6] A 1975 article in the *New York Times* tells of an Aymara Indian in a coca-growing region of Bolivia who owns a plot that produces 300 pounds of leaves a year, worth about $250. This will yield about 1 kg of cocaine, sold for $75,000 at retail prices in New York. First coca paste or crude cocaine is extracted; this may be transported to Chile or Colombia for further refining and conversion to cocaine hydrochloride. The cocaine hydrochloride crystal or powder is then smuggled by way of Central America to California or directly into San Antonio, Miami, or New York, where it is sold especially in Latin bars and restaurants. Large-scale smugglers often use bulk freight like coffee. Smaller independent operators may use private boats and planes or employ individual carriers called "mules" or "camels" who use such ingenious methods as soaking their clothes in a cocaine solution or swallowing condoms that contain the powder. (Occasional stories appear in the newspapers about smugglers who die when the condoms burst in their stomachs.) In the United States cocaine may be sold to middlemen who supervise cutting (dilution) and packaging in secret "coke mills" (where workers can earn as much as $300 a day) and then distribute the drug for retail sale. At retail, it is often sold by the "spoon," a variable amount of powder averaging about one-quarter of a measuring teaspoon, or half a gram; the cocaine content, of course, may be much lower.[7]

The dimensions of this trade are naturally obscure, but it is clear that the quantity of illicit cocaine on the market has been increasing rapidly since about 1970. In 1970 the federal government seized 305 pounds; in 1971 it seized 787 pounds. Seizures by the Berkeley, California, police department went up one-third in the same period. In New York City there were 874 arrests for sale or possession of cocaine in 1970 and 1,100 arrests in 1971; the amount seized by police had already gone up five times from 1969 to 1970.[8] Bureau of Customs seizures rose from 11 pounds in 1960 to 199 pounds in 1969 and 619 pounds in 1972. The amount of cocaine taken by federal agents first exceeded the amount of heroin in 1970, and in 1974 a Bronx prosecuting attorney declared that it had supplanted heroin as the main "drug of abuse" in New York City. The increase in total police seizures from 1969 to 1974 was sevenfold, and the Drug Enforcement Administration is now devoting half its efforts to cocaine.[9] It goes without saying that police seizures are only a tiny fraction of the amount actually on the market.

Prices in the illicit drug trade are subject to wild disparities and irregularities, but there is no doubt about the great financial cost of using cocaine and the enormous profits to be made from dealing in it. A man we interviewed who was familiar with the local market in the United States spoke of "people who like being involved with big cash" and "the hook of making that easy money," which he regarded as more dangerous than the hook of the drug itself. In late 1973 *Newsweek* quoted a Drug Enforcement Administration official as saying, "In the last three years, the coke traffic has gone right through the roof. Right now, *anybody* can go down there, turn a kilo for $4,000, and sell it back here for $20,000." One Mexican citizen arrested in the United States in 1972 for selling cocaine posted a cash bond of $250,000 and then disappeared.[10] This kind of money entices a growing number of people into the business. An article in the *Journal of Psychedelic Drugs* describes the advantages of cocaine dealing from the point of view of a former heroin addict and heroin dealer who had stopped using and selling opiates with the help of a methadone maintenance program. At the time of the interview he was using three spoons (a gram or two) of cocaine a day and claimed a net income of $3,000 to $4,000 a week:

I deal coke because you associate with a better class of people. I sell to executives, teachers, businessmen, pimps. I don't have to fuck around with low-life junkies, always sniffin' and scratchin' and always a dollar short beggin' to get a ping off your cotton. Shit, who ever heard of a skag [heroin] dealer workin' the X [a luxury apartment house]. Some of my best customers live in there. . . . I've got what I always wanted, I got my Mark IV, I've got my pick of any bitch on the strip and I've got a piece of a legitimate business in "A" [a resort city]. I'm OK now.[11]

Figures from various sources will give some idea of the current prices and profits. The cocaine dealer "Jimmy," portrayed in Richard Woodley's study *Dealer: Portrait of a Cocaine Merchant,* bought 425 grams of pure cocaine in New York for $4,400 in 1970. By 1972, according to a *New York Times* article, a "key" (kilogram) of cocaine worth $1,000 in Arica, Chile, might sell for $23,000 in New York. In reporting the capture of a ring of smugglers in late 1974, the *New York Times* gave the estimated wholesale price in New York as $10,000 a pound, or $24,000 a kilogram. Couriers were said to get $1,500 a kilogram. Recently a New York policeman was indicted for pocketing $15,000 in "buy money" that was supposed to be exchanged for 500 grams of cocaine.[12]

Retail prices are even more outrageously inflated. The heavily cut cocaine sold by "Jimmy" in 1970 was priced at $150 a tablespoon (not to be confused with a "spoon"). A "two-and-two" (two "lines" or snorts for each nostril) was worth $10, and it would cost "Jimmy" about $50 to "coke up" for an evening's pleasure. In 1972 cut cocaine was said to be selling for $500 to $1,500 an ounce, or $50 a gram. The PharmChem Research Foundation reported that the street price of samples assayed in late 1975 averaged in the range of $1,000 to $1,800 an ounce, or $45 to $85 a gram. The prices were not necessarily correlated with the degree of purity, which varied greatly. If price is measured by the cost in dollars per minute spent under the drug's influence, cocaine is probably ten times as expensive as heroin. It is cheaper in South American cities, and North Americans sometimes go there simply to get it. Several of the people we interviewed had made a trip to Colombia mainly for this purpose; one woman bought an ounce of pure cocaine there in 1973 for $200. Inflation and rising demand may shoot all these prices even higher; the opening of new smuggling networks through Mexico would tend to lower them. The fact that hospitals and pharmacies were buying the drug legally as an anesthetic for $31.50 an ounce (about a dollar a gram) in 1975 should give some idea of the risks and profits in the illicit trade.[13]

The large-scale traffic is controlled by Latin Americans, especially Colombians and Cubans, and to a lesser extent North American blacks;

white North American small-time independent dealers who consider themselves part of the counterculture are apparently being put out of business. But people in many different social circumstances are involved at one level or another. One recent arrest broke up a ring transporting the drug from Bolivia to New York that included a New York physician and the former president of a pharmaceuticals company in New Jersey. Columbia University undergraduates have been arrested for selling it, and professional football players are suspected of being involved in smuggling. A former chief of the narcotics police in Chile (1966 to 1969) was extradited to New York in 1974 for his alleged part in a cocaine smuggling operation that included four other former high Chilean drug officials who are now fugitives. Military aircraft and diplomatic pouches, which are not subject to regular customs inspection, are said to be common means of transportation. This is another indication that many members of South American governing elites are involved in the trade. Meanwhile, in August 1974, 435 North American couriers—many of them students or other tourists persuaded to try to earn a share of that easy money—were in Mexican prisons, some for marihuana or hashish smuggling, but many on cocaine charges.[14]

A particularly interesting and important element is the Cuban gangs, reported to be running most of the Miami and New York operations, which have been smuggling cocaine since the days when it was a small business instead of a big one. After Castro's revolution, exiles resumed operations in Florida. They include former Cuban government officials; one was arrested on cocaine charges in July 1973. Some of the desperadoes once trained by the CIA for what proved to be abortive invasions of Cuba are also apparently involved. A Drug Enforcement Administration official has been quoted as saying, "Cuban brigades seeking political asylum in Miami . . . most are dope peddlers. Not some, not a few. But most." [15] "Dope" here refers mainly to cocaine; the Cubans have not yet entered the heroin business to any great extent. Miami is probably the biggest center of cocaine traffic, and many small businesses in southern Florida are said to be financed by cocaine money.

The United States has been increasing its drug police force in South America, but these agents obtain insufficient cooperation from the governments and people of the countries they are assigned to. Over half of the 200 major Colombian cocaine and marihuana traffickers have been indicted in the United States, but under present treaties they cannot be extradited and they are not prosecuted at home. The $22,000,000 a year provided in grants to Latin American countries to combat the drug trade in the last few years has had limited effect. From July 1973 to May 1975,

457 cocaine traffickers were arrested and 73 laboratories destroyed, but the trade keeps growing. An apparent exception to the rule of insufficient cooperation is Chile, where an ingenious Drug Enforcement Administration official persuaded the *junta* to extradite 19 important traffickers on the admittedly spurious ground that they threatened state security because radical groups might use money from cocaine to buy arms. Shipments from Chile are estimated to have dropped from 200 kg to 11 kg a month since the expulsion.[16] Some arrests by the United States and Chilean governments suggest that cocaine charges are being used to persecute political opponents of the military regime in Chile. For example, a professor at the University of Wisconsin was acquitted on charges of cocaine smuggling in a federal district court in Brooklyn in July 1975, after contending that he was being prosecuted because he had worked in an organization to help Chilean radicals after the coup d'état.[17] Cocaine dealers in general are anything but radical, and, like Castro, other Latin American radicals are opposed to the traffic.

A United States diplomat is quoted in a 1975 *New York Times* article as saying, "These countries don't have a drug problem themselves. There's no mutual interest to work with."[18] What this means is not that South Americans do not use cocaine but that they do not regard it (or cannabis) with the horror that North American drug enforcement officials consider appropriate. It is hard for them to take the menace of cocaine seriously while the coca leaf serves as the ordinary daily drug of millions in Peru and Bolivia—even if they are poor and often despised Indians. (It is the same with opiates in Southeast Asia.) Cocaine itself has always been relatively easy to buy, too. A 1949 article in *Time* tells how it is sold openly, in spite of the laws, at market stalls in Callao, Peru. It quotes a dealer: "If you're poor, you're hungry. *Pichicato* [cocaine] fixes that. If you're rich, you want an aphrodisiac. *Pichicato* fixes that too. It's a sure cure for everything."[19] Although the drug may no longer appear quite so openly in marketplaces, reports we have heard from travelers in South America suggest that North Americans with a little money or the right friends can obtain it easily. As for the Cubans, an exile journalist in Miami told a *New York Times* reporter: "Some Cubans think that the cocaine is like the violation of the tax. But the heroin is a vice."[20] Most of the people who deal in cocaine, unlike big heroin operators and like illicit alcohol refiners, use the drug they sell. The game of evading the cocaine laws is like the game of evading income taxes commonly played in many countries, or like the methods once adopted in the United States in the face of alcohol prohibition. Cocaine traffickers, as we have seen, have government connections in Latin America, and police are easily bribed. In the circumstances, it is unlikely that narcotics officials can keep the traffic from increasing.

Cutting

A problem (and an argument for legalization) that always arises with illegal commodities is adulteration. In the 1920s no one could be sure what was in a bottle labeled whiskey; today, heroin addicts do not know the proportion of quinine or lactose (milk sugar) in their drug. Cocaine, too, is almost always cut or diluted with various inert substances or other drugs before it reaches the retail market. Many people are getting little more from their black-market cocaine than a numb palate and the social status of having sniffed it. Since dealers and police cooperate in a kind of de facto conspiracy to prevent knowledge of the kinds and quantity of adulterants from reaching the customers, it is especially hard to find information on the subject. Aspirin, boric acid, and bicarbonate of soda were mentioned as common adulterants in the 1920s. According to "Jimmy," the subject of Richard Woodley's study, cocaine could be distinguished from the synthetics benzocaine and procaine because it numbed the tongue but did not "freeze" it. He told Woodley, referring to the number of parts of sugar to each part of cocaine that the drug he bought would tolerate without losing its effect, "Best superfine dope will stand a twelve. My stuff can take a three." He actually claimed to prefer cut or "stepped on" to pure cocaine; some of our interview subjects agreed. Marc Olden estimates that the highest quality generally available on the street in New York City is 25 percent.[21]

The result is that in using cocaine people usually do not know what or how much they are taking into their bodies. "Jimmy," who buys and sells only cut cocaine himself, is quoted as saying, "Drugstore coke [i.e., pure] is so strong it'll wreck you." In Bruce Jay Friedman's novel, *About Harry Towns* (1974), the apparently autobiographical hero, if asked about the effects of cocaine, "would say it was subtle and leave it at that." But he and his friends know that less subtle stuff is available too: " . . . they would tell each other stories about coke they had either heard about or tried personally, coke that was like a blow on the head, coke that came untouched from the drug companies, coke so strong it was used in cataract eye operations." [22] One of our interview subjects told us, "Cuts can sometimes be total highs in themselves. One person I know sells a cut composed of procaine, lidocaine, caffeine, and something else. There's no coke in it. But you get a freeze from the procaine, a buzz from the lidocaine, and the caffeine keeps you going . . . that's one of the pitfalls of an illegal product." Although it is hard to tell, most cocaine users have the impression that quality has been declining as the popularity of the drug rises. As one man who had used cocaine long before it became fashionable told us in an interview, "The reason why I curtailed, cut down, was that it wasn't *coke* any more." The use of cuts may be increasing because organized crime is becoming more involved in the traffic. The problem of quality control that plagued Freud and his predecessors has returned for entirely different reasons.

But too much of the information about cuts is only rumor, some of it put about by narcotics officials with an interest in frightening or discouraging the public. That is why the work of the PharmChem Research Foundation, a chemical laboratory in Palo Alto, California, that analyzes samples of street drugs, is so interesting. An administrative regulation promulgated by the Drug Enforcement Administration in mid-1974 has prevented this group from publishing quantitative data on individual samples of illicit drugs contributed by anonymous donors. The ostensible purpose is to keep dealers from using it to check on quality. But the effect is to keep everyone in the dark and make using the drugs as dangerous as possible, while dealers continue to sell whatever they have on hand. The samples analyzed by PharmChem are not necessarily typical, and they come almost exclusively from the West Coast; but they provide one of the few sources of reliable information on this important subject.

In an issue entitled "Heroin and Cocaine Adulteration," based on analyses made in 1973, *The PharmChem Newsletter* reported that 73 percent of the 40 samples of alleged cocaine contained cocaine as the only *drug*. Twenty-one percent also contained synthetic local anesthetics, and 6 percent had amphetamine, caffeine, phencyclidine, some other drug, or no drug at all. The newsletter points out that procaine, the most commonly found local anesthetic, is also the least toxic—about one-third as toxic as cocaine or lidocaine. Benzocaine is particularly dangerous if injected because it is relatively insoluble and may cause blood clots. Three percent of the samples analyzed contained it. The cocaine content of samples containing no other drug varied from 14 percent to 89 percent (pure cocaine hydrochloride), with an average of about 60 percent. Price was unrelated to purity. The most common inert adulterants were mannitol (an alcohol derived by reduction from fructose) and lactose. Later lists compiled under the new regulations can give no percentages for individual samples, but they reveal that glucose, sodium bicarbonate, and especially inositol (a member of the vitamin B-2 complex) are also being used as adulterants. In a 1974 issue the editors indicate that the average proportion of cocaine alkaloid in the samples analyzed was 65 percent when the sample contained no other drug (as was usually the case) and 39 percent when it contained other drugs. Inositol, mannitol, and lactose were the most common cuts. Nothing as dangerous as quinine, strychnine, or asbestos was found, and there were no amphetamines, either. The percentage of pure cocaine in these samples actually seems high. Maybe people who know enough to send what they have bought to the Pharm-Chem Research Foundation also have access to particularly reliable sources of drugs; maybe, as the Drug Enforcement Administration fears, some of them are dealers who have nearly pure supplies that they intend to adulterate further and sell. According to Richard Ashley, however, West Coast cocaine is purer than the drug found in the East because the local traffic has not yet been monopolized by Cuban exiles.[23]

Several methods of assaying a substance for cocaine are used in the illicit traffic. Cobalt thiocyanate is sold as part of a kit for testing the quality of illicit cocaine. It produces a blue precipitate that dissolves in hydrochloric acid; the deeper the blue, the purer the cocaine.* Another test is to drop the powder into a dark-glass jar containing sodium hypochlorite (Clorox). If it is cocaine, a white halo will appear as it drops to the bottom. If any red appears, it has been cut with a synthetic local anesthetic. Pure cocaine hydrochloride (the form in which cocaine is usually used) dissolves instantly in water and readily in methyl alcohol; any powder that does not dissolve is probably a sugar. Burned on aluminum foil, nearly pure cocaine hydrochloride produces a gray or red-brown stain; a black residue means sugar, popping sounds mean amphetamine, and sizzling means procaine; salts do not burn, but remain as a residue. Most cuts, and especially the most common ones—sugars—dull the sparkle of the crystalline powder. Sugar makes the taste sweeter, salt makes it more bitter, and synthetic local anesthetics have a stronger numbing effect.

* Drug-enforcement agents use a version of the cobalt thiocyanate method as a field test for cocaine hydrochloride. They add a 2 percent solution of cobalt thiocyanate in water, diluted in the field with glycerine, to the suspect substance. If it contains cocaine, a blue precipitate will form; concentrated hydrochloric acid will redissolve the precipitate; and chloroform will cause the blue color to reappear in the lower layer. This assay is said to distinguish cocaine from any other substance except the rare coca alkaloid tropacocaine. See "Cocaine Identification," *Drug Enforcement* 1 (Spring 1974):26–27. A recent article points out that the test is adequate for pure drugs but does not distinguish cocaine from certain mixtures of lidocaine (a synthetic local anesthetic) with other drugs; it recommends supplementary confirmation by thin-layer chromatography that the drug is an alkaloid and not a mixture of synthetic substances. See Charles I. Winek and Timothy Eastly, "Cocaine Identification," *Clinical Toxicology* 8 (1975): 205–210. For a test for cocaine in urine, see F. Fish and W. D. C. Wilson, "Gas Chromatographic Determination of Morphine and Cocaine in Urine," *Journal of Chromatography* 40 (1969): 164–168.

Methamphetamine burns the nasal passage when it is sniffed, and Epsom salts may cause diarrhea.[24]

Cocaine Use and Users

On the question of who is buying and using whatever passes for cocaine on the retail market, there are plenty of anecdotes and rumors and few quantifiable data. We begin with statistical samples, accepting their possible bias or unreliability. In 1973 the National Commission on Marihuana and Drug Abuse estimated that 4,800,000 people in the United States had used cocaine at least once, about 3.2 percent of all adults and 1.5 percent of adolescents; this was slightly higher than the number using heroin. Eighty percent of the people surveyed knew no one who had used cocaine. A 1972 survey suggested that 10.2 percent of all college students had tried it. A questionnaire distributed in 1973 to 1,629 students on the Chicago Circle and Urbana-Champaign campuses of the University of Illinois revealed that 4.6 percent of them had used cocaine, 0.8 in the preceding month. Of the main psychoactive drugs, only heroin was used less. Proportionately more blacks (8.5 percent) than whites (4.4 percent), more liberal arts majors than education, engineering, and business majors, and slightly more higher- and middle-income than lower-income students had used it. There was no significant correlation between the use of cocaine (or any other drug) and grades. A survey of over 50,000 high school students in 1971 indicated that 5 percent of them had used cocaine at least once, and 2 percent in the preceding week. About 0.6 percent had used it ten or more times. The corresponding figure for amphetamines was 3 percent, for marihuana 8 percent, for tobacco 23 percent, and for alcohol 43 percent.[25] Even so, the numbers seem too high for 1971. Some of the students must have been boasting, and others must have confused cocaine with cocoa, Coca-Cola, or other drugs.

A recent analysis by the Drug Enforcement Administration based on reports from hospital emergency rooms, inpatient units, county medical examiners and coroners, and drug crisis intervention centers provides further evidence that cocaine is not yet commonly used. The analysis covers about two dozen cities for the period July 1973 to March 1974; the data concern only drug abuse episodes that caused the person involved or someone in contact with him to seek help at one of the listed places. Cocaine alone or in combination with other drugs was mentioned 1,764 times, 1.5 percent of the total; about one-third of the cocaine episodes involved blacks, and 70 percent involved males. According to the study, perhaps half of them were caused by cocaine alone. The distribution between crisis centers (69 percent) and emergency rooms (25 percent) resembles the pattern for amphetamine (60 percent and 34 percent), LSD (69 percent and 29 percent), and marihuana (83 percent and 14 percent), as opposed to heroin (49 percent and 42 percent), phenobarbital (13 percent and 74 percent), or Valium (10 percent and 85 percent). Cocaine is twentieth on the list of drugs causing crises reported in this way. It might be suggested that the study indicates only that cocaine is relatively innocuous, not that it is rarely used. This interpretation cannot be discounted, but in any case it would not exclude the one we have made. In fact, it looks as though for most drugs the number of reported crises is related to frequency of use. Drugs that are relatively safe but commonly used appear high on the list: Valium (diazepam) is first, marihuana fifth, and aspirin seventh. Nothing is said about the *kinds* of episodes involved, but it is reasonable to assume that most of the crisis intervention center cases (for all drugs) were anxiety or panic reactions that required mainly support and reassurance rather than medical attention. Half of all mentions of cocaine came from three cities: Miami, Los Angeles, and New York.[26]

A woman we interviewed wrote an account for us of how she became involved in using cocaine. It began in the spring of 1973, when she went to work for a radio station:

> I was aware that many of the people working for the same radio station were into coke, but I had to become a trusted person before I knew the extent of it. At first I was just offered coke at work, a kind of pick-me-up during the day. Or at a party someone might put some out. I considered it an act of generosity. There's no denying that whoever put out the coke became a bit more prestigious to those who were snorting it up. I found myself attracted to people who had coke. I can even say I formed friendships on that basis. There was something nice in sneaking into the bathroom or a closet at the station, snorting some coke with a friend, coming out high and together with our secret. After snorting coke I came back to my work with a freshness and vitality that I enjoyed. Considering the long hours I put into preparing news stories, writing and producing them in the studio, coke became an aid I counted on. Three months or so after I started using it, it became evident that I could not rely on being offered coke. I wanted to have it accessible to me. So I considered buying some. I also felt that it was my turn to lay out some coke to the friends who had given me some, and I wanted that element of distinction.
>
> It was not difficult finding someone to buy coke from since others who worked at the station had formed a kind of cooperative coke pool. One person bought quantity, a couple of ounces, and then sold it to us for the price he got it

for. He made a gram selling it and we were able to get good quality coke (75 percent or more) for what was then a reasonable price—$55 to $65 a gram.

The same woman told us, "Cocaine is something you expect to see at promotion parties; it has become almost a way of payment. . . . Musicians expect to get it before a concert. They come into town and the first thing their road managers do is try to locate a source of cocaine. . . . People stuck in a job that's forcing them to put out a lot of energy, cocaine is getting them through." Although there are no surveys that provide a reliable indication of who uses cocaine, clearly workers in high-pressure elite environments of this kind—the entertainment business, radio and television, advertising, and so on—are most heavily involved. It is unusual, for example, to leaf through a couple of issues of a so-called underground newspaper or a publication devoted to popular music without finding several casual references to cocaine. According to *Rolling Stone* magazine, "Already it is responsible for wasting a number of top musical names." One singer calls it a "temporary cure-all." Another refers to it as "vitamin C" and uses it in the same quantities that others use the real vitamin C. Rock songs make open references to it; the musicians and their entourages use it for rehearsals and performances as well as at parties. Several rock stars are supposed to have refused to attack cocaine on television because it would be hypocritical; they were more willing to warn against amphetamines and barbiturates. Occasionally, the situation is publicly acknowledged for one reason or another. For example, a member of the rock group Grateful Dead was arrested for possession of cocaine in 1973. And during one of the periodic scandals in the recording business it transpired that an assistant to the president of Columbia records was regularly buying $100 worth of cocaine for visiting musicians and writing it off as a business expense; he declared that the drug was part of the ordinary "lifestyle" of the pop music world. Nicholas von Hoffman, in an article entitled "The Cocaine Culture: New Wave for the Rich and Hip," reports that when one rock musician last went on tour, "one of the members of his astonishing, rocking, rolling, and rollicking entourage was paid $20,000 to do nothing but the holding. He was given extra money, maybe another $20,000, to score the cocaine, but his main job was to take the fall in case of a bust." According to von Hoffman, "There are people around here [Los Angeles] who'll tell you that Mr. Kung Fu, Bruce Lee, died of overcoking, that it is wrecking music, if not the industry: 'The drug culture has lost its taste buds. They assault you. Help! I've just been mugged by a hit record.' " [27]

Other places where cocaine dealing has come to the attention of the press through arrests or scandal are the Columbia and Indiana University campuses, a Via Veneto nightclub, and an apartment in a fashionable district of London. After a cocaine arrest at the Playboy mansion in Chicago, Hugh Hefner was investigated (and cleared) by the DEA on charges of distributing the drug. In 1974 the DEA broke up a distribution ring in Aspen, Colorado, in a project humorously designated "Operation Snowflake." The conjunction of high altitude, high life, and vigorous physical exercise occurring at fashionable ski resorts appears to make them an ideal geographical and cultural setting for cocaine.[28]

In the 1920s Joël and Fränkel described the group of cocaine users in each city as a kind of secret society with its own meeting places, jargon, poetry, and songs. However it was then in Europe, now in America the secret is open, and there is no cocaine set, apart from the various smart and jet sets who indulge in the drug. Most of the people whose lives revolve around cocaine are those who make their living from it. But the cocaine scene can sometimes be identified by physical as well as sociological location. Von Hoffman quotes a Los Angeles record producer: "There are certain clubs in this town where you feel out of place if you aren't wired to the teeth. Whole clubs are based on being wired. Even the waitresses are wired. That fast cocaine tempo. You feel you've got to eat, drink, and get out of the place in an hour." [29] In any case, cocaine is still largely restricted to groups that are privileged or socially marginal or both. Outside of these circles it is known more by reputation and occasional encounters than by intimate acquaintance.

Snob appeal is important in the spread of cocaine use. It may be inducing people to pay $100 a gram for a substance that is often largely sugar. The high cost of cocaine and its powerful reputation as an elite drug tend to persuade people that it is desirable. It is an ideal mode of conspicuous consumption. As von Hoffman puts it, "It has come home to you that you can buy anything and do anything, and you've got to do it. How do you solve your problem? You get into cocaine." As one man we interviewed said, "I felt I'd better enjoy the experience because it was so hard to get hold of." Marc Olden quotes a rock musician on his reasons for using the drug: "Getting high, getting laid, getting a lot of work done, being up and being 'on,' being strong, being creative and superaware, being hip and being 'in.' " [30] Being hip and being "in" are probably not the least of these. Buying cocaine is like buying expensive cosmetics: the outlay is partly for glamour and romance. Cocaine is called icing drug, stardust, gold dust, rich man's drug, pimp's drug, and by one wit "the thinking man's Dristan." To wear a T-shirt imprinted with the legend "Things Go Better with Coke" or "It's the Real Thing," or to carry a gold snuff spoon

around one's neck is (or was until recently) to advertise membership in
an elite club. "Cocaine is a luxury, it's not a lifestyle," a man we interviewed told us. But for anyone who can afford to buy it continually or
knows how to obtain it from people who can, this luxury, like others, may
easily become an essential part of a certain way of living. As cocaine invests this way of living with its psychopharmacological qualities, the way
of living in turn affects the perception of what cocaine is—its social definition.

Nothing provides better proof of cocaine's new social status than its
reappearance as a topic of public reference in the media of popular culture. Several national magazines, including *Playboy* and the *New York
Times Magazine*, have carried articles about it in the last few years. The
films *Easy Rider* and *Superfly*, in which the heroes were cocaine dealers,
have helped to make the drug's reputation among, respectively, white
cultural rebels and young blacks. (In *Easy Rider* the Mexican who sells
the drug to the heroes declares, "It is life.") Robert Woodley's book-length
portrayal of a Harlem cocaine dealer (*Dealer: Portrait of a Cocaine Merchant*) appeared in 1971; a novel that devotes much attention to its hero's
dealings with and ambivalent attitudes toward cocaine (*About Harry
Towns*, by Bruce Jay Friedman) was published in 1974. The black come-
dian Richard Pryor, formerly a heavy cocaine user, has a recorded comedy routine about the drug and its effects ("He told me, 'It's not habit
forming. I've been using it every day for 15 years and I don't have a
habit.' ") Cocaine has also been conspicuous as an inspiration for the
lyrics of popular songs. In the 1920s there was a "Ballad of Cocaine Lil"
and also a song about a man who shot his woman while high on cocaine.
Huddie Ledbetter ("Leadbelly") wrote a song recorded in 1933 called
"Take a Whiff on Me":

> *Whiffaree and whiffarye*
> *Gonna keep on whiffing till I die*
> *An' ho, ho, baby, take a whiff on me.*
>
> *Cocaine for horses an' not for men.*
> *Doctors say it kill you but they don't say when.*
> *An' ho, ho, baby, take a whiff on me.*

Dave Van Ronk's "Cocaine Blues" runs, in part, "Come here mama/
Come here quick/ This old cocaine/ 'bout to make me sick./ Cocaine,
cocaine/ running around my brain." Another balladeer, Hoyt Axton, sings
of a friend buying a "one-way ticket" to heaven on "an airline made of
snow" and ultimately "Flyin' low/ dyin' slow/ blinded by snow." The ideal
large-scale cocaine consumer, for financial, professional, and cultural

reasons, is a rock musician; the lyrics of rock songs have been attesting to familiarity with the drug since the mid-1960s. Richard Ashley cites references by Laura Nyro in 1967 (it's even better to get off the "poverty train" than get high on "sweet cocaine"), by the Rolling Stones in 1970 ("And there will always be a space in my parking lot/ when you need a little coke and sympathy," from "Let It Bleed"), the Grateful Dead, also in 1970 ("Driving that train/High on cocaine/ Casey Jones you'd better/ watch your speed," from "Casey Jones"), the Jefferson Airplane in 1971 ("Earth Mother your children are here/ high and feeling dandy/ Earth Mother your children are here/ ripped on coke and candy"), and the Stones again in 1971 ("Sweet cousin cocaine laid his cool, cool hands on my head," from "Sister Morphine").[31] A hit song recorded in 1975 by Ringo Star, the former Beatle, is less laudatory; its doggerel concerns a series of drugs, including cocaine, that the singer has given up because he is "tired of waking up on the floor."

Today cocaine—also known as coke, snow, flake, blow, leaf, medicine, candy, happy dust, freeze, C, Carrie, Dama Blanca, lady, and girl (heroin is "boy")—is usually taken through a process of vigorous sniffing called snorting or blowing. It is more powerful injected under the skin or intravenously, but this method seemed to be largely confined to heroin addicts; most people find it more difficult to overcome the barrier against using hypodermic needles than to overcome the barrier against using illicit drugs. Today cocaine is rarely eaten or drunk, as in the days of Mariani's wine, because for most users the mild stimulation does not justify the high price. Since it is absorbed through mucous membranes anywhere in the body, it can also be applied to the genitals—a method that may produce some tingling caused by partial local anesthesia but is otherwise valued more for its symbolic licentiousness than for any strengthening of the drug effect. Cocaine is usually chopped up with a safety razor and sniffed from a measuring spoon, a soda straw, or a tightly rolled cigarette paper or dollar bill (higher denominations for those who want to display their wealth). But recently there has been a proliferation of expensive accessories that confirm and promote its *premier cru* status among illicit drugs. Advertisements can be found for "shotgun spoons" ($6), gold-capped safety razors ($30), gold straws ($75), gold-jacketed glass storage bottles ($200), mirrors and jade blocks for chopping, spring-loaded snorting devices or "toots," and an elaborate "dessert service tray" ($69), pictured with a glass of liqueur.[32] Everything suggests that cocaine is becoming one of the routine luxuries provided by our consumer society.

The Future

The most significant sociological fact about cocaine today is that it is
rapidly attaining unofficial respectability in the same way as marihuana
in the 1960s. It is accepted as a relatively innocuous stimulant, casually
used by those who can afford it to brighten the day or the evening. Several serious books and articles on it have appeared in the last few years,
none of them displaying the kind of horror that was once standard in the
literature. Use of cocaine is gradually spreading in the upper middle
class. College students, young professional men and women, and middle-
class radicals have begun to experiment with it. A new elite mainstream
has been created by the convergence of the old Hollywood-Broadway,
black, and Latin cultural practices and the drug culture of the 1960s; the
present status of cocaine is both a product of this convergence and the
perfect symbol of it. A glance at *High Times* magazine, the highly successful new publication that serves as the *Playboy* of the drug culture,
reveals that cocaine is placed on a par with marihuana as a pleasure drug
and accorded none of the ambivalence or distrust associated with alcohol,
amphetamines, and barbiturates or the somewhat awed respect reserved
for psychedelics. Given the price of illicit cocaine, this implies that *High
Times*, in spite of its vague gestures of radicalism and its necessary opposition to the activities of the drug police, does not represent anything
like a counterculture; no unworldly hippie is going to buy the expensive
cocaine-sniffing apparatus advertised in its pages. The cocaine situation
is symbolic of the death of the counterculture, to the extent that it was
more than a creation of the mass media (underground *and* orthodox)—or
at least it constitutes a sign that the counterculture has been absorbed
into American society as a whole, the former adversary relationship between the two reduced to the practical issue of skirmishes with the law.

Whether the unofficial acceptability of cocaine to an increasingly large
section of the middle class can be translated into official respectability is
doubtful, but there are bound to be some changes. Once the drug was
used mainly by the rich and privileged, who were sheltered from the law
and public disapproval, or by blacks, small-time criminals, and others
who had no resources for public opposition. But now the kind of person
who is both a potential victim and a potential manipulator of the legal machinery has arrived on the scene. These people are articulate and have
access to legal skills and sources of publicity; for example, a 1974 article
in the *New York Times Magazine* was practically an endorsement of
cocaine.[33] The new cocaine users are often veterans of the cultural civil

war of the 1960s and conscious of the irrationality of our drug laws and policies. They not only know what they want but have ideologues, social philosophers, or gurus to tell them why they should want it. Some victories have already been won in the marihuana campaigns, and now, with the help of public defenders, radical law firms, and affidavits from prominent physicians and psychiatrists, the cocaine issue is being brought to attention.

The most celebrated case of this kind is the arrest of the radical leader Abbie Hoffman in 1973 for selling cocaine to police in New York City, but there have been several other similar legal actions. We have examined briefs and affidavits from some of these cases, and they all follow the same pattern. Apparently the first step in making a drug respectable, for radicals and cultural rebels as well as for physicians, police, and government officials, is to establish that it is not an opiate, is not like the opiates, is not a steppingstone to the opiates, and is not for the most part used by the same people who use opiates. As we shall explain under "A Note on Cocaine and Opiates," at the end of Chapter 6, this devil theory of heroin and morphine is misleading. Nevertheless, the approach is very effective and sensible, not only because most people believe the theory but also because laws that classify cocaine as a narcotic along with the opiates are particularly irrational and therefore subject to Constitutional challenge on due process and equal protection grounds. The legal briefs and affidavits we have read properly point out that cocaine should be classed with amphetamines as a stimulant and also insist that it is actually not so dangerous as amphetamines, alcohol, barbiturates, and other less restricted drugs.

The immediate effect of legal challenges is likely to be educational: judges may learn to mete out more lenient punishments whenever they are not required by statute to impose absurdly harsh ones. But, as the difficulties encountered in the struggle to reform the laws on marihuana or sexual activity show, nothing moves more glacially than official attitudes toward punitive statutes based on a formerly entrenched popular prejudice—even when the men who make and enforce the laws no longer really believe in them. Nevertheless, there are already signs of change, at least at the level of government commissions that operate at some distance from the effects of popular prejudice. In September 1975 a group consisting of representatives of 11 federal departments and agencies concerned with drugs presented to President Ford a White Paper on drug abuse which classifies cocaine as a minor problem compared to all other illicit pleasure drugs except marihuana, much less serious than barbiturates, amphetamines, and heroin. They add the reservation that cocaine might become a bigger problem if it were used more extensively, but they conclude that drug-enforcement officials should devote less attention and resources to it—for example, working more often in Mexico, where heroin is being smuggled, and less often in Miami, where they are more likely to find only cocaine or marihuana.[34] The recommendations of presidential commissions are often considered radical and not put into effect until long after they are made, if ever. But in this case President Ford has actually issued a general endorsement of the White Paper's conclusions in a statement for the press.[35] How soon this change in attitude will be translated into action remains in doubt. If the number of cocaine arrests rises to the level already reached by marihuana arrests (according to the Federal Bureau of Investigation, there were 445,000 arrests for marihuana in 1974 and only 101,000 arrests for cocaine and opium derivatives together), the strain on government time and financial resources may become so great that serious policy changes will be necessary. But the chances are against substantial change in the laws themselves in the near future; even if statutes declaring cocaine to be a narcotic were voided as unconstitutional, they could be reenacted with the same penalties but with the offending terminology removed. At most, the prospects are for an attenuation but not a cessation of hostilities, on the pattern supplied by the recent history of marihuana.

Price is now the biggest obstacle to a higher level of cocaine consumption. It depends to some extent on the operations of drug-enforcement agents, but also on the efficiency of the refining and smuggling network, and, of course, on the level of demand. In spite of the increased supply in recent years, the cost of cocaine has been rising fast and the quality declining. If the economic recession deepens, cocaine use may decline as it did in the Depression of the 1930s. It is also possible that South American governments will become serious about suppressing the traffic. There are no signs of this right now, in spite of the draconian penalties occasionally imposed on individual cocaine dealers and some pretense of coordinating antidrug operations at the request of the United States government; cocaine is too much a part of the social fabric in Latin America and too useful in too many ways, especially as a source of income for corrupt officials and a means of prosecuting those whom they consider undesirable for one reason or another. But truly revolutionary governments with puritan ideas about drugs might effectively suppress the cocaine traffic, which is associated with decadent high life among the old ruling class; that is what Castro apparently did in Cuba. If that ever happened, and if the coca fields of Peru and Bolivia were destroyed, as the United Nations demands, cultivation would have to be resumed somewhere else in the tropics where no mass demand in the form of a traditional coca-chewing population exists. At the moment, there is no prospect of laboratory synthesis on a commercial scale. Meanwhile, however, new smuggling networks are being opened, and the demand is rising. So many variables are involved that it is hard to predict the future of cocaine as a pleasure drug.

PART II

4

FROM PLANT
TO INTOXICANT

Botany and Cultivation

Coca is cultivated principally in warm valleys on the eastern slope of
the Andes, at altitudes of 1,500 to 6,000 feet. The region of cultivation extends from Colombia in the north to the Cochabamba region of Bolivia in
the south and eastward into the Amazon basin. Some coca is also grown
on the Pacific slope of the Andes in Peru. In the past, it was also grown
commercially in British Guiana, Jamaica, Madagascar, the Cameroons,
India, Ceylon, and especially Java. For best yields the shrub requires continuous high humidity and a uniform mean temperature of about 65°F
throughout the year. Frost kills plants grown at high altitudes, and the
excessive heat of the Amazon lowlands apparently lowers the content of
cocaine and related compounds. The plant needs a soil free of limestone
and thrives in the red clay of the tropical Andes, often on land that is too
poor for other crops.

The domesticated species of coca are identified botanically as members
of the family Erythroxylaceae and the genus *Erythroxylum* (sometimes
written *Erythroxylon*).* Otherwise their classification is somewhat un-

---

* Patrick Browne first described the genus in 1756 and gave it the name *Erythroxylum*. But most botanists doubted that his description was technically adequate and therefore accepted instead the form
Erythroxylon used by Linnaeus when he redescribed the genus in 1759. The consensus is now shifting
back to recognition of Browne's priority and the name *Erythroxylum*.

certain. It appears that one species, *E. coca,* predominates in the Amazon basin, in Bolivia, and in most parts of Peru. Two variants of another species, *E. novogranatense* (after "Nueva Granada," the old Spanish name for Colombia), are also cultivated, one in Colombia and the other on the Pacific slope of the Andes in Peru. *E. novogranatense* was once grown more extensively throughout the Caribbean and, more recently, in Java and Ceylon. Further research now being conducted on the taxonomy of the genus *Erythroxylum* may modify these species designations and also identify the wild ancestors of the cultivated forms of coca.

In Peru and Bolivia, *cocales,* or coca farms, cover whole terraced mountainsides. Some are large plantations; others are small family holdings farmed by Indians. The basic methods of cultivation have been the same for hundreds of years. In a nursery with a thatched roof, young plants are started from seed or from cuttings. After six months or a year, when they are about 18 inches high, they are set out in rows separated by small earthen walls. After five years the shrub reaches full height, 12 feet, or higher in some varieties; it is usually pruned back to six to nine feet for convenience in harvesting. The coca plant produces a creamy white flower with tonguelike appendages on the petal, and it forms a deep red, small, egg-shaped fruit with a pit. The fruit is used only for seeds; it is the leaves that are harvested. They vary considerably in size and shape but tend to be oval, about an inch or two long and half an inch wide. They are considered mature when they are faintly yellow or when they tend to break off in response to bending. At this stage, which is achieved about a year or 18 months after sowing, the first harvest (*quita calzón,* or "taking off the underpants") begins; it is more a trimming than a harvest. Thereafter the leaves are harvested three or four times a year. The regular harvests (*mittas*) have traditional seasonal names, less often used today: *mitta de marzo* in spring; *mitta de San Juan* at the festival of St. John, late in June; and *mitta de Todos Santos,* the All Saints' harvest, in October or November.

The leaves are picked, often by families working together, emptied into sacks, and conveyed to drying sheds that open into closed courts. There they are spread in thin layers on a slate or concrete pavement, or on the ground, and dried in the sun. (Coca that has dried fast is known as *coca del día* and commands a high price.) The crisp leaves are now thrown in a heap for several days to absorb moisture from the air so that they will not be too brittle to pack. Then they are dried in the sun for another half hour and packed. The well-cured leaf is olive green, pliable, clean, smooth, and slightly glossy; it retains some moisture and the subtle coca aroma, which resembles new-mown hay. According to Mortimer, the leaf can be preserved for several years if it is properly stored and handled. He estimates that each bush produces four ounces of fresh leaves or 1.6 ounces of dried leaves at each harvest and that the bushes are planted 7,000 to the acre. With three harvests a year, that means 2,200 pounds of dried leaves per acre each year. The leaves are commonly packed for shipping in bales that weigh 20 to 25 pounds and then transported—formerly by mule or llama, now by truck—to the point of sale.[1]

Chemistry and Extraction

Cocaine is one of the group of alkaloids, substances occurring naturally in plants. Alkaloids are basic (alkaline) compounds synthesized by living organisms from amino acids; they have a closed carbon ring that contains nitrogen and are pharmacologically active in animals and man. Some examples are caffeine, from the coffee and other plants; morphine and codeine, from the opium poppy; quinine, from cinchona bark; strychnine, from *strychnos nux-vomica;* and nicotine, from tobacco. The term is used somewhat loosely; for example, substances lacking some of the defining characteristics may be called alkaloids if they are pharmacologically active and their distribution in nature is restricted to a few plants; a compound like the B vitamin thiamine is not usually classified as an alkaloid simply because its distribution is nearly universal. Some compounds that do not have the correct ring structure, like mescaline, the derivative of the peyote cactus, may be called either protoalkaloids or alkaloids.

The alkaloids of *Erythroxylum coca* belong to six closely related groups: 1) cocaine and some variants, including cinnamylcocaine and truxillines; 2) methylecgonine and methylecgonidine; 3) benzoylecgonine; 4) tropeines, e.g., tropacocaine; 5) dihydroxytropane; 6) hygrines, including hygrine, beta-hygrine, cuscohygrine, and hygroline. From 50 to 90 percent of the alkaloid content of the coca leaf is cocaine. Leaves from *E. novogranatense* are richer in alkaloids than *E. coca*— about 2 percent by weight as opposed to 0.5–1 percent—but the latter has a greater proportion of cocaine to other alkaloids.[2] Cocaine, like several other coca alkaloids, is a tropane, chemically related to the psychoactive tropane alkaloids produced by plants of the Solanaceae family, which includes belladonna, or deadly nightshade; henbane; jimsonweed, or thorn apple (*Datura stramonium*); and mandrake. Although these drugs—atropine, hyoscyamine, and scopolamine (hyoscine)—resemble

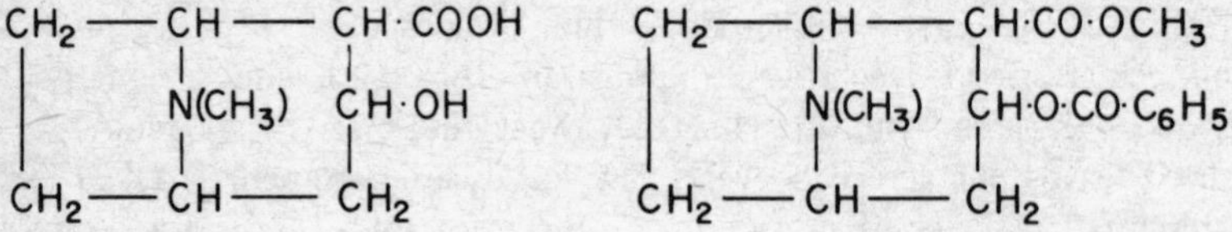

$$CH_2 - CH - CH \cdot COOH$$
$$| \quad\quad N(CH_3) \quad CH \cdot OH$$
$$CH_2 - CH - CH_2$$

ECGONINE

$$CH_2 - CH - CH \cdot CO \cdot OCH_3$$
$$| \quad\quad N(CH_3) \quad CH \cdot O \cdot CO \cdot C_6H_5$$
$$CH_2 - CH - CH_2$$

COCAINE

FIGURE 1
*Structural formulas of ecgonine and
its methylbenzoyl derivative, cocaine.*

cocaine structurally, their pharmacological mode of action and observed effects are quite different.

Chemically, cocaine is designated as 2-beta-carbomethoxy-3-beta-benzoxytropane. It is also described as the benzoyl ester of methylecgonine. Pure cocaine, or cocaine base, is a colorless, odorless transparent crystalline substance, almost insoluble in water and freely soluble in ether. It also dissolves in dilute acids, forming salts that are soluble in water and therefore more commonly used in medicine and recreation than cocaine base itself. Cocaine hydrochloride, also a transparent crystalline substance, is by far the most popular of these salts; it is 89 percent cocaine by weight. Both cocaine base and the hydrochloride are normally chopped into a white powder for use. Cocaine hydrochloride must be kept in a closed container or the water in the air will dissolve it, but it is resistant to the effects of sunlight and room temperature heat. Cuts do not have any effect on the potency of the remaining cocaine; but on one account, cocaine can be decomposed by molds, and the presence of sugars might enhance that process.[3]

There are several methods for obtaining refined cocaine hydrochloride on a large scale, some of them commercial secrets, but they are all variations on two patterns: extraction directly from the leaf, or semisynthetic production from ecgonine (more precisely, methylecgonine) another coca alkaloid. Crude cocaine or coca paste, called *masa* in Peru, is a mixture of alkaloids that is about two-thirds cocaine. It is produced by dissolving dried leaves in sulfuric acid and precipitating the alkaloids with sodium carbonate. A dull white or brownish powder with a sweet smell, it is sometimes smoked in pipes with marihuana. In the direct extraction process, coca paste is converted to "rock" cocaine by solution in hydrochloric acid and further treatment to remove the other alkaloids. This is the form of cocaine hydrochloride usually sold in the illicit traffic. Production of pharmaceutical-quality, or crystalline, cocaine, popularly known as "flake," requires repeated further purification. One technique employs petroleum ether, methanol, and hydrochloric acid; another employs a mixture of acetone and benzene. The semisynthetic process was commonly used before World War II with coca leaves from Java that contained a relatively small proportion of cocaine. The mixed coca alkaloids are converted to ecgonine, and cocaine hydrochloride is generated by the addition of methanol and benzoyl chloride. Direct synthesis of cocaine, without recourse to the coca leaf, is a lengthy and difficult process. Willstätter first achieved it in 1902, but neither his original method nor others introduced later have so far proved commercially practical even at the present inflated price.[4]

Unlike amphetamine and opium, for which many congeners have been synthesized and manufactured, cocaine has comparatively few analogues and substitutes with similar psychopharmacological effects. One of the other natural coca alkaloids, tropacocaine, has anesthetic and stimulant properties, and it may be chemically more accessible to structural modification. Cocaine diethylamide, a synthetic derivative, related to cocaine as LSD is related to lysergic acid, is about four times more powerful as a stimulant in mice and rabbits than cocaine itself.[5] Other alterations of the cocaine molecule to yield compounds that retain the stimulant effect would be useful to experimenters.

Pharmacology

When cocaine first came into use in the late nineteenth century, experimental pharmacology was in its beginnings and ideas about drug action and the nervous system were uncertain. Freud discussed the theory that cocaine is a "source of savings," an agent enabling the body to make more efficient use of energy and so to survive with less food. Experiments by Moreno y Maíz, von Anrep, Gazeau, and others had shown, he said, that animals given cocaine succumbed to starvation just as fast as or even faster than control animals. But he pointed out that a historical "experiment," the use of coca in the siege of La Paz in 1774 as described by Unánue, suggested the contrary. Freud concluded that the savings theory was probably false, but that "the human nervous system has an undoubted, if somewhat obscure, influence on the nourishment of tissues; psychological factors can, after all, cause a healthy man to lose weight." Freud at least understood how little he knew. Mortimer, more than ten years later, maintained that coca frees the blood from uric acid and enables it to repair tissue more effectively: "Coca simply makes better blood and healthy blood makes healthy tissue." He also thought that cocaine might diminish the consumption of carbohydrates by the muscles during exertion and so economize on oxygen.[6] The unlimited possibilities that seem to exist when there are no generally accepted models of drug action create conditions in which the idea of a wonder drug or panacea thrives.

Although there are still many mysteries in neurophysiology and neuropharmacology, we now know enough to describe the mechanism of cocaine's effect with more plausibility and precision. It has two principal actions; either may be made to predominate by varying the dose and method of application. Cocaine produces anesthesia at the site of application, and this is often the most obvious effect when it is applied to the skin or mucous membranes. Swallowed, inhaled, or injected, it enters the nervous system from the bloodstream and produces the pleasant stimulation that makes it a recreational drug. The local anesthesia is a result of its capacity, at relatively high concentrations, to block the conduction of electrical impulses within nerve cells. The stimulant effect, apparent at much lower concentrations, is achieved by interference with communication between nerve cells; cocaine does this by modifying the chemical signals passed through the synapse or neural junction. At the synapse, the electrical impulse generated in a nerve cell causes the release of a transmitter substance that diffuses across a short space to a receptor site on the adjoining cell and generates another electrical impulse; in this way signals are broadcast through a nerve network. The signal-carrying chemicals, called neurotransmitters or neurohormones, are different in different parts of the nervous system; the distinctive powers of a drug like cocaine therefore depend on the kind of neurotransmitter it affects.

Cocaine acts mainly on norepinephrine (also called noradrenalin), a neurotransmitter found in both the central nervous system (the brain and spinal cord) and the peripheral nervous system, which carries messages in the body outside of the brain. Most central and peripheral nerve cells that use norepinephrine are associated with the sympathetic nervous system, which is part of the network governing involuntary functions—muscles and glands that normally operate independently of one's intentions. Its general purpose is to put the body onto an emergency basis for coping rapidly with environmental changes. Activity of the sympathetic system produces the bodily reactions characteristic of emotional excitement, especially fear and anger; it mobilizes the organism for flight, defense, or aggression. Specific physiological effects include dilation of the pupils, increased heart rate, higher blood pressure, constriction of blood vessels in the skin and mucous membranes, increased blood sugar, rise in body temperature, and inhibition of the digestive process with increased tone of the sphincter muscles; these are all largely ways of preparing for sudden physical and mental exertion. The hormone epinephrine (adrenalin), released by the adrenal glands in times of stress, is a close chemical relative of norepinephrine and reinforces its effects. Because cocaine augments the effect of norepinephrine and therefore activates the sympathetic system, it is described as a sympathomimetic drug.

Another network governing involuntary functions is the parasympathetic nervous system, which generally produces physiological effects opposite to those of the sympathetic system: it narrows the pupils, slows the heart, lowers blood pressure, stimulates digestion, relaxes the sphincters, and so on. The transmitter substance in these nerves is acetylcholine, which is chemically unlike norepinephrine; cocaine affects parasympathetic nerves only indirectly, when they respond to sympathetic activity.

The interest in cocaine as a recreational drug derives not from its effect on such organs as the heart and digestive tract but from its direct action on the central nervous system (CNS), and in particular on its most important part, the brain. It has proved difficult to identify and locate the neurotransmitters used by the brain because of its complexity and relative inaccessibility for delicate chemical manipulation. Nevertheless, it is generally agreed that about 1 or 2 percent of CNS neurons (nerve cells) use either norepinephrine or dopamine, a related substance, as a transmitter. Dopamine is also the biosynthetic precursor of norepinephrine; that is, the body normally synthesizes norepinephrine by a chemical transformation of dopamine. Both substances belong to a class of compounds called catecholamines that also includes epinephrine; cocaine seems to augment the effect of catecholamines in the CNS just as it does in peripheral nerves.

The neurons that use catecholamines as transmitters are concentrated in a few specific areas of the brain, most of them functionally related to the sympathetic cells of the periphery. Norepinephrine is found in the hypothalamus, which regulates appetite, thirst, body temperature, sleep, and sexual arousal, as well as orchestrating anger, fear, and other emotional responses. One region of the hypothalamus, the median forebrain bundle, contains areas where electrical stimulation produces a sensation of pleasure. Norepinephrine is also an important neurotransmitter in the ascending reticular activating system (RAS), which controls mechanisms of arousal and attention. External stimuli normally reach consciousness only through the RAS, and when it is inactive the organism is asleep. Norepinephrine is also used in nerve pathways leading from the hypothalamus and RAS to the cerebellum, which controls fine motor coordination, and the cerebral cortex, which governs higher mental activity like concept formation and memory. Dopamine is found in the corpus striatum, which is part of the network governing motor functions, and also in the region of the hypothalamus regulating appetite and thirst.

Cocaine is only one of several drugs classified as stimulants that exert their influence by modifying catecholamine transmission in the CNS; caffeine and the amphetamines are others. The precise mechanism by which cocaine acts is uncertain, although there are several plausible theories. One way a stimulant effect can be produced is by preventing normal destruction or inactivation of the neurotransmitter after it has carried its message at the synapse. In most nerve cells, including all those that use catecholamines, the major inactivation process is reuptake: the neurotransmitter is partially reabsorbed by the nerve cell that released it and held in storage, so that it need not be continually resynthesized at the same rate that it is used. A commonly accepted theory is that cocaine acts primarily by blocking reuptake of catecholamines, leaving an excess of the neurotransmitter in the synapse to stimulate receptors. Imipramine and other drugs used in treating depressed patients apparently work in this way. But some experiments suggest that cocaine has no more effect on the rate of norepinephrine inactivation than procaine (Novocain), a local anesthetic without any stimulant effect; if that is so, the stimulant power of cocaine cannot be attributed solely to its effect on reuptake.[7]

A stimulant may also work by displacing neurotransmitters from their storage pools in the nerve terminal and provoking their release. Amphetamines almost certainly do this; cocaine probably does not. The effect can be demonstrated by preventing normal synthesis of catecholamines and therefore artificially depleting neurohormone stores. A substance that does this is AMPT (alpha-methyl-para-tyrosine); if animals are pretreated with AMPT and then given amphetamines, the usual amphetamine effects are not observed because there is too little of the neurohormone left to be released. By contrast, cocaine effects, in some experiments, seem to be diminished but not eliminated; this suggests that cocaine does not work predominantly by forcing release of stored catecholamines. But such conclusions are problematic because the experiments are often hard to interpret; for example, in order to release a neurotransmitter, the drug may also have to block reuptake, since it takes over the uptake mechanism to gain access to the storage pools.[8]

Because it is difficult to explain the effect of cocaine fully by release of catecholamines or reuptake blockade, some researchers now favor the theory that it acts primarily by heightening the sensitivity of the receptor sites that combine with catecholamines in nerve cells receiving their chemical message. If (as the experiments with AMPT suggest) cocaine can be active even when very small quantities of neurotransmitter are present, this hypothesis has some support. According to one version of the theory, cocaine does not act directly at the receptor site where catecholamines generate the nerve impulse but rather at nearby sites; its effect is to change the shape of the receptor molecule so that the neurohormone is more likely to generate a signal. The process is known as allosteric interaction.[9] One reason for suggesting that cocaine acts in this indirect fashion is that, unlike amphetamines, it bears little molecular resemblance to the catecholamines; this may also be the reason why it does not displace catecholamines from nerve storage pools.

There is also evidence that cocaine does not affect the transmission of dopamine, as opposed to norepinephrine, so strongly as amphetamines. For example, the drug haloperidol, which blocks receptor sites for dopamine, but not norepinephrine, seems to reduce the effect of amphetamine more than that of cocaine. (Chlorpromazine, which blocks both dopamine and norepinephrine receptor sites, affects cocaine and amphetamine in the same way).[10] Either of these stimulants may cause in animals a kind of stereotyped motion—repetitive head turning, gnawing, sniffing, and so on—which is probably controlled by dopamine-using nerve cells in the corpus striatum. But it takes much higher doses of cocaine to produce such behavior. In one recent experiment injections of cocaine even in doses as high as 10 mg per kg of body weight (the equivalent of 700 mg in a 150-pound man) caused no stereotyped behavior in rats, although amphetamine began to produce the effect at less than 1 mg per kg. It is possible that amphetamines are more active than cocaine at dopamine synapses because they actually release the neurotransmitter from storage sites, while cocaine only blocks its reuptake or makes receptor sites more sensitive.[11]

This apparent difference between cocaine and at least some of the amphetamine compounds is significant for two reasons. First, it suggests an explanation of why cocaine is somewhat different in its subjective effects from amphetamines (as we will see) and apparently tolerable and desirable at higher doses. In experiments on rats, injections of cocaine in doses up to 10 mg per kg increased the rate at which they stimulated themselves by means of electrodes placed in the median forebrain bundle (pleasure center) of the hypothalamus. Amphetamine increased the rate of self-stimulation most at .63 mg per kg and actually depressed it at higher doses as stereotyped behavior became more prominent.[12] The release of dopamine was apparently interfering with or superseding the effect of the norepinephrine-using nerve cells in the pleasure center. A second reason why differences between norepinephrine and dopamine effects are so important is that dopamine mechanisms may be the cause of the "model psychosis" produced by both cocaine and amphetamines, but more often by amphetamines. We will eventually discuss these matters in more detail.

A reasonable conclusion is that cocaine affects the nervous system in somewhat the same way as amphetamines, but with subtle differences that may turn out to be significant. It would be particularly useful for any hypothesis to account for an important difference between the organism's response to cocaine and its response to amphetamines, which we discuss more fully later: that amphetamines often produce tolerance, and cocaine apparently does not. Needless to say, no theory has been able to fully explain this or other unique features of the pharmacology of cocaine.*

Detailed information on the processes by which the body alters, inactivates, and eliminates cocaine is lacking, but some facts are known. Recently a study of monkeys receiving cocaine by injection showed that after half an hour about one-quarter of the drug is found in the brain as norcocaine, a breakdown product which blocks reuptake of norepinephrine in the test tube and may be active in the living brain as well. Eventually cocaine is inactivated when it is broken down by enzymes in either the bloodstream or the liver, where it is converted to the related alkaloids ecgonine and benzoylecgonine; the liver can detoxify the equivalent of one lethal dose an hour. Some of the cocaine absorbed by the body is also excreted unchanged in the urine. Various studies have reported values ranging from 0.5 percent to 21 percent; in one case of overdose, as high as 54 percent. Studies on coca chewers showed that about

---

* Amphetamines and cocaine both produce anorexia (loss of appetite), presumably by interfering with appetite regulation in the hypothalamus. Amphetamine anorexia, like other amphetamine effects, is nullified by AMPT (see, for example, A. Groppetti, "Amphetamines and Cocaine on Amine Turnover," *Life Sciences* 13 [1973]: lxii–lxiii). If cocaine anorexia is less affected by AMPT, it too may be caused by heightening of receptor sensitivity rather than release of catecholamines. Tolerance to amphetamine anorexia notoriously develops rather quickly (see, for example, Lester Grinspoon and Peter Hedblom, *The Speed Culture: Amphetamine Use and Abuse in America* [Cambridge: Harvard University Press, 1975], pp. 210–211). If cocaine anorexia has a different mechanism, tolerance to it may arise much less easily and cocaine may be a more effective diet drug. We know of no experiments on this.

10 percent was eliminated unchanged in the urine if the chewers did not use alkali. If they did, the proportion rose to 22 percent to 35 percent. Some researchers believe that when cocaine is absorbed through the digestive tract, as in coca chewing, it is partly broken down to ecgonine and benzoylecgonine even before it reaches the bloodstream and is carried to the brain. But this is still in doubt.[13]

5

THE ACUTE
INTOXICATION

IN RECOUNTING the history and pharmacology of cocaine, we have necessarily mentioned most of the effects it can produce. Before discussing these effects more systematically and in more detail, we must distinguish several topics or dimensions of analysis. One of these is physiological versus psychological: roughly, effects on the involuntary functions of the organism versus effects on mood, perceptions, voluntary action, and higher mental activity. It must be remembered that the physiological and the psychological are merely ends of a continuum: many effects, like rise or decline in sexual interest, changes in power of endurance, or tendency toward restlessness or calm, are not easily identified as one or the other. Another important dimension is acute versus chronic intoxication: the effect of a single dose, large or small, as opposed to the effect of long continued use. Still another is the distinction between smaller and larger doses or "moderate" and "excessive" use. We put these words in quotation marks because they tend to be vague, misleading, and emotionally fraught. For many people there is simply no such thing as moderation in the use of certain drugs, and cocaine is one of them. We will return to this question when we contribute our opinion on the question of when and how cocaine is harmful or beneficent. Meanwhile we speak of moderate and excessive use of cocaine, or small and large doses, in the same common-sense way that most people—not, of course, prohibitionists—speak of small and large amounts of alcohol or moderate and excessive drinking. There are two other relevant distinctions. One is coca versus cocaine; the other is the difference among various methods of administration: subcutaneous or intravenous injection; sniffing; oral ingestion by eating, drinking, or chewing; and, more rarely, rubbing into the gums, smoking, and topical application to the genitals.

The three main continua or modes of distinction, then, are physiological-psychological, acute-chronic, and small dose-large dose or moderate-excessive. We shall try to evaluate the effects of cocaine in all these categories, paying attention where necessary to the differences between coca and cocaine and among various routes of access to the body. A classification like this cannot determine how and how much the drug actually is or will be used. That depends on conditions of availability and other rather indeterminate psychological and social factors. Although this statement seems obvious, we introduce it here because what it emphasizes is so often ignored; for example, the effect of an overdose of very strong hashish or pure tetrahydrocannabinol (the main active principle of the hemp plant) is discussed as though it were the typical effect of marihuana. Since cannabis is not often used that way, the picture of the situation is distorted even when the description is technically accurate. To judge a drug in context, we need more than an enumeration, even an orderly enumeration, of all the effects it has ever produced or might possibly produce; we must have some idea of how much it is going to be used, how, and by whom. This question involves the important and obscure subject of dependence, with its mixture of pharmacological, psychological, and social issues.

Individual variability and a shortage of information make it difficult to specify the dose of cocaine that is likely to produce a given effect. An amount that is barely a stimulant for one person may cause a mild paranoid reaction in another; in surgery an amount that produced only local anesthesia in most people was fatal to a few. Statistical or quantitative studies on the frequency of different reactions at different doses in human beings are simply not available, and we have only more or less anecdotal material—clinical case histories, interviews, and literary descriptions—and a few small-scale experiments on animals and human beings. Fortunately, new work is in progress, and much more information should be available in a few years.

Acute Effects of Cocaine

The euphoria produced by cocaine probably resembles amphetamine euphoria more than the effect of any other drug. But it is more transient: a single moderate dose (the average in recreational use is about 20 mg) might work for a half hour to an hour, as opposed to several hours for dextroamphetamine. It also tends to be subtler or milder than amphetamine, at least when taken orally or sniffed. Often the user feels little the first few times and has to learn to appreciate it; this is rare with amphetamines. A man we interviewed said, "Of course when I first snorted some coke I didn't get off on it, like most people, I've noticed. . . . I figured, well, there's got to be something to this, everybody really likes it . . . and after trial and error I just sort of started to perceive the subtle areas." Another statement: "The first time it was nothin' at all, didn't get off. It's very subtle, in the small amounts." Ernst Joël and F. Fränkel in their book *Der Cocainismus* confirm this. Gutiérrez-Noriega and Zapata Ortiz say that it often takes a few weeks to become sensitive to the effects of the coca leaf; they believe this implies an increasing failure of the liver to metabolize cocaine.[1] But probably the sensitization is mostly a matter of learning to perceive the effects, a phenomenon well known in connection with cannabis, opiates, and alcohol. People apparently take longer to learn to "read" cocaine than amphetamines, but less time than they need for opiates or cannabis. Many do report strong effects the first time they use the drug, especially if they have the opportunity to repeat the dose at intervals for a couple of hours. It is also worth mentioning again that the unreliable quality of illicit cocaine magnifies the apparent variability of its effect and sometimes gives a delusive impression of mildness or subtlety.

Arno Offerman, in an early (1926) controlled experiment testing the CNS effects of cocaine and other local anesthetics, administered cocaine in subcutaneous injections of 40 mg to 100 mg to a number of normal subjects, mostly physicians, as well as victims of various forms of schizophrenia and brain disease. In normal subjects the more or less consistent result was increased talkativeness and confidence, fast pulse and pounding heart, reduction of fatigue, and restlessness. A phlegmatic psychiatrist felt "as though he had drunk strong coffee" and talked more freely. The effect lasted 50 minutes. Another psychiatrist began to pace up and down and talk incessantly without ending his sentences or listening for replies. For about an hour, beginning 15 minutes after the injection, this state alternated with one in which he sat still in a corner staring into space, grimacing and chomping on his cigar butt. Still another psychiatrist said he saw everything more clearly, "as through a magnifying glass," and felt both physically strong and happy in such a way that he did not know which phenomenon was the cause of the other. A woman doctor began to rock back and forth on her chair, made little chewing and tongue motions, felt pleasantly restless, talked fast and continuously, sometimes in unfinished sentences, felt coquettish and turned her eyes with enlarged pupils on the experimenter, wanted to dance, and regarded her own behavior as affected and slightly ridiculous. The effect lasted two hours. A male physician felt lightness and a sense of well-being, a heightened emotional state, as though he could cry for joy. Offerman himself felt animated as after drinking alcohol but without the same giddiness or numbness, strong, and talkative. After half an hour he was tired and apathetic. He had no desire to eat for ten hours afterward. Except in the case of the woman doctor, the subjective effects began ten or fifteen minutes after injection and lasted from a half hour to an hour. Injections of saline solution administered to the same subjects under the same circumstances had no effect, and injections of synthetic local anesthetics had very little effect.[2]

Reports from literary sources and interviews may be less sober and the descriptive terms not always consistent, but the general tendency is the same: "It gives me a hilarious, exhilarating feeling"; "It's something like drinking but does not make me feel fuzzy like whiskey"; " 'Rich' is the only word to describe how he felt"; "that familiar snow feeling . . . I began to want to talk. Cocaine produces . . . an illusion of supreme well-being and a soaring overconfidence in both physical and mental ability. . . . There was also that feeling of timelessness"; "that pale cocaine edge, pale like acetylene flame . . . that voluminous hollow rush inside, that slippage of control systems, that cocaine express"; "It's good to do any job with a good attitude, with a happy frame of mind, positive energy"; "It's nothing more than stand-up-and-move energy. It's nothing that stands out as anything more than a good deep breath will do you"; "One thing you do, you get a lot of energy. So naturally you're gonna be almost like a racehorse at the gate, quivering inside"; "Depending on my mood, different aspects will be accentuated. There's obviously the speed quality . . . the first experience is a burning of the nose; then there's a kind of rush that goes straight to your brain, sort of a giddiness, a mild elation. After the rush, not exactly a speed high, but a feeling of being able to concentrate . . . I'd find myself smoking a lot more, just living at a faster, more intense pace . . . talking is a tremendous amount of fun"; "If I do some coke, I want to draw, write a poem, talk to people—it makes me verbal—I want to work things out, go places and do things. . . . I feel like a vegetable without it. With it, I'm more in touch with my feelings"; "It's the biggest sense-of-humor dampener I've ever seen. . . . If you sit around and blow joints all night and everybody gets real tired or something, and somebody pulls out some cocaine, everyone starts to talk and go again, but the smiles go away. It's an edge, cocaine."

Here are some more comments. "I've taken a couple of snorts of heroin, and all I can say, heroin is like the opposite. Because when you do cocaine, everything's *not* okay. There's a whole lot to do, the house is a mess all of a sudden." "If I had to describe it in physical terms it's like sitting in a field in Switzerland full of flowers, fresh spring grass, and the wind is blowing, and it's chilly but it's beautiful, a brisk day. Dynamite." "It has a positive effect on communication. . . . You can sit and talk for hours and hours, and feel a total rapport with the person you're talking to." "I do better with something which is beautiful in a subtle and quieting way [marihuana], rather than something which is beautiful in a racy and flashy way, like coke . . . and sometimes . . . you pay for it . . . you feel too much energy or something, just generally feel a little bit nervous or anxious. . . . I still don't know if maybe there's something that's not meant for this world about that product, you know, like just being so spiritual and so pleasurable that it may not be real, and life isn't meant to be lived in absolutes like that. . . . I would consider it in a category totally by itself, different from hallucinogens, different from other things like speed, because coke is much more integratable with your normal life. If you take acid, like you're just *gone* for 15 hours, and that's the way it is, an isolated experience. But this is *not* an isolated experience, it's an integrated experience." "Rapport, ease of communication . . . I started wanting to talk again." "It's the main ego-inflating drug there is." "Under cocaine I feel like a king. I make plans of great ingenuity. I can move about for hours." "You wouldn't know it, but when I don't take coke I'm rather bashful." "There never was any elixir so instant magic as cocaine. To one the drug may bring liveliness, to another languor, to another creative force, to another tireless energy, to another lust. But each in his way is happy." Freud described his own experience this way: "The psychic effect of cocainum muriaticum in doses of .05 to .10 gram consists of exhilaration and lasting euphoria, which does not differ in any way from the normal euphoria of a healthy person. . . . One is simply normal, and soon finds it difficult to believe that one is under the influence of any drug at all." [3]

These remarks, the purple passages as well as the sober ones, are about cocaine eaten, sniffed, or taken subcutaneously. Intravenous injection produces even more immediate and striking effects. "An intoxicating

wave of the highest rapture . . . at this moment the world seems perfect." "The first time I ever did it, I shot it. . . . It's very remarkable, it
doesn't last very long. It's a very intense rush, a very *very* intense rush. If
you're standing up, you can conceivably fall over. Very very intense rush.
It just sits you in your seat, for ten or fifteen minutes, blissed out." "Just
started as a *zoom* up . . . after the zoom up, it was like a thing that we
call, you hear bells, your ears started ringin' and you dealt with that for
two or three minutes, and then . . . you level off, and then things was
just bright . . . that strange, magical feeling . . . I used coke for that
zoom and that strange feeling." William Burroughs describes it more art-
fully in *Naked Lunch:*

> Ever pop coke in the mainline? It hits you right in the brain, activating con-
> nections of pure pleasure. The pleasure of morphine is in the viscera. You lis-
> ten down into yourself after a shot. But C is electricity through the brain, and
> the C yen is of the brain alone, a need without body and without feeling. The C-
> charged brain is a berserk pinball machine, flashing blue and pink lights in
> electric orgasm. C pleasure could be felt by a thinking machine, the first hid-
> eous stirrings of insect life.

And in an article on drugs for the *British Journal of Addiction* written in
1956 and later appended to the Grove Press edition of the novel,
Burroughs wrote:

> Cocaine is the most exhilarating drug I have ever used. The euphoria
> centers in the head. Perhaps the drug activates pleasure connections directly
> in the brain. I suspect that an electric current in the right place would produce
> the same effect. The full exhilaration of cocaine can only be realized by an in-
> travenous injection. The pleasurable effects do not last more than five or ten
> minutes. If the drug is injected into the skin, rapid elimination vitiates the ef-
> fects. The same goes double for sniffing.[4]

Acute Effects of Coca

At the other end of the emotional and rhetorical scale, the solemn, taci-
turn, introverted people of the Peruvian and Bolivian highlands rarely
admit that coca does anything except reduce hunger, thirst, and fatigue.
There is in fact some doubt about how much cocaine is actually entering
their digestive tracts and bloodstreams. According to the report of the
United Nations Commission of 1950, the average daily consumption of
coca leaf is 50 to 100 grams, with a few old *coqueros* chewing as much as
250 grams. A 1948 study found that in each *cocada* an average of 34

grams of leaves was chewed, with 180 mg of alkaloids including 112 mg of cocaine; the study concluded that coca chewers ingest 200 to 500 mg of cocaine daily. In a recent study of ten families of coca chewers in a town in southern Peru, Joel M. Hanna found the average consumption to be 60 grams of coca leaves, or 250 mg of cocaine, daily. The estimates of the total production of coca leaves and number of coca users given in Chapter 1 suggest an average of 30 to 80 grams a day; more coca illegally converted to cocaine would lower the mean, and a larger number of occasional or intermittent users would imply a higher mean consumption among those who take the drug daily. This is a substantial amount (the 250 mg of cocaine a day used by Hanna's subjects might cost $20 in the United States), and it is sometimes questioned why the apparent effect is not greater.[5]

One possibility suggested by experiments is that oral cocaine is not absorbed well. For example, in a 1971 study subcutaneous cocaine increased the motor activity of mice at 10 mg per kg, while oral cocaine did not affect it even at 50 mg per kg. The drug improved learning of a flight reaction when administered by either route. Parenteral and oral cocaine were equally effective in reducing fatigue as measured by reflex reaction time. Oral caffeine and amphetamines, in contrast with oral cocaine, do increase spontaneous motor activity in mice at relatively low doses. Recent experiments by Robert M. Post and his colleagues on depressed patients suggest that cocaine taken orally up to 200 mg a day produces no consistent mood change, although it modifies sleep patterns. These experiments were controlled and double-blind, but the subjects were of course not normal. Freud experienced some euphoria when he drank 50 to 100 mg in a water solution. The apparent relief of neurasthenia and other disturbances by coca wine or coca tea also implies that oral cocaine is effective. Gutiérrez-Noriega and Zapata Ortiz believe that cocaine is absorbed well by the digestive system and even contend that the difference between the lethal oral dose and the lethal subcutaneous dose in dogs is *less* than it is for other drugs of the same type: a ratio of 2:1 or 1.5:1. One and a half mg per kg of oral cocaine (110 mg in a 150-pound man), they state, usually produce some euphoria and faster reaction times in human beings. More thorough experimental studies are needed to determine whether and in what respects the effect of cocaine is different or weaker when it is taken by mouth.[6]

A question that arises in this connection is the function of the *llipta* or *tocra*—the alkaline substance used in chewing coca. It may be activating oral cocaine that would otherwise be ineffective. Analysis of its chemical composition shows that potassium and calcium compounds are the effective agents.[7] Two main explanations for the use of alkali have been proposed: 1) it helps to extract the alkaloids from the organic acids that bind them in the leaf; 2) it facilitates absorption of the extracted alkaloids into the bloodstream from the digestive tract. The second seems more plausible on the available evidence. In one experiment, 14 men chewed 20 grams of leaves each; eight used alkali (bicarbonate of soda) and six did not. The subjects who did not use alkali eliminated an average of 11.8 mg of cocaine in the urine in the first six hours after they began to chew; the others eliminated an average of 19.8 mg. The amount of alkaloids (mostly cocaine) extracted from the leaf, as measured by chemical analysis of samples before and after chewing, was about the same in both groups. Without alkali, subjects eliminated in the urine 10 to 20 percent of the cocaine they ingested; with alkali, urinary elimination of unchanged cocaine reached 22 to 35 percent. The experimenters concluded that the alkali speeds the absorption of cocaine into the bloodstream and therefore its elimination.[8]

Emilio Ciuffardi made the most thorough study of the effect of alkaline substances. He had 124 *coqueros* chew their accustomed amount, 71 of them with *tocra* (*quinoa* ashes) and 53 without it. (He notes that those who were not given *tocra* protested and asked permission to use the *tocra* they had brought along). The proportion of alkaloids by weight in the leaves varied from 0.5 percent to 0.7 percent, of which about 80 percent was cocaine. The subjects who did not have *tocra* chewed for an average of 95 minutes; those who had it chewed for an average of 113 minutes. The average total of alkaloids ingested with *tocra* was 213 mg; the average without *tocra* was 272 mg. With *tocra* 87 percent of the alkaloids in the leaves chewed were extracted (ingested) and without *tocra* 83 percent. The smaller total amount of cocaine ingested with *tocra* had considerably more effect on pulse rate, body temperature, muscular vigor, and mood than the larger amount ingested without the alkali. Obviously the main function of the alkali was not to extract cocaine from the leaf but to facilitate absorption into the bloodstream, since the coca chewers who did not use *tocra* actually ingested more cocaine and other alkaloids in a shorter time but still failed to obtain the accustomed effect. In other experiments it appears that the lethal dose of cocaine is lower in an alkaline solution and higher in an acid one. *Tocra* apparently somehow prevents cocaine from being neutralized or destroyed in the stomach before it is absorbed into the blood.[9]

A related problem is the contribution of the associate alkaloids to the effect of coca. This is a matter of great interest to admirers of the drug like Mortimer and Richard T. Martin, who believe that the coca alkaloids work synergistically or that ecgonine, benzoylecgonine, and hygrines act directly on muscle tissues rather than through the central nervous system like cocaine.[10] Gutiérrez-Noriega, on the other hand, contends that coca has basically the same effect as cocaine. Freud agrees; a large dose of ecgonine, incidentally, had little effect on him.[11] Ecgonine has the same legal status as cocaine, but that is presumably because cocaine is easily synthesized from it. There is no experimental evidence on the volatile aromatic hygrine alkaloids that Mortimer considers so important, but one study is available comparing the effects of ecgonine, benzoylecgonine, and cocaine in mice. The lethal dose of ecgonine proves to be very high, about 4 grams per kg intravenously; 1 gram per kg has no apparent effect at all. (These doses are the equivalent of 280 grams and 70 grams respectively in a 150-pound man). On a test of the grasping reflex used to measure fatigue and a test of conditioned learning, ecgonine at suitably high doses improved performance, although to a lesser degree than cocaine. Benzoylecgonine, which is chemically intermediate between cocaine and ecgonine, is also intermediate in its effect on learning and fatigue.[12]

The most interesting result of this experiment was that cocaine administered orally at low doses had effects like those of ecgonine; it was concluded that digestive juices turn most of the cocaine into benzoylecgonine and ecgonine before it is absorbed.[13] The function of *llipta* or *tocra,* then, may be to create an alkaline solution that permits absorption of the cocaine into the bloodstream before it is partly broken down into the relatively inactive benzoylecgonine and ecgonine.

But there is little evidence that the associate alkaloids in the coca leaf itself either moderate the effect of cocaine considerably or produce substantial effects of their own. It has not been shown that any common use of the coca leaf cannot be duplicated by cocaine in suitable doses and methods of administration; therefore there may not be any phenomena that need to be explained by the synergistic action of several alkaloids. Cocaine is not only the most active alkaloid but the one present in by far the greatest amount; it constitutes about 80 percent of the alkaloid content. As for the possible moderating effects of this relatively small proportion of other alkaloids, none have been found. A purified alkaloid mixture from the leaf was administered subcutaneously to 42 rats in a dose of 300 mg per kg, the amount of cocaine hydrochloride previously determined to be lethal 50 percent of the time in that form. The effects were qualitatively the same as those of pure cocaine hydrochloride, and 40.4 percent of the rats died.[14] Further research would be useful, but we must conclude for now that the associate alkaloids in the leaf do not have an important pharmacological role.

The defenders of coca are correct in stating that it rarely produces the more lurid symptoms of acute or chronic cocaine abuse. The obvious reason is that it is impossible to absorb enough cocaine into the bloodstream fast enough that way. Pure chemicals taken by injection, sniffing, or smoking have more immediate, powerful, and dangerous effects than plant matter containing only small quantities of these chemicals and entering the bloodstream by the normal digestive route for the absorption of alien substances. (The defenses of the gastrointestinal tract against cocaine, as we saw, seem to create the need for alkali in chewing coca.) Eating opium does not give even the same effect as smoking it, much less the sudden, strong euphoria of a morphine injection; most people drink coffee, beer, and wine, but no one injects caffeine or ethyl alcohol into his arm. These distinctions may have something to do with the modifying effects of subsidiary alkaloids and other plant substances. Andrew Weil observes:

> There is no question that if the only form of sugar available to us were the whole raw kind, our total sugar intake would be a fraction of what it is now. The immediate sensory signals coming from the secondary compounds of the whole plant tell you not to overdo it. They provide a kind of insulation from the highly reinforcing properties of sucrose. I suspect that the secondary alkaloids in opium likewise help people not to overdo the use of morphine, and I feel sure that the many coca alkaloids play a similar role with respect to cocaine.[15]

But this remains speculative; the distinct effects are more obviously explained by differences in the concentration of the active compound and the speed of its absorption. Probably the social context of the use of a drug in one form rather than another and the fears and expectations surrounding it (for example, attitude toward hypodermic needles) are also important. The relation between these factors and the concentration and route of absorption is, of course, a sociological and historical problem, not a pharmacological one.

The reticence of Peruvian and Bolivian Indians about the effect of coca on their mood cannot be regarded as evidence that it has no effect. Even the early Spanish chroniclers noticed the difference between their reserved, solemn character and the ebullience and (at first) friendliness of the coastal natives. This passivity, or shyness, or dignity, or apathy, or melancholy, or aloofness—how one decides to describe it probably depends on cultural bias—is heightened by the natural suspicion among oppressed people about the intentions of outsiders who come to investigate their habits. Gutiérrez-Noriega, who made the most extensive studies of the effects of coca on its habitual users, confesses that they almost uniformly deny any psychological effects at first, so that it is difficult to get information from them. But he persisted and, with the help of auxiliary experimental studies and investigations carried out among inhabitants of the coastal areas, obtained testimony to some of the same effects that cocaine users describe.

In one study Gutiérrez-Noriega examined 30 men, all poor peasants. One-third of them used 70 to 120 grams of coca leaf a day, and two-thirds used 30 to 70 grams. They all said it suppressed hunger, thirst, and fatigue. Many said it made them feel stronger and took away their worries: "With coca I forget my troubles, while with liquor it's the opposite." It tended to intensify the association of ideas, and it produced optimism and confidence. A few reported changes in the visual field—mostly, objects seemed to stand out in relief and space became larger. Time sometimes seemed to be speeded up. The effect lasted one to two hours and was often followed by fatigue.[16]

Gutiérrez-Noriega also made a study of 25 men in Lima, 20 of them prisoners in the local jails, who used coca at the rate of 50 to 100 grams a day. The difference in temperaments and social setting obviously mattered a great deal. The prisoners produced descriptions that resemble Mantegazza's in some ways more closely than they resemble the Andean peasants'. They often found colors brighter and had a sense of proximity to the faraway. There might be a sense of a presence without visual content; for example, a feeling that an enemy was at one's side. Many of them reported fantastic visions or hallucinations, which they never confused with actual perceptions. One described his visions as "beautiful landscapes and golden castles." Street noises might turn into voices or melody or the steps of a pursuing policeman. The body seemed bigger, stronger, more agile—some used the word *superhombre* ("superman"). The slang term these men used for the intoxicated state, incidentally, is *armado,* which means literally "armed" but might be translated "loaded" or "charged up." The corresponding term in the United States is "high" or, at a more advanced stage of intoxication, "wired" or "cranked up." Coca produced quick but uncontrolled association of ideas and illusions of superior intelligence. It also sometimes induced fantasies of wealth or sexual fantasies. But one *coquero* is also quoted as saying, "Coca increases self-control, alcohol makes you lose it." [17]

Some skepticism about these reports is necessary. Gutiérrez-Noriega tended to exaggerate the effects of coca, which he wanted to abolish; the Lima criminals were probably not the most reliable of men and may have been telling him what they thought he wanted to hear. Unfortunately, he does not specify dosages after he has mentioned the average (moderate) consumption of 50 to 100 grams a day. Some of his descriptions sound like the effects not of a few grams of coca leaf but of a fairly large dose of pure cocaine working on highly impressionable personalities. Not even the heaviest cocaine users we interviewed reported anything like fantastic visions. Still, the possibility that they might occur is confirmed by Mantegazza's testimony and by comments like this one from the nineteenth-century traveler Poeppig about his *mestizo* guide: "His description of the beautiful images that came to him at night in the woods and the depiction of his glorious feelings at such times had something truly terrifying about them." [18]

In another study of coca users, Francisco Risemberg Mendizábal describes the coca effect much more soberly. He finds a period of extraversion and euphoria at the start of chewing followed by a desire to be alone and withdrawal into oneself because of the multitude of thoughts passing through the mind. The *coqueros* only appear to be gloomy, he writes; their imagination is soaring. They can work without boredom for hours at a mechanical task because, in the words of one of them, they "are not there." When Risemberg Mendizábal asked coca users what they would most like to do when in that state, many of them answered: drink alcohol. None of the men he interviewed had experienced any symptoms of intoxication from excess chewing.[19]

For a firsthand report from a coca chewer who is not a South American Indian, here is an account provided to us by Andrew Weil:

> Next to my typewriter is a bag of dry coca leaves sent to me by a friend who purchased them in the marketplace of Huanuco, a town in Central Peru. I take a small handful of them and place them in my mouth, moistening them and working them into a wad. Over the next few minutes I add more leaves until I have a walnut-sized mass. Now I add a small chunk of grayish-black material also from the Huanuco market. It is called *tocra* and consists of the ashes of the stalks of a cereal plant cultivated in the Andes. The salty taste of tocra blends well with the distinctive flavor of coca, and its alkalinity will increase the effects of the leaves.
>
> I keep my quid of coca between cheek and teeth, working it from time to time to suck the juices from it. I do not swallow the leaves. I savor the taste for I have good associations to it. Within a few minutes a pleasant tingling and numbness permeate the inside of my mouth and spread down my throat, a sign that the leaves are releasing their activity. In another few minutes I am aware of a delicious feeling in my stomach, a sort of warm contentment not wholly unlike the sensation that persists in my mouth. Now I feel a subtle energy moving from my stomach through my muscles. My mind is clear and alert. My mood picks up. I turn my attention to my typewriter and enthusiastically begin

to write. In half an hour or so I will have sucked my quid of leaves dry of their virtues. Then I will spit them out and rinse my mouth. The subtle stimulation I now feel will continue for perhaps an hour, then will fade gradually and imperceptibly without leaving me different from normal. My mouth will feel clean and refreshed. I will not chew coca again until I wish to use its stimulation for some other specific purpose.[20]

Summary of the Cocaine Feeling

Gutiérrez-Noriega is probably exaggerating when he declares that coca produces an "illusory satisfaction of desires . . . giving an impression of vivid reality so perfect that it provides genuine relief in the most adverse circumstances of life." It is easier to agree when he writes, "The drug does not . . . display any predetermined tendency to develop any particular type of feeling. The latter depends . . . on the content and quality of the thoughts and fantasies through which the subject passes. . . . It also depends on the company in which he finds himself and other social factors." In this respect moderate doses of cocaine are no different from moderate doses of opium, alcohol, or other psychoactive drugs. One person may become garrulous, another contemplative. One may feel rapture, another nervousness. One may want to sit still, another to pace or dance. Offerman in fact described the effect of cocaine as "an almost comically heightened amplification of one's character and individual predispositions," as though the drug turned a man into a caricature of himself.[21]

But in spite of all ambiguity and variability, the cocaine feeling cannot be confused in the long run with the physical relaxation and emotional disinhibition induced by alcohol or the placid, drowsy opiate sensation, or even the relatively long-lasting and physically intense excitement produced by amphetamines. It is characterized most by heightened self-confidence and a feeling of mastery. One writer put it this way: "With cocaine, one is indeed master of everything, but everything matters intensely. With heroin, the feeling of mastery increases to such a point that nothing matters at all." [22] Another way to put it is that opiates tend to cause a loss of interest in the self that makes mastery of the external environment irrelevant: the feeling is a kind of Nirvana. In contrast, a stimulant like cocaine heightens the sensory and emotional brightness and distinctness of the self against its environmental ground. A man we interviewed described it this way: "I knew that everyone knew *I* was

there; I thought that everyone could *see* me. Everything became alive."
This strong focal awareness may degenerate into paranoia with delusions
of reference: everything and everybody seem to be threatening "me," the
object of greatest interest in the universe. The heightening of the sense
of self—"the greatest ego-inflating drug there is"—may be a correlate in
the mind of the sympathomimetic action of stimulants. For in moments
of stress and danger the organism has to concentrate its forces to preserve itself against a suddenly hostile environment. It must *realize* its
own distinctness from the world and discourage any tendency toward a
relaxed merging with it. In the mind this may produce a strong sense of
individuality and of power and control, and a drug that simulates the central as well as the peripheral effects of sympathetic arousal will do the
same.

This phenomenon may be accompanied by joy, anger, even anxiety or
fear, or a mixture of all these: "It [cocaine] makes my mind ten times
quicker than normal, but also makes me anxious"; "It makes me jumpy
and scary, but I like it." [23] In a famous experiment, the effect of injections
of epinephrine on a group of subjects proved to be dependent on the behavior of a confederate of the experimenter who was in the room with
them. If he seemed happy, they felt happy; if he acted angry or annoyed,
they became ill-tempered too. In a control group that received placebos,
there was no such contagion of emotions.[24] The moral about the importance of the social component in the experience induced by psychoactive
drugs is obvious. But the experiment also shows that even the definition
of a physiological phenomenon, especially one that mimics the effect of
stress, as this or that emotion depends on personality and setting. The
coward defines the coursing of epinephrine and norepinephrine through
his body as fear, and the brave man defines it as the joy of battle or the exhilaration of facing danger. The fighter pilot, the racing driver, and the
mountain climber are often seeking the kind of exhilaration, the sense of
living more intensely, that comes from putting one's life in danger. Central stimulant drugs are less equivocal in their effects than injected epinephrine (which does not easily cross the physiological barrier between
the bloodstream and the brain) or stress, because they apparently influence the arousal and pleasure centers of the brain before they affect rage
and fear mechanisms. But the exhilaration they produce has the same
source as the pleasures of adventure (not, of course, that people think
they are risking their lives when they use cocaine). This effect, combined
with the further adventure of performing an illicit act, may contribute to
the interest in cocaine, especially among adolescents. The various emotions, or rather interpretations, associated with catecholamines (and

therefore with drugs that affect them) may also coexist in the same person at successive moments or even at the same moment. André Gide once wrote, "The hare or deer pursued by another animal takes pleasure in its running, its leaps, and its feints." This beautifully expresses the emotional ambiguity of sympathetic nervous system activity by referring it to the condition of creatures that have no words for joy or fear.

Effects on Intellectual and Physical Performance

Like other stimulants, cocaine gives a feeling of improved intellectual and physical capacity that is often of doubtful validity. It "makes you feel superior to other people," as one man we interviewed said. Another reported:

> It's efficient, it's an effector. On all levels. A potentiator. . . . If I'm going to do a large amount of coke, then I should be in a position in which I am actively communicating, or receiving. . . . Like if you're going to get into writing, or performing, or something like that . . . it's a level on which you become much more fluid, you have a whole lot more psychic poise and mental agility and physical agility. . . . It's not something that you don't know what's happening. You *do* know what's happening . . . and you're in control. . . . Cocaine gives me a feeling of mental, psychic comprehension, a level of confidence, of competence, a unity of my own work, my own validity.

On the other hand, a male cocaine user commented about attracting women, "You snort so you can pull your rap on her. My rap wasn't any better on cocaine, but see, I thought it was." Malcolm X, in his *Autobiography,* recounts how a foolish overconfidence induced by cocaine caused him to walk up to his white mistress in a public place and talk to her while she was with a friend of her husband's. He thinks the husband would have killed him if he had not been arrested the same day. And William Burroughs, who has certain artistic and intellectual achievements to his credit as well as a broad acquaintance with psychoactive drugs, considers the strong CNS stimulants useless for serious work: "Amphetamines and cocaine are quite worthless for writing and nothing of value remains." [25]

Almost all the evidence about cocaine's effect on performance is anecdotal; but it seems to be qualitatively similar to that of other stimulants like amphetamines and caffeine, which have been studied in controlled experiments and are relatively well understood. The experiments show that both drugs, and especially amphetamines, often improve performance on fairly simple intellectual and physical tasks. They do not enhance the quality of more complicated intellectual work (although they may speed it up) and may even lower it by inducing anxiety, restlessness, or overestimation of one's capacities. Amphetamines in particular can increase the endurance of athletes and enhance their performance in the short run, at the possible cost of eventually overstraining their physical capacities. Amphetamines reduce reaction time, hasten conditioned learning, and increase the rate of learning a motor skill. They also improve arithmetic calculation. But they have no effect on general intellectual capacity as measured by IQ and other intelligence tests.[26]

What little is known about cocaine in this area confirms the expected resemblances. It improves attention, reaction time, and speed in certain simple intellectual performances. For example, Johannes Lange in 1920 found that as little as 20 mg by mouth reduced the time taken on an arithmetic calculating test and on a word-association test produced more associations (but also more superficial ones—for example, sound rather than meaning). Hans W. Maier, also working in the 1920s, tested a habitual cocaine user on addition problems after giving her 400 mg at one time and 650 mg at another time (apparently by mouth). Cocaine greatly increased the number of additions completed, from 33.5 per minute without the drug, to 42.2 per minute with 400 mg, to 57.4 per minute with 650 mg. It also doubled the number of mistakes, from four to eight out of a total of about 700; this suggests that cocaine, like amphetamines, can give speed at the expense of accuracy. In two other experiments, also on habitual cocaine users, cocaine (20 mg sniffed) produced better scores than no drug or cocoa, but worse than coffee or even nearly caffeine-free coffee. Maier concluded that the nervousness caused by cocaine sometimes impaired performance, but he believed that it directly excited associative activity in the mind.[27] Twenty years later Vicente Zapata Ortiz studied sensory reaction time and attention under the influence of oral cocaine in 13 coca chewers and 13 controls. Sensory reaction time slowed in both groups but on a test of attention both groups worked faster and the coca chewers also made fewer errors than they had made without the drug. Zapata Ortiz suggests that the divergence between the two tests may be caused by a tendency to distraction by fantasizing that worked more strongly on the former. Incidentally, coca chewers were slow on both tests compared to nonusers, even under the influence of cocaine.[28]

For mental work that requires wakefulness, a free flow of associations, or the suppression of boredom and fatigue, a drug like cocaine is un-

doubtedly useful. The secretary who drinks several cups of coffee a day at
the office in order not to fall asleep over the typewriter, Balzac sustaining
himself on caffeine for the production of novel after novel, and John Ken-
nedy enhancing his "vigor" with amphetamines were making use of this
property of stimulants. In the late nineteenth and early twentieth centuries a number of writers and intellectuals obviously used cocaine in this
way. If Freud wrote his cocaine papers, General Grant his *Memoirs*, or old
Leo XIII his papal encyclicals under the influence of cocaine, it was serving the same function for them that coffee served for Balzac and many
others. Freud wrote that cocaine "steels one to intellectual effort." [29] It
was for this reason that students sometimes used cocaine, as they now
use amphetamines, to get through examinations. An interview subject
told us:

> SUBJECT: I wrote a whole entire master's thesis, a hundred pages on Samuel
> Beckett, and got an honorable mention on the thing. And I wrote it in seven
> days, start to finish, with about a gram of coke. . . . And I never could have
> written it otherwise, because I couldn't understand how all these various
> ideas fit together. . . . I never wrote so fast in my life. . . . I didn't eat and I
> didn't sleep. I like broke the thing up into three parts, maybe two days each,
> and after two days of just constant work, I would like take a Valium, and just
> sleep till I woke up. . . .
>
> INTERVIEWER: It aided your creativity?
>
> SUBJECT: Maybe it reduced it, maybe it reduced all the thoughts in my head to
> a clear line of action, which gave me a theme, which I didn't have
> before. . . .

But no drug will provide intellectual capacity, talent, or genius where it
is lacking; that is where the illusory aspect of the mental stimulant effect
becomes obvious. Euphoria and confidence can make the user attribute
to the drug effects that have nothing to do with it and cause him to
overestimate the value of the changes it does produce. Conan Doyle
seems to have been conscious of this when he had Sherlock Holmes say,
after calling cocaine "transcendentally stimulating and clarifying to the
mind," that he never used the drug when he was working on a case.
William Burroughs, as we noted, has also testified against the use of
cocaine and amphetamines for intellectual work. The delusive overestimation of one's capacities that stimulants can produce has been amply
documented in the case of amphetamines.[30] After chronic stimulant
abuse, as William Halsted's own later estimate of his condition when he
was taking large doses of cocaine shows, intellectual faculties may be directly disrupted. In Chapter 2 we quoted the first sentence of one of his
papers written at that time. It suggests a man in a mentally disjointed,

confused, distracted state. When we discuss chronic abuse of cocaine more fully in Chapter 6, we will provide other examples of this effect.

Cocaine may improve performance in several kinds of tasks that contain elements of both the mental and the physical. One of these is routine, unvarying, repetitious labor that nevertheless requires mental concentration. A man we interviewed said, "I know people who have shot coke and cleaned their whole house, you know. They get that up feeling, and then they start emptyin' ashtrays. Next thing you know they's cleaning the tables, next thing you know they's washing windows. . . . I've gotten into that, when it's good coke." It is possible, as we have mentioned, to induce incessant stereotyped turning, gnawing, or sniffing in rats, cats, and dogs by means of amphetamines or cocaine. The fact that amphetamines in large doses can sometimes cause absorption in routine activities that seem meaningless to the outsider is so notorious that a Swedish psychiatrist has suggested a special name for the phenomenon: "punding." Animal experiments indicate that cocaine produces less of this stereotyped behavior than amphetamines, but some remarks made in our interviews suggest at least a mild form of it. We shall discuss this further in another context.

A strong CNS stimulant can also fortify body and soul, at least in the short run, for the kind of performance in which boldness, confidence, and a feeling of mastery are themselves a large part of the achievement. A stimulant may eliminate the stage fright of the actor or musician, the timidity that suppresses the potential eloquence of a public speaker, the hesitancy that keeps the salesman from making his sale or the criminal from going through with his crime. Cocaine, like amphetamines, is often found useful when a performance in the colloquial sense—a good showing or a good show—is needed. This has been one of the greatest merits or temptations of the drug for its users ever since the days of advertisements recommending it "for young persons afflicted with timidity in society." A former actor told us, "You have at your command more of yourself. Say you're an actor, and you're going over a speech, and you've worked on it, and then you take a blow, and you go through the same speech. You will do it better." Another interview subject said, "To realize your own powers depends 90 percent on confidence. . . . When I ski, I don't consider confidence or energy derived from use of cocaine as misleading. Because the confidence that I need to get through a slalom course . . . the more aggressive I am, the better I ski. And the less hassle I have—I'm not going any faster. . . . It's just that my attitude is aggressive." No one recognizes the virtues of a performance in this sense better than actors

and musicians, and from Adelina Patti and Sarah Bernhardt to the rock
singers and Hollywood stars of today they have been prominent among
the users of CNS stimulants. So have the producers, directors, talent
agents, writers, and other auxiliary personnel of what is called show business. As we have mentioned, cocaine is now the stimulant they prize
most, for reasons of fashion as well as its intrinsic psychopharmacological
properties.

Effects like these are hard to define and almost impossible to test; the
fact that cocaine eliminates fatigue and permits the user to continue
physical activity more intensely or for a longer time is more obvious and
universally recognized. Freud wrote:

> The main use of coca will undoubtedly remain that which the Indians have
> made of it for centuries: it is of value in all cases where the primary aim is to
> increase the physical capacity of the body for a given short period of time and to
> hold strength in reserve to meet further demands—especially when outward
> circumstances exclude the possibility of obtaining the rest and nourishment
> normally necessary for great exertion. . . . Coca is a far more potent and far
> less harmful stimulant than alcohol.[31]

Freud did not distinguish between coca and cocaine. Today cocaine is
rarely used merely to stay awake or increase physical capacity; for those
who are willing to chance the dangers of strong drugs for this purpose,
amphetamines are more effective and certainly cheaper. Only if staying
awake implies going directly from work to an all-night party, or increasing physical capacity suggests permitting the user to dance for hours, can
it be said that cocaine is commonly used in this way. For example, although professional athletes undoubtedly use the drug during off hours,
we have seen no reports of their sniffing it before games. The nineteenth-
century example of the Toronto Lacrosse Club and the French bicyclists
sipping "Velo-Coca" is not (or not yet) being followed today, although a
man wearing a T-shirt with the legend "Enjoy Cocaine" was recently
seen running in the Boston Marathon.

But, as Freud indicates, for centuries cocaine (in the form of coca) was
recognized mainly as a physical stimulant. Before the invention of the
amphetamines, it was possibly the most potent stimulant drug available.
We have mentioned a number of references to the physical capacities of
men under the influence of coca. J. J. von Tschudi, the archaeologist who
visted Peru in 1838–1842 and left us some of the most substantial information on coca chewing from the nineteenth century, mentioned a 62-
year-old *coquero* who worked on diggings with him for five days and
nights with only two hours of sleep a night and no food, ran tirelessly
alongside a briskly stepping mule for 23 leagues (stopping only to chew

coca), and then announced that he would do more of the same work if Tschudi would give him more coca.[32] We have also mentioned the experiment undertaken by the 78-year-old Sir Robert Christison in 1876 and reported in the *British Medical Journal* and later by Freud. Christison tired himself to the point of exhaustion by walking 15 miles without food. When he took the same walk after chewing coca, he arrived home after nine hours without food or drink neither tired nor suffering from thirst and hunger. He woke the next morning refreshed. Further testimony comes from a Bolivian writing in the United Nations *Bulletin on Narcotics,* a publication not ordinarily favorable to the coca leaf: "During the Chaco War [1929–1932], coca chewing among the *mestizo* and white Bolivian soldiers was widespread. I have experienced its beneficial effects myself on forced marches." Even Gutiérrez-Noriega, who was hostile toward the coca habit, concluded after self-experimentation that the drug could relieve fatigue temporarily without any serious depression afterward.[33]

Freud was one of the first to test quantitatively the effect of cocaine on physical capacity. He used a dynamometer, a machine that measures the motor power of the arm, and served as his own experimental subject. Fifty to 100 mg of cocaine hydrochloride (apparently taken by mouth, although he does not say) produced a considerable increase in motor power that began 15 minutes after the drug was taken and lasted in diminishing degree for some hours. Cocaine also shortened the time it took him to react by a hand motion to a sound signal. A few other experiments have confirmed the anecdotal evidence and Freud's work. Gutiérrez-Noriega found that resistance to fatigue in dogs, as measured by the length of time they would keep swimming in order to avoid drowning, was increased from 69 percent to 150 percent by cocaine at 4 mg per kg subcutaneously (the equivalent of 280 mg in a 150-pound man), a result similar to the effect of caffeine at 8 mg per kg. Testing athletes with a bicycle ergometer, Dore Thiel and Bertha Essig found that cocaine increased endurance. Robert Herbst and Paul Schellenberg found that cocaine speeded recuperation after hard labor.[34] More carefully controlled experiments are needed, especially comparisons of cocaine and amphetamines in various doses and methods of administration. The little evidence that is available, including most notably the reports of people who have taken both drugs, suggests that amphetamines have stronger physical effects.

The question of whether stimulants improve performance directly or by indirect action that primarily affects confidence and interest has proved obscure. For example, Freud noted that although his strength under the influence of cocaine might exceed the maximum it could attain under normal conditions, the increase was greatest when his general conditon was poor. He concluded that the action of the drug was not "a direct one—possibly on the motor nerve substance or on the muscles—but indirect, effected by an improvement of the general state of well-being." But C. Jacobi, writing in 1931, observed that doses of coca or cocaine too small to produce perceptible central stimulation nevertheless made work easier; he concluded that the effect was related to the peripheral nerves and not to the brain. Joel M. Hanna, in a recent experiment using a bicycle ergometer, found no significant endurance differences between men who were chewing coca and men who were not, except that the coca chewers performed slightly longer at the very highest level of effort. He concluded that coca does not actually improve work performance but only reduces fatigue sensations and makes labor seem less burdensome.[35] Amphetamines work particularly well on experimental subjects who are bored, which suggests a primary effect on mood; but they also improve performance in an athletic competition like the shot put, where neither boredom nor physical fatigue in the ordinary sense is likely to be decisive; yet there is no evidence that they have any significant direct effect on skeletal muscles.

One way of resolving these issues is to refer to the connection between stimulant drug effects and those of an emergency situation. In moments of danger or stress people sometimes perform feats otherwise beyond their physical and mental capacity—whether or not they claim afterward to have felt unusual euphoria or confidence. It would be absurd to ask someone whether the catecholamines released in his body and brain made him "more favorably disposed" toward, say, saving his life in a moment of mortal danger, as though it were a matter for decision. In this case the attitude and the action are simply two aspects of the same process, a psychological-physiological unity. The verbal elaborations produced by the cerebral cortex that are usually used in experiments as evidence of attitude or disposition are related to the attitude expressed *in* the action performed only in a complex and indirect fashion. There is no distinct performance-improving effect separable from attitude in every sense; one cannot do something unusual without in some sense feeling different. On the other hand, euphoria, interest, or confidence as indicated by words or thoughts need have no *causal* relationship with improvement in performances that are mostly the work of nonlinguistic and often noncortical brain areas. Short-term increases in physical and mental capacity under the influence of cocaine probably resemble the similar increases that sometimes occur under the influence of stress and danger;

they are neither more nor less genuine, they are achieved neither more nor less directly, and they have the same limitations.

The long-term effects of cocaine on endurance and work performance are less clear. However, it is known that stimulants cannot save energy but only redistribute its expenditure. Eventually the body has to "come down" at best or "crash" at worst. It is possible to regulate one's intake of the drug judiciously so that one works harder or feels more vigorous while taking it and sleeps a little longer or feels slightly tired when one is not. But it is impossible to avoid resting one's body for a long time, with or without stimulants, and not pay the price in physical exhaustion. This is well documented in the case of amphetamine abuse, where the crash is often associated with extreme fatigue. Cocaine is apparently milder, but fatigue after excessive use is common enough. For example, "Jimmy," the cocaine dealer portrayed by Richard A. Woodley, preferred to come down while sleeping but was willing to tolerate a little depression. But toward the end of his association with Woodley, he was telling the writer that he was run down, tired, worn out. A doctor finally insisted that he get some rest, because he was exhausted from insomnia brought on by cocaine.[36]

Studies of the effects of habitual use of coca on the work performance of South American Indians, like all attempts to elucidate the unspectacular long-range consequences of a common practice, are plagued by the difficulty of isolating relevant variables. For example, if the men who use coca are also undernourished and underpaid, or alcoholics, they may work poorly even though the coca does them no harm. Opinion on the question in Peru and Bolivia varies greatly. The armies of these countries have forbidden coca to the troops (it was part of their rations until the 1930s), and the officers believe that this makes them stronger because they eat better. One wonders whether they eat better because they are not using coca or rather simply because for the first time in their lives they have enough food. The UN Commission report quotes a Peruvian Director of Mines as saying that coca chewers are not employed in the mines at any but the simplest tasks, and it reports that the president of the Agricultural Society at Cuzco believes men who do not chew to be more efficient and intelligent. Some mine foremen consider *coqueros* to be dull, inattentive, accident-prone workmen. But many priests, physicians, and other professional men consider coca harmless.[37] In general the quality and quanitity of labor produced by *coqueros* is not impressive (according to Gutiérrez-Noriega, the average working day is only five hours), but there are so many plausible reasons for this—we will discuss them in another context—that it seems dubious to single out the coca leaf.

Besides relieving fatigue, cocaine "satisfies the hungry": that is, it tends to make hunger easier to bear and also to deaden appetite, partly by anesthetizing the palate and tongue but mainly by acting on the hypothalamic hunger center. A man we interviewed said, "You don't eat, you don't sleep enough. . . . I was down to like a hundred pounds. People were really freaked by it." Richard Schultes of Harvard University, an authority on psychoactive plant substances who chewed the coca leaf on and off for eight years while he was in South America in the 1940s, recounted to us two incidents in which his canoe tipped over in rapids and all his food was lost. In one case he had coca to chew, and four days of paddling without food but with coca was much less of an ordeal than three days of paddling without food or coca—more, he believes, because he felt less hunger than because he felt less fatigue. People in the coca-chewing regions of the Andes do not eat much, and there has been acrimonious controversy about whether their low calorie intake is a cause or a consequence of the coca, or neither, or both, and in any case how harmful it is. We will discuss this in Chapter 6. The appetite-deadening effect is another area where experiment is badly needed. It would be particularly interesting if tolerance to it proved to develop less quickly than tolerance to the anorectic action of amphetamines.

Sexual Effects

Sexuality is notoriously a playground of legend and rumor, because interest is nearly universal and the amount of reliable information most people have about other people's sexual activity never seems to be enough. The relationship between sex and psychoactive drugs is particularly obscure because the central nervous system is connected with sexual functions in such a complex and indirect way. Nowhere is the great individual variability in drug effects greater, and nowhere is the influence of temperament, culture, social circumstances, and, above all, expectations more important; this is familiar in the case of alcohol but less recognized for other drugs. Animal experiments are usually inadequate testing devices; for example, cannabis is anaphrodisiac in animals but many human users of the drug consider it aphrodisiac. The situation is further complicated and obscured when the use of a substance is illegal, as in the case of cocaine. Illegality invests a drug with the glamour of sinfulness, and insofar as sex is still regarded as sinful both the opponents of the drug and its users may associate it with what they may call excessive or better sex, depending on whether they approve of the sexual activity or the drug or the law. No doubt there are many people who wrongly believe that heroin is an aphrodisiac simply because they have been taught that it is an illicit pleasure that is dangerous, evil, and enjoyed by blacks.

If a drug is not almost unequivocally anaphrodisiac, like the opiates, expectations of sexual activity when it is used are likely to be self-ful-filling; that is often the case with cocaine. It is important that the drug is a luxury, used by many people only on special occasions. One can be-come intoxicated almost as fast on cheap red wine as on champagne, but champagne drinking is the form of alcohol consumption particularly as-sociated with romance. The use of cocaine by fashionable rich people and entertainers who are assumed to have particularly interesting sex lives has enhanced its reputation as an aphrodisiac.

Keeping all these important qualifications in mind, we can say that cocaine, like other stimulants, often heightens sexual interest and sexual powers in the short run. Interviews on drug use and sexual activity among patients at the Haight-Ashbury Free Medical Clinic in San Fran-cisco showed that these young people who used many different psy-choactive drugs often saved cocaine as a luxury for sexual occasions. Ten of the twenty men interviewed who had injected cocaine intravenously reported erections. Some also reported painful priapism, and one had ex-perienced multiple orgasms. Cocaine might also be used to anesthetize the penis and lengthen the time before ejaculation, and several women came to the clinic with the mucous membrane of the vagina inflamed from sexual intercourse under these conditions. The arterial vasodilator amyl nitrite ("poppers"), often used to heighten sensation at the moment of orgasm, was considered similar to cocaine but briefer and more intense in its effect. In another study, sailors at a navy drug rehabilitation center gave heightened sexual pleasure as their main reason for using cocaine: about a third of those described as light users and almost half of those de-scribed as heavy users took it to "get a sexual feeling" or to "improve sex-ual pleasure." Among other drugs, only cannabis was used for these pur-poses (alcohol was not included) and then by only a fifth of both light and heavy users. So, at least among dwellers on naval bases and in the Haight-Ashbury district, cocaine is a highly "sexual" drug. In Haight-Ashbury, but not on the naval base, amphetamines were also used for sexual purposes. They were explicitly given credit for greatly augmenting sex drive, "making chicks nymphos," causing men to "go all day and not

come," and instigating group sex. They also induced a fantasy state satisfied by casual sex contacts and occasionally caused erections in men and
orgasms in women when injected. Gay and Sheppard call amphetamine
"a true aphrodisiac." E. H. Ellinwood, in a study of amphetamine psychosis, also noted that the victims were often hypersexual: "The greatest
increase in libido was often noted in women and especially those who had
been relatively frigid prior to abusing amphetamines." [38]

The cocaine users we interviewed also reported sexual stimulation
commonly, though by no means universally. One said, "It's probably the
greatest aphrodisiac known." Another added, "If you're in a sexual situation . . . everything is delayed, so prior to orgasm, it may take three times
as long, and an orgasm is, like, expanded over a long period of time . . .
cocaine definitely not only heightens the act but increases the desire for
it, and generally does have all the qualities of a classical, good aphrodisiac." An opposite opinion was, "I heard other guys' stories, you know,
you shoot some coke, your sex desires go right to the top. It never affected
me that way." Still another interview subject commented, "You spend all
day snorting and making love and listening to music. . . . But I think you
can do some damage, because the body's like an astronaut, going on a
blastoff. 'Cause making love on cocaine, you don't know when to stop."
When asked whether he enjoyed sex more on cocaine, he said he did not.
Others said cocaine in small doses was somewhat aphrodisiac, but large
doses tended to make them lose interest. One man reported that "alcohol
makes you lustful, but sort of cloudy mentally, and cocaine makes you
lustful, but clear mentally." Another said that with large doses, "You can't
keep your mind on sex. You get beyond it. You get beyond most everything except the cocaine." Still another man said: "My senses were stimulated, but not sexually particularly." But we were also told that "If it is
an ego-boost for some people, it may be an ego-boost sexually as well."

Cocaine, like amphetamine, may apparently encourage a certain kind
of hard-driving sexuality in which the sexual act is a performance requiring power and mastery. A recent doctoral thesis confirms, in a sense, the
reports of Gay's and Sheppard's subjects by showing experimentally that
cocaine and methamphetamine are the only ones among a wide variety of
psychoactive drugs that "improve" sexual performance in male rats as
measured by the number of ejaculations achieved in a given time. (*High*
doses of these drugs, however, are shown to inhibit sexual activity.) In a
recent Harold Robbins epic, a tireless black man, a lust-crazed white
woman, and an endless supply of cocaine and amyl nitrite are assumed to
be the perfect combination for best-selling sex.[39] Some Latin Americans
who use cocaine, and especially those who deal in it, are reported to

regard it as a *macho* drug that readies them for battle or sex at will. Cocaine is also supposedly used at what journalists call jet set orgies (where no doubt its sheer expense has almost as much allure as its psychopharmacological powers). The prolongation of sexual activity in men by the delay of ejaculation can also give a sense of power and control.

Many people consider cocaine to have the same special aphrodisiac effect on women that Ellinwood saw in amphetamine abusers. "Girl" and "lady" are a couple of the drug's aliases, possibly because men (and some women) believe that women love it. Early writers speak of the "nymphomania of cocainist women" [40] as some of the men interviewed by Gay and Sheppard spoke of the "nympho chicks" produced by amphetamines. One woman we interviewed said, "I've blown cocaine and had ecstatic sex. It's an aphrodisiac. . . . It's called 'the lady'; it brings out the most lady qualities in women. It makes you really feel incredibly feminine . . . coquetry . . . immediately, you do a little blow and you go and you put on your best clothes. . . . It brings out your artistry. In terms of lovemaking, very artistic lovemaking." Prostitutes commonly use it both to relieve fatigue and to make their work tolerable, if not enjoyable; it is said to be a popular accessory in high-priced brothels. One writer estimated that in 1913 half the prostitutes in Montmartre used it.[41] (At that time, of course, both prostitution and cocaine were legal in France.) Pimps have a reputation for keeping large supplies on hand. Like belladonna ("beautiful lady"), the source of the chemically related drug atropine, cocaine also dilates the pupils of the eyes and so makes women momentarily more attractive.

One of the men treated at Maier's clinic in Zürich in the 1920s gave him a particularly lurid account of the effect of cocaine on women. He believed that among habitual users it excited them sexually much more than it excited men, who were often psychologically aroused but unable to sustain an erection or come to a climax. He said that a little bit of cocaine made women insatiable and described orgies at "cocaine clubs" after which the participants felt remorseful and ashamed to look at one another. It is safe to assume that Maier's patient wanted to tell a good story and that Maier himself was prepared to associate an illicit drug with illicit sexual activity. Another student of cocaine use in the 1920s remarks that abusers' accounts of their sexual prowess and exploits are as dubious as their resolutions to renounce the drug and start a new life. A writer in *Playboy*, one of our own culturally approved sources of sexual connoisseurship, asserts, "Men consistently told me that women turn on for coke, but the women I talked to were vague on the subject." [42] Nevertheless, the testimony of Gay's and Sheppard's patients and the observations of Ellinwood on amphetamines suggest that the association of hypersexuality in women with the use of stimulants is more than just an expression of male fears or desires.

But people in general, and especially drug users, may have a tendency to magnify verbally the importance of sexuality and the sexual effects of drugs. Any drug, from amphetamine to barbiturate, that removes inhibitions may heighten erotic desires and fantasies. One of our society's standing joke topics is the use of alcohol to seduce women. Stimulants in particular may prolong and intensify sexual activity. But it is interesting to note that most of the people interviewed by Gay and Sheppard thought marihuana was the drug that most enhanced sexual pleasure, largely by reducing inhibitions and heightening sensibility without actually increasing sexual drive as stimulants may. And in the end, even among these drug sophisticates, "an almost invariable return to 'sex on the natch' [natural] is described." [43] Obviously the more spectacular sexual effects may come to seem undesirable, and once the novelty wears off, cannabis and other drugs often lose their apparent sexual powers.

Enhancement of sexual interest and capacity by cocaine, insofar as it exists, is in any case unreliable. L. Vervaeck, writing in 1923, found that only 20 to 30 percent of cocaine users became sexually excited and that stimulation often gave way to impotence and frigidity after a while. A number of the people we interviewed, including women, also reported no or only occasional sexual excitement. In spite of unreliable anecdotes about old *coqueros* with great sexual powers and some reports of erotic fantasies, there is little evidence that coca chewing has any significant aphrodisiac effect. Gutiérrez-Noriega says that the Indians rarely find it to be sexually stimulating. The women of the Kogi tribe, as we have mentioned, regard coca as a rival for the interest of the men rather than an ally of their own sexuality and desire for children. In large doses or after chronic consumption of substantial quantities, cocaine prevents at first ejaculation and later erection in men, even when erotic fantasy remains strong. In most of the cases of chronic cocaine abuse studied by Maier, the victims lost interest in sex after a few months. [44] Since people often go on using cocaine, as they go on using alcohol, long after the drug ceases to increase sexual interest or activity, the sexual effect is evidently not one of the most important reasons for using it. The effect of cocaine in this as in other respects may resemble that of an emergency situation. A suggestion of adventure or novelty may make sex more exciting for some members of the uncomfortably domesticated human species, but chronic stress makes it nearly impossible.

Acute Psychological Effects—Large Doses

The acute psychological effects of large or excessive doses of cocaine are strikingly similar to those of large doses of amphetamine. We will leave most of the discussion of the more severe symptoms of cocaine intoxication for Chapter 6, which deals with chronic use, because many of the phenomena associated with acute abuse do not appear until there has already been considerable chronic abuse, and the symptoms are often the same. But there may sometimes be serious acute effects, especially from injection, even before any chronic abuse. We have already spoken of the stereotyped behavior produced by both cocaine and the amphetamines. (Apparently its onset is slower with cocaine.) [45] This can take various forms, from minor chewing or teeth-grinding movements and unusual interest in tasks like sewing or typing to the absorption in apparently meaningless repetitive activities known as punding. Cocaine abuse, like amphetamine abuse, may also cause paranoia, ideas of reference, and delusions of grandeur.

A man we interviewed who had injected both cocaine and heroin intravenously described some paranoid episodes: "Cocaine will scare you, if it's strong enough. . . . I've had that experience, and the only way to bring that down is to shoot some heroin. . . . When I did get paranoid, I got paranoid about the police. I thought the police was after me. . . . You could walk down the street and turn the corner, and walk down the block half a block, and swear you hear somebody around the corner calling you, and you'll walk back, and turn the corner, and go back up the block . . . and there's nobody there." A man who sniffed rather than injected the drug reported this incident: "I was driving home, I stopped my car, I swear, and got out, walked to the back of the car, and peeped up the exhaust pipes, 'cause I thought somebody was in the car with me." Another comment from an interview was: "When I've shot a lot of coke—now I've shot coke for two or three hours, all night—I wake up in the morning and I come out—hey! think somebody behind me. . . . I know people that had been riding around town, thinkin' the police was after them . . . very seldom, I felt that way, paranoia. Only when I was overdoin' it, and *knew* I was overdoin' it." But none of the cocaine users we interviewed knew anyone who had experienced the kind of serious psychotic episode that, as we shall see in Chapter 6, is possible although not common.

In an 1889 article J. Chalmers da Costa describes "Four Cases of Cocaine Delirium" produced by application to the urethra or prepuce for surgical purposes. One man became pale and immobile, apparently ceased to breathe, and had an imperceptible pulse. His lips were pale and his face bathed in sweat. After 15 seconds shallow and slow breathing with a light pulse began, as unconsciousness and general insensibility to pain continued. Then his facial muscles twitched convulsively and he tossed his arms and legs. He began to talk incoherently, to laugh and sing, seemingly oblivious to his surroundings. As the delirium passed, he concentrated on his own ideas, which flowed torrentially and with intellectual brilliance. About an hour after the onset of the symptoms he fell asleep, waking two hours later with complaints of headache, giddiness, numbness of the extremities, and dryness of the throat. He repeated the urethral application on his own one night, and suffered severe after-effects including prostration, numbness, and dimness of vision. In another case the patient cried out after five minutes that he could not breathe, brushed imaginary bugs off his coat, and said that the physician was his brother. He staggered about like a drunken man, upset chairs, aimed blows at the doctor, and declared with distinct articulation that the doctor wanted to kill him. He was taken home and put to sleep with sodium bromide and morphine. In a third case the patient began to mutter a few minutes after application to the urethra and moved his hand as if sending a telegram (he was a telegraph operator). He talked about the need to send the message and "telegraphed" with great energy. After ten minutes he recovered, complained of difficulty in breathing, and had no memory of the immediately preceding period. The next morning he felt dull and heavy for a few hours.[46]

Acute Physiological Effects—Moderate and Excessive Doses

The unequivocally physiological effects of cocaine are as a rule not nearly so important as the psychological ones. We have already mentioned most of them or referred to them by implication. If cocaine is sniffed it produces a cold or numb sensation in the nose and palate; taken nasally or orally it anesthetizes the taste buds. Otherwise, the physiological phenomena it causes are characteristic of sympathetic excitement. It increases the basal metabolic rate and produces hyperglycemia (a heightened level of blood sugar), an increase in muscle tone, and mydriasis (dilation of the pupil of the eye); it also constricts peripheral blood vessels. The throat often becomes dry, and stomach and intestinal activity is usually reduced. (Freud, drinking 50 mg in a 1 percent solution, reports a "cooling eructation.") Cocaine also makes breathing faster and deeper and increases the heart rate, although it may sometimes slow the heart momentarily at first by central stimulation of the vagus nerve. It raises body temperature and sometimes induces sweating; men who chew coca often say that it makes them feel warm. In a group experimentally exposed to cold, those who were chewing coca showed a slower core temperature decline and a quicker decline of temperature in the extremities (because of peripheral vasoconstriction). The hyperthermia (temperature elevation) is primarily a central rather than a peripheral effect, since it does not occur in animals under general anesthesia. It is not related to muscular activity, since it appears even in animals prevented from moving. Probably the thermal regulator mechanism in the hypothalamus is affected directly.[47]

At larger doses, especially intravenous or subcutaneous, cocaine can produce headache, pallor, cold sweat, rapid and weak pulse, tremors, Cheyne-Stokes respiration (fast, irregular, and shallow), nausea, vertigo, convulsions, unconsciousness, and death. Most people who take large doses, including several men we interviewed, are frightened by the pounding of their hearts and likely to believe that death, if it occurred, would be from cardiovascular collapse. It is true that if a large amount of cocaine enters the body very rapidly (i.e., by intravenous injection) there may be an idiosyncratic direct toxic effect on the cardiac muscle that causes it to stop beating, usually after ventricular fibrillation (rapid twitching of individual fibers or small bundles of fibers in the ventricle of the heart). But paralysis of the medullary brain center that controls respiration, often preceded by convulsions, is the most common cause of death.[48]

Incidence of Death and Acute Poisoning

The lethal dose is uncertain and variable. According to Gutiérrez-Noriega and Zapata Ortiz, in dogs it is 20 mg per kg orally (about 1.4 grams for a 150-pound man) or 10 to 12 mg per kg subcutaneously or intravenously (about 700 to 850 mg in a 150-pound man). They claim that the safety margin is low, because the lethal dose may be only twice the optimum stimulant dose. But William Hammond, as we shall see, took

800 mg and 1.2 grams subcutaneously on two occasions (the latter, admittedly, in four doses over a 20-minute period). Although both these quantities exceed the dose estimated to be lethal by Gutiérrez-Noriega and Zapata Ortiz, he suffered no permanent ill effects. Reliable information on this subject is very hard to come by, but as little as 20 mg applied to the nasal mucous membrane or 1.2 grams taken orally is supposed to have caused death in unusual cases; this is probably a rare anaphylactoid effect.[49]

In the first 40 or 50 years of cocaine use there were many reports of acute poisoning and sometimes death, usually in surgery. One of the earliest of these is J. B. Mattison's 1891 article, "Cocaine Poisoning." From the period 1888 to 1891 he recounts six deaths and a number of other cases of acute poisoning. He refers to other fatalities and poisonings totaling over 200 recorded cases and suggests that still more have gone unreported. He advises great caution in the use of "this peerless drug" with its highly uncertain lethal dose. To cite one of his cases:

> Male, age 29, one drachm [a teaspoonful] of a 20 per cent solution injected in urethra, prior to urethrotomy. Instrument was scarcely removed, when patient made a foolish remark, facial muscles twitched, eyes staring, frothed at mouth, and face was congested, breathing embarrassed and a violent epileptiform convulsion, lasting several seconds, ensued. This, increasingly severe, continued several times a minute; the whole muscular system was involved, requiring force to keep him on table. Lung action first failed, then the heart irregular and slow, breathing more and more disturbed, face and entire body deeply congested, and twenty minutes from the first convulsion, patient was dead.

This was a large dose, nearly 800 mg. Other deaths came from injections into the eyelid (three-quarters of a syringe [of unspecified size] of 5 percent solution), breast (225 mg), and gums (1 gram), or large amounts—about a gram and a half—taken by mouth. In an incident that was not fatal, "Would-be suicide took 22 grains [about 1.5 grams] in beer, causing great belly pain, intense dyspnoea and vertigo, and urine suppression for 24 hours." Another example: "J. Chalmers da Costa reported to me the case of woman, age 22, in whose forehead he injected 10 min. [a minim is one-sixtieth of a teaspoon, or about one drop] of a 6 percent solution [only about 60 mg of cocaine], causing shallow, rapid breathing, quick, weak pulse, great tremor, temperature 102, with delirium for several hours, and complete analgesia."[50]

In a report in the *Journal of the American Medical Association* for 1924 on "The Toxic Effects Following the Use of Local Anesthetics," Emil Mayer analyzes 43 fatalities. Although by that time procaine was being used far more than any other local anesthetic, cocaine caused 26 of the deaths, usually by producing convulsions and respiratory failure a few minutes after its application. The most dangerous sites and techniques of administration were applications to the inflamed urethra and to the tonsils (mostly for tonsillectomy). The accidental substitution of cocaine for procaine caused two deaths. The committee headed by Mayer recommended that cocaine not be injected into the submucous tissue or subcutaneously and also that only solutions of low concentration be used on skin and mucous surfaces. It opposed the use of cocaine paste or "mud" as a preoperative measure.[51]

It is as difficult to say how common acute poisoning or death from cocaine was or is as to say what dose is lethal. Deaths in surgery now occur very seldom, since surgeons now use the drug less often and more cautiously. One exception is a case recently reported in the journal *Anesthesia and Analgesia*. In preparation for a bronchogram to determine lung pathology, a physician in a medical clinic administered a 10 percent solution of cocaine to the throat and trachea by mistake instead of the correct 1 percent solution. The patient, a 22-year-old woman, suffered a cardiac arrest which produced permanent brain damage and has been in a coma since June 1972. She has been awarded $2,000,000 in damages from the clinic and the pharmacy that supplied the drug, an amount said to be the largest ever awarded to one person in a malpractice suit. Deaths from recreational use have probably always been rare, and sniffing in particular rarely causes serious acute poisoning. For example, in a 1920 article based on the records of the New York City Narcotic Clinic, S. Dana Hubbard cites only one recorded death in 1919 from cocaine and 51 from opiates. Statistics from the City and County Coroner's Office of San Francisco in 1973 show 80 deaths from heroin, 137 deaths from barbiturates, 553 deaths from alcohol, 10 deaths from amphetamines, and none at all from cocaine. (There were two in 1971–1972.) A 1974 article in the *New York Times* on the East Side singles scene reported the death of a young woman in a Times Square hotel from what appeared on autopsy to be an overdose of cocaine. But the article states that medical examiners and police believe cocaine deaths to be rare and therefore suspect murder in that case.[52]

These very low numbers may be misleading. Deaths from cocaine may simply not have been recognized in the past, because large-scale cocaine users almost always take other drugs that are more familiar to the authorities, and cocaine use comes to the attention of the law and physicians less often than, say, opiates, because of the absence of a physiological need for the drug. Also, cocaine is metabolized quickly and therefore difficult to detect in the blood or urine. Official attributions of cause of death are often influenced by socially accepted myths; for example, it is now recognized that many of the deaths once attributed to heroin overdose were probably caused by a combination of heroin with alcohol or barbiturates, by alcohol and barbiturates alone, or even by the quinine used to cut most street heroin. A recent experiment on mice is especially interesting in this respect. It indicates that the LD 50 (dose that kills half the experimental population) of heroin is 57 mg per kg, of quinine 138 mg per kg, and of cocaine 31 mg per kg, all intravenously. The LD 50 of heroin is raised by small proportions of cocaine but actually lowered by large proportions. R. D. Pickett, who conducted the experiment, believes that in the doses usually used by heroin addicts (equal proportions by weight of the two drugs) cocaine potentiates heroin's lethal effect.[53] The tendency of a large amount of cocaine to paralyze the respiratory center in the brain may be supplementing the respiratory depression produced by opiates. The LD 50 of cocaine alone in this experiment was the equivalent of about two grams intravenously in a 150-pound man, far more than most people ever use; but human beings are probably more susceptible than mice to the drug, and may become more sensitive to its toxic effects as they continue to use it. Whatever we make of these results, it is possible that as cocaine is used more often and becomes more prominent in the public consciousness, more deaths will be attributed to it, either truly or falsely. Certainly as long as the law continues to treat the drug as it does now, practically everyone will have an interest in obscuring the truth and confusing the issue in one way or another.

Deaths from recreational use, then, may not be so rare as they seem, at least if the drug is injected rather than sniffed. As Maier observed in 1926, death by cocaine poisoning may be attributed to "heart attack" if no tests are made for chemicals in the blood. Autopsy after an overdose usually shows hyperemia (congestion with blood) of the brain, lungs, liver, and kidneys. There may also be fatty degeneration of liver cells and lung infarctions (necrosis of tissue) produced by embolisms (obstructions of blood flow) formed because of the slowing of capillary circulation·by peripheral vasoconstriction. But these phenomena are not diagnostically specific for cocaine poisoning. Maier was convinced that three of his former patients' "heart attacks" were deaths from cocaine. If fatalities occurred even in surgery, when immediate help was usually available, they must have occurred in other situations too and gone unreported or misreported out of fear of the law.[54]

In the 1920s it was commonly accepted that an overdose of cocaine could be lethal, and sometimes the drug was used in suicide attempts like the one described by Mattison. Vervaeck recounts several cases of suicide by cocaine; one was a London dancer who died after swallowing at least 500 mg. In Jean Cocteau's novel *Le grand·écart* (*The Big Split*), published in 1923, the hero, after hearing about a man who died from "sniffing too much powder," tries to commit suicide by drinking whiskey containing ten grams of cocaine. He survives only because the bartender who sells him the drug swindles him by diluting it. Cocteau had used cocaine and other drugs himself and probably knew what he was talking about when he implied that one could die from sniffing cocaine, although he may have been thinking of the complications of chronic abuse rather than acute overdose.[55]

Nonfatal incidents of acute poisoning from recreational use of cocaine are easier to document. A man we interviewed recounted one: "I stopped shooting it. I've shot it once since I was 20, and it was a disaster. . . . I took too much, and I felt as though my head was going to explode. I was sick for about 12 hours, and I puked for about three hours. My friends kept telling me to go to bed. Later they told me I was moaning all night, loudly, and I didn't know it. My body was in complete agony, and there was no relief. That was the last time I shot it." Another interview subject described ill effects from excessive snorting: "I've had too much coke, and I've felt, you know, for like an hour, all the traditional things you're supposed to feel when overdosing, crawling like bugs in your skin, really hot sweat, nausea, and just feeling like you're almost dying, for almost 15 minutes." In another incident, an interview subject who had sniffed two-thirds of a gram in a short period of time (she is not sure whether the cocaine was cut with other drugs) felt this effect: "It seemed like I was going to pass out any second . . . a very dead feeling in my whole body, but my heart racing. . . . I was afraid of my heart giving out. . . . I was aware that taking that much was self-destructive; I felt it was sort of suicidal. I had no idea how much it took to o.d. . . . Since that experience, I've dropped off. I became more conscious of the heartbeat speeding, paranoia kinds of effects." Although incidents like these may become more common as cocaine becomes more popular, they are probably unusual now. If a recent DEA survey can be trusted, cocaine overdoses rarely bring people to the emergency rooms of hospitals.[56]

The basic treatment for cocaine overdose is revival of breathing by artificial respiration or administration of oxygen under pressure, facilitated if necessary by a muscle relaxant like diazepam or even succinylcholine. If convulsions occur, a short-acting barbiturate like sodium pentothal intravenously in small doses (25–50 mg) is recommended. One clinical toxicologist suggests 1 to 2 mg intravenously of propanolol (a drug that blocks catecholamine receptors in the peripheral nervous system) and 1 mg physostigmine intramuscularly, or else 200 mg oral secobarbital. A heart stimulant like phenylephrine or cardiac massage may be necessary. Action must be taken very quickly, because death usually arrives in less than five minutes and rarely takes as much as half an hour. If the patient survives this early period he or she will probably recover fully.[57]

Postscript: Report of an Experiment

William A. Hammond (1828–1900) was Surgeon General of the United States Army during the Civil War and later became an enthusiastic advocate of the use of cocaine. He left an eloquent and detailed report of an experiment on himself that he undertook in 1885 to determine the effects of different doses of the drug.[58] Of course, his testimony lacks all statistical virtues and even the experimental virtues of controls and double-blindness, but in the absence of any substantial information of this kind about the effects of cocaine on human beings (Offerman's experiment is a partial exception), what Hammond has to say is of great interest:

> I began by injecting a grain [65 mg] of the substance under the skin of the forearm, the operation being performed at 8 o'clock P.M. The first effect ensued in about five minutes, and consisted of a pleasant thrill which seemed to pass through the whole body. . . . On feeling the pulse five minutes after making the injection, it was found to be 94, while immediately before the operation it was only 82. With these physical phenomena there was a sense of exhilaration and an increase of mental activity that were marked, and not unlike in character those that ordinarily follow a glass or two of champagne. I was writing at the time, and I found that my thoughts flowed with increased freedom and were unusually well expressed. The influence was felt for two hours, when it gradually began to fade. At 12 o'clock (four hours after the injection) I went to bed, feeling, however, no disposition to sleep. I lay awake till daylight, my mind actively going over all the events of the previous day. When I at last fell asleep it was only for two or three hours, and then I awoke with a severe frontal headache. This passed off after breakfast.
>
> On the second night following, at 7 o'clock, I injected *two grains*. . . . All the phenomena attendant on the first experiment were present in this, and to an increased degree. In addition there were twitching of the muscles of the face, and a slight tremor of the hands noticed especially in writing. . . . I felt a great desire to write, and did so with a freedom and apparent clearness that astonished me . . . when I came to peruse it . . . it was entirely coherent, logical, and as good if not better in general character as anything I had previously written. The effects of this dose did not disappear till the middle of the next

day, nor until I had drunk two or three cups of strong coffee. I slept little or none at all, the night being passed in tossing from side to side of the bed, and in thinking of the most preposterous subjects. . . . Four nights subsequently I injected *four grains* [260 mg] of the hydrochlorate of cocaine into the skin of the left forearm. The effects were similar in almost every respect with those of the other experiments except that they were much more intense. . . . I wrote page after page, throwing the sheets on the floor without stopping to gather them together. When, however, I came to look them over on the following morning, I found that I had written a series of high-flown sentences altogether different from my usual style, and bearing upon matters in which I was not in the least interested . . . and yet it appeared to me at the time that what I was writing consisted of ideas of very superior character. . . .

The disturbance of the action of the heart was also exceedingly well marked, and may be described best by the word "tumultuous." At times, beginning within three minutes after the injection, and continuing with more or less intensity all through the night, the heart beat so rapidly that its pulsations could not be counted, and then its action would suddenly fall to a rate not exceeding 60 in a minute, every now and then dropping a beat. This irregularity was accompanied by a disturbance of respiration of a similar character, and by a sense of oppression in the chest that added greatly to my discomfort.

On subsequent nights I took *six, eight, ten, and twelve grains* of the cocaine at a dose. . . . The effects . . . were similar in general characteristics though of gradually increasing intensity. . . . In one, that in which *twelve grains* [780 mg] were taken, I was conscious of a tendency to talk, and as far as my recollection extends, I believe I did make a long speech on some subject of which I had no remembrance the next day. . . . Insomnia was a marked characteristic, and there was invariably a headache the following morning. In all cases, however, the effects passed off about midday. . . . A consideration of the phenomena observed appeared to show that the effects produced by twelve grains were not very much more pronounced than those following six grains. I determined, therefore, to make one more experiment, and to inject *eighteen grains* [1170 mg]. . . .

I had taken the doses of eight, ten, and twelve grains in divided quantities, and this dose of eighteen grains I took in four portions within five minutes of each other. At once an effect was produced upon the heart, and before I had taken the last injection the pulsations were 140 to the minute and characteristically irregular. In all the former experiments, although there was great mental exaltation, amounting at times almost to delirium, it was nevertheless distinctly under my control. . . . But in this instance, within five minutes after taking the last injection, I felt that my mind was passing beyond my control, and that I was becoming an irresponsible agent. I did not feel exactly in a reckless mood, but I was in such a frame of mind as to be utterly regardless of any calamity or danger that might be impending over me. . . . I lost consciousness of all my acts within, I think, half an hour after finishing the administration of the dose. Probably, however, other moods supervened, for the next day when I came downstairs, three hours after my usual time, I found the floor of my library strewn with encyclopaedias, dictionaries, and other books of reference, and one or two chairs overturned. I certainly was possessed of the power of mental and physical action in accordance with the ideas by which I was governed, for I had turned out the gas in the room and gone upstairs to my bedchamber and lighted the gas, and put the match used in a safe place, and undressed, laying my clothes in their usual place, had cleaned my teeth and gone to bed. Doubtless these acts were all automatic, for I had done them all in pretty much the same way for a number of years. During the night the condition which existed was, judging from the previous experiments, certainly not sleep; and yet I remained entirely unconscious until 9 o'clock the following morning, when I found myself in bed with a splitting headache and a good deal of cardiac and respiratory disturbance. For several days afterward I felt the effects of this extreme dose in a certain degree of languor and indisposition to mental or physical exertion; there was also a difficulty in concentrating the attention, but I slept soundly every night without any notable disturbance from dreams. . . .

Certainly in this instance I came very near taking a fatal dose, and I would not advise anybody to repeat the experiment. . . .

It is surprising that no marked influence appeared to be exercised upon the spinal cord or upon the ganglia at the base of the brain. Thus there were no disturbances of sensibility (no anaesthesia) and no interference with motility, except that some of the muscles, especially those of the face, were subjected to slight twitchings. In regard to sight and hearing, I noticed that both were affected, but that while the sharpness of vision was decidedly lessened, the hearing was increased in acuteness. At no time were there any hallucinations.

6

EFFECTS OF CHRONIC USE

THE JUSTIFICATION for outlawing cocaine was mainly the supposed psychological and physiological consequences of prolonged use. But the law does not distinguish, as we must, between moderate and excessive doses. The more spectacular consequences of cocaine abuse are not typical of the drug's effects as it is normally used any more than the phenomena associated with alcoholism are typical of the ordinary consumption of that drug. We insist on this here because we may seem to be overemphasizing the most harmful effects, and we do not want to imply that it is impossible to use cocaine (or any other drug) in moderation. If we seem to speak about severe abuse at too great length, it is because this is more important to the user and society than the occasional Saturday night "blow," just as alcoholism is more important than ordinary social drinking. The question of how often cocaine or any other drug will in fact be abused, or used at all, is quite different from the issue of the effects of its abuse. It involves a number of conditions, including above all availability, that conspire to create what is called drug dependence or a drug habit. This is largely a social and cultural phenomenon, which we shall discuss more fully in Chapter 8.

Effects of Chronic Coca Use

The best source of information, inadequate as it is, on the long-term effects of moderate doses of cocaine is studies of the coca-chewing regions of South America. The anecdotal evidence, from the time of the Spanish conquest on, is unfortunately variable, unreliable, contradictory, and heavily colored by the biases of the observers. We are told on the one hand that *coqueros* are liars, depraved, pickpockets, indolent, submissive, depressed, stupid, and subject to muscular degeneration, anemia, jaundiced skin, digestive complaints, and other diseases; and on the other hand that coca is a harmless stimulant and tonic which has never caused any nervous or physical disease. Poeppig believed that habitual use of coca caused mental and physical decadence, but Mantegazza, Tschudi, and others considered it healthful in moderation and rarely overused. Richard E. Schultes told us that the members of the Yucuña tribe in the Amazon region of Colombia, the largest consumers of coca he had ever encountered, were also remarkably strong and healthy. He doubts that misery and malnutrition among the highland Indians has much to do with coca.

Gutiérrez-Noriega and Zapata Ortiz undertook the first serious systematic studies of the chronic psychological and physiological effects of coca chewing. In spite of their conviction that the habit was dangerous, the results have to be called inconclusive. In a 1947 study, "Mental Alterations Produced by Coca," Gutiérrez-Noriega asserts that *coqueros* who chew 20 to 50 grams of leaves a day are very much like non-*coqueros* matched for relevant variables. But those who chew over 100 grams a day, he says, are very different: they are dull and torpid, sit silent and motionless for hours, have dry skin and bad posture, lack sexual interest, answer questions vaguely, make contradictory statements, and cannot handle abstract ideas. Gutiérrez-Noriega believes that the women, who do not chew coca, are more intelligent and energetic. He also comments that most people in the area he studied believe that coca is good for health. He admits that the apathy, timidity, and introversion found in some *coqueros* is an accentuation of the Andean temperament that apparently existed even before the Spanish arrived and consumption of coca became common. But he concludes that "There is no medical problem of greater importance in Peru." In another article Gutiérrez-Noriega and Zapata Ortiz note that illiteracy is highest where coca chewing is most common, but they also note that these are the areas where Quechua is spoken instead of Spanish. They admit that most *coqueros* show no obvious personality or mental deterioration, and the only abstinence symptoms they found were occasional depression and irritability.[1]

Gutiérrez-Noriega and Zapata Ortiz also studied "Intelligence and Personality in Persons Habituated to Coca." They found subnormal intelligence on the Binet and Porteus Maze tests and a longer reaction time to sound stimuli than in non-*coqueros*. The longer a *coquero* had been chewing, the lower his scores were. On the Rohrschach test *coqueros* produced fewer total responses, fewer global responses, and fewer original ideas than controls. The Rohrschach results revealed personalities described as apathetic, indolent, and hypoaffective. Gutiérrez-Noriega and Zapata Ortiz concluded that chronic intoxication by coca was a major cause of intellectual deterioration.[2] This study was conducted in Lima, where *coqueros* are a deviant minority, and it has been criticized for this reason.

J. C. Negrete and H. B. M. Murphy have made the most carefully controlled study of this topic. In an analysis, "Psychological Deficit in Chewers of Coca Leaf," they chose as their subjects 50 *coqueros* and 42 controls who worked in the fields on a sugar plantation in northern Argentina. All were men between the ages of 25 and 49; about half were Bolivian migrant workers and half local residents; half were literate and half illiterate; and half were older than 36. This location was chosen because neither men who chewed coca (as in Lima) nor those who did not chew it (as in some areas of the Peruvian and Bolivian Andes) were considered a deviant minority, and the population was more or less equally divided between chewers and nonchewers. Negrete and Murphy eliminated anyone with a history of mental disorder, head injury, epilepsy, physical illnesses known to affect mental functioning, "erratic or antisocial work practices," or excessive drinking. The exclusion of heavy drinkers, they say, biased the sample so that the controls were slightly younger and more literate than the chewers; but they believe the sample remained satisfactory. A chewer was defined as someone who had chewed 200 grams a week for at least ten years. The men used as controls were never seen to use coca while working and denied having taken more than an average of 10 grams a week.

Noting the fact that chronic users were often said to look dull and apathetic, Negrete and Murphy used a battery of tests designed to measure psychological deficit caused by organic brain damage. They admit that there is no single reliable method of measuring this and that the concept of brain damage is a very loose one. They also point out that there is no evidence of damage by the coca alkaloids to any specific part of the brain. The battery of tests included a verbal intelligence scale, auditory and visual memory tests, figure completion and similarity recognition tests from the Army Beta scale, an attention test (the Knox cubes), a test of manual ability, tactile and spatial memory, and learning (the Seguin Form Board test), and a block design test adapted from Wechsler. Although the examiner knew whether each individual was a chewer or a control, Negrete and Murphy believe that examiner bias had no influence, since the researchers did not expect to find a difference between the two groups.

When allowances were made for age, literacy, and background (local or immigrant), the *coqueros* performed worse than the controls in almost all respects. Only the verbal intelligence test yielded no significant differences; the researchers expected this from previous work on brain damage. The differences were greatest on the Army Beta tests and the learning measure of the sorting test (Seguin Form Board). The researchers asked about diet but found no reason to suspect greater dietary deficiencies among the chewers. (It is not clear how thoroughly they examined this question.) They reject the suggestion that men who take up the habitual use of coca are deficient to start with. First, they believe, a lower native intelligence would have shown up most clearly on the verbal intelligence scale. Second, they found a relationship between duration of chewing, apart from age, and four of the test scores (auditory memory, manual learning, and immediate and delayed attention). Temporary intoxication did not seem to affect the results: there was no correlation between lower test scores and recent large doses. Negrete and Murphy reject the hypothesis that chronic use of coca induces retardation or depression by exciting inhibitory reflexes, because chewers took no longer on the tests than controls. They conclude, however, that "coca must be assumed to have an adverse effect on the brain until it is proved otherwise." Still, the deficiencies revealed by the tests were not observable on casual inspection or even during the test taking; the authors believed they would remain unimportant as long as *coqueros* lacked social opportunities to make use of their intelligence.[3]

In a later study Murphy, O. Rios, and Negrete tried to determine whether long-term coca use would in fact prevent *coqueros* from learning new skills if the occasion arose. Working in the same sugar-growing area of northern Argentina and with a similar population, they admitted 20 chewers and 10 controls to the empty wing of a plantation hospital for ten days. Of the chewers, 10 were supplied with their customary amount of coca and 10 were required to abstain. Abstainers tolerated the absence of the drug easily, and some of them said they would not have used it in such circumstances anyway, since its purpose was to suppress hunger and fatigue. They were more annoyed about the enforced abstention from alcohol. All the subjects took repeatedly, at intervals of three days, a single battery of tests consisting of the items in the previous study by Negrete and Murphy that had distinguished best between coca users and controls: the auditory memory test, the Army Beta tests, the attention test, the Seguin Form Board, and some of the block design tests.

Assuming a chronic psychological deficit of some kind, one could expect *coqueros* to score worse initially and perhaps improve less than controls. If the immediate effect of coca compensated for some of the chronic damage, the continuing users should score better than the abstainers. Finally, if there was a strong withdrawal reaction, the scores of the abstainers, especially on attention tests, might be expected to decline. The results showed no substantial effects of recent consumption or withdrawal. All three groups improved on each test in approximately equal degrees from the first to the third trial. The subtests on which the continuing users improved most compared with the other groups were ones that required manual dexterity rather than memorizing or abstract thinking. On the memory and attention tests, literate controls scored much higher than illiterate ones, while literate *coqueros* performed no better than illiterate ones. The authors suggest that there is a kind of memory developed by formal training in school and destroyed by chronic consumption of coca.

From a detailed analysis of the tests, the authors conclude that chronic coca use does not affect simple untrained memory, hand-mind coordination, or elementary concept formation. *Coqueros* showed no clinical signs of ataxia or poor muscle control, and the tests revealed no language difficulties or perception disturbances that suggested damage to a particular area of the brain. The coca users fell behind most on memorizing associated with literacy and on the more difficult abstract sections of the Army Beta subtests. They found it easy to compare figures to which they could give a name but hard to compare, say, geometric designs. The authors believe that the failures of the *coqueros* have a common source in a deficiency in abstract thinking of the kind found in some lobotomized patients and patients with lesions of the frontal lobes and possibly related to extensive loss of brain tissue in any lobe. Kurt Goldstein's description of this kind of patient, according to the authors, resembles the accounts that have sometimes been given of heavy coca users in the high Andes. They conclude that older *coqueros* might find it difficult to grasp the principles behind innovations in social organization but would have no trouble understanding and using new techniques in farming and industry presented in sufficiently concrete form. Since the present conditions of their lives do not permit abstract thinking, coca is probably doing as much good as harm. For the time being, the authors believe, efforts should be devoted to social reform rather than eradicating the habit.[4]

The test results may in fact be more ambiguous and inconclusive than Negrete, Murphy, and Rios think. The basic question, of course, is whether the relationship between coca use and test scores is causal. A subsidiary issue is the precise meaning of the capacities supposedly measured by the tests. First, there are problems of bias in the sample. By excluding excessive drinkers and men with erratic work practices, the authors may have weighted the coca-using group with subjects whose prior intellectual or psychological deficiencies would have caused alcoholism or inability to work regularly if they had not taken to chewing coca instead. The question of diet was not examined closely, either; other research, as we shall see, suggests that malnutrition is a major cause of coca chewing. As for the tests themselves, the authors rely heavily on the hypothesis that verbal intelligence (on which *coqueros* and controls scored the same) is less easily impaired by organic brain damage than the capacities measured by the Army Beta figure completion and similarity recognition tests (on which *coqueros* scored lower). In rejecting the contention that the coca users in their sample were less intelligent than the controls to begin with, Negrete and Murphy mention an association between duration of chewing and certain test scores. But these scores do *not* include the abstract sections of the Army Beta tests, supposed to be the best measure of a deficiency in abstract thinking associated with chronic brain damage. In any case, it is also possible that whatever environmental or physiological conditions cause a man to keep chewing coca also cause progressive deterioration in performance on some of the tests. In short, confusing environmental variables and difficulties in interpreting the tests make the results of the work by Negrete, Murphy, and Rios less conclusive than they might hope.

Working in conjunction with Murphy and Negrete, D. Goddard, S. N. de Goddard, and P. C. Whitehead conducted a study, "Social Factors Associated with Coca Use in the Andean Region," in the same area and using the same criteria for admission to the sample. The subjects were 58 of Negrete's and Murphy's respondents and 20 others: 40 users and 38 controls. In the region studied it was regarded as natural to use coca but no stigma was attached to not using it. Coca was chewed "to avoid sleep, to get a will to work, to have willingness to work, not to work better," and also sometimes to provide physical strength for heavy labor and courage to face dangers like snakebite and accidents. It was said to make the step lighter and to quench thirst in hot weather. The workers rarely invested coca with curative or magical powers, although some thought it good for stomach pains. Some subjects in both the *coquero* and control groups considered coca a mild vice, but the drug had no effect on the organization of social life. The attitudes of the nonusers toward coca were classified as one-third negative, one-third indifferent, and one-third favorable. But no one felt very strongly about it. The use of coca by women was not disapproved, and 28 percent of the total sample had wives who sometimes chewed it, mostly in gatherings with friends or while doing the laundry or other household tasks.

In this study the effect of coca on children, family life, and social activity was also examined. From fragmentary evidence, the authors conclude that coca chewing by parents has no adverse effect on children's school performance. Teachers thought that coca use among older children made little difference except that the chewers might be slightly more alert. Adult *coqueros* did not have lower aspirations for themselves or their children than adult controls, and *coqueros* showed neither more nor less social isolation than controls. The authors conclude that coca chewing in the Andes does not define or restrict social relationships; it is very much like the use of gum, tobacco, betel nut, or alcohol elsewhere. They do not mention the resemblance to the use of coffee and tea, which is even greater.[5]

The methodological difficulties in showing a causal relationship between coca chewing and organic deterioration are almost as great as those of showing a relationship to psychological damage. Gutiérrez-Noriega and Zapata Ortiz, studying 500 *coqueros* in 1948, found many constitutional disturbances and signs of degeneration: anemia, eye disease, caries, hepatomegaly, muscle weakness, hyperthyroidism, and so on.[6] But very little can be proved without controls; it is hardly surprising that these conditions afflict undernourished peasants with practically no access to physicians. In 1968 Alfred A. Buck and his colleagues made a controlled study of the correlation between coca chewing and various measures of physical health among residents of a Peruvian village. The village, population 492, was at an altitude of 2,400 feet and its climate was humid and tropical. Quechua Indians, migrants from the high Andes, made up 23 percent of the population; the rest were *mestizos*. Of the 53 coca chewers in the village, 28 were Quechuas and 25 *mestizos*. Fifty-one of the coca chewers were matched with controls of the same sex, ethnic group, and approximate age; medical information was obtained by interviews, physical examinations, laboratory tests, and skin tests. Three hypotheses were tested: that coca causes malnutrition by diminishing the sensation of hunger; that it produces indifference to personal hygiene; and that the work performance of *coqueros* is inferior (presumably for reasons of organic dysfunction).

As measures of nutritional state, weight-height ratio, skinfold thickness on the back and upper arms (a measure of subcutaneous fat), and serum albumin level were examined. There were statistically significant differences favoring the controls on the last two measures. The indicators of personal hygiene used were prevalence of scabies and pyodermia (skin infections), prevalence of intestinal parasites in stool specimens, and frequency of reported rat bites. Only in the prevalence of pyodermia was there a statistically significant difference in favor of the controls. Fewer coca chewers were infested with amoebae, and the authors wonder whether coca leaves have amoebicidal properties. (Cocaine does have a paralytic effect on lower organisms, and in one instance reported to us it seemed to relieve amoebic dysentery.) To test work performance effects, the researchers asked each subject how many work days he had lost because of illness in the preceding month. The average was almost twice as many days for the coca chewers as for the controls. Coca chewers had more often suffered severe anemia, and their hemoglobin levels were significantly lower in all weight-height categories. Hepatomegaly (enlarged liver) was twice as common among *coqueros* as among controls. The authors conclude that under the conditions of their study, habitual coca chewing is associated with poor health.

But they admit: "The directions and sequences of causes and effects cannot be identified clearly, because the conditions recognized by the study as possible disease determinants are arranged in a vicious circle." Difficulties were also created by the choice of a village in which only a small minority of the population chewed coca and the attempt to compensate by matching techniques. As the authors note, there were significantly more Protestants in the control groups, since the Evangelical mission in the village did not condone the use of coca, tobacco, or alcohol. In fact, two-thirds of the Quechua controls were Protestant (and, of course, none of the *coqueros* were). But conversion from traditional Catholicism to Protestant Evangelism in such a village must imply differences in native disposition or changes in attitudes toward work and hygiene that would be overwhelmingly more important than the effects of chewing coca. In this case, giving up coca was a symbolic act representing a decision for a new way of life. In general, matching techniques are unreliable, since matching for some variables may cause the groups to be dissimilar in respect to more important ones. The very fact that each control was of the same age, sex, and ethnic group as the corresponding coca chewer makes us wonder why one used coca and the other did not; the answer to this question, as we have pointed out in the case of the Protestants, may be far more significant than any consequence of the use of coca itself.[7]

In a further study of possible physiological effects, James E. Hamner III and Oscar L. Villegas examined the cheeks of coca chewers for signs of cancer. They did biopsies of 36 tin miners and found that the mucous membranes were swollen, gray-white, and opaque, but without signs of carcinoma or chronic ulcer. They noted the betel-nut chewing in southern Asia, with or without added tobacco, was highly correlated with oral cancer, but that oral cancer was rare in Bolivia and Peru. They concluded that the condition they observed was not premalignant.[8]

The evidence associating coca use with minor deficiencies in mental and physical health or intellect is convincing (at least for the Andes and their foothills—anecdotal reports from Colombia and the Amazon give a different impression), but the causal connection has not been established. If coca users are thin, or their standards of hygiene are low, or they suffer more from minor illnesses and accidents, or they cannot handle abstractions and sometimes seem demoralized, that is easy to understand from many circumstances of their lives. It is likely that the men who suffer most from personal inadequacies and oppressive social conditions turn to the drug for solace. They also turn to alcohol. The connection between coca use and alcoholism is especially interesting. Gutiérrez-Noriega asserts that "the alcoholism that generally accompanies cocaism" makes it difficult to isolate the effects of coca. The United Nations Commission notes that observers in the Andes often say that alcohol is a far more serious problem than coca.[9] In all the studies we have discussed, adjusting for the probable effects of alcohol has been a problem. Any study that includes heavy or excessive drinkers is likely to confound the effects of alcohol with those of coca, and any study that deliberately excludes them is unavoidably creating a biased sample. Since some of the same conditions that drive men to use alcohol may drive them to use coca, *coqueros* deprived of their drug might replace it with alcohol.

Alcohol is only one of the many variables in this situation which may or may not be mere nuisances to be factored out. Probably the most important is poor nutrition, which is obviously a source of physical deficiencies and probably also associated with low scores on intelligence tests, poor work performance, and carelessness about personal hygiene. The trouble is that it has not been determined whether the use of coca is a cause or an effect of dietary deficiency, or neither, or both. According to a report by Zapata Ortiz on a brief experiment in the Peruvian Andes, *coqueros* gave up the drug when their nutrition was improved. Zapata Ortiz also points out that Indians tend to stop using coca when they join the Peruvian army, which provides them with proper food, and go back to the drug when they return to their homes.[10] Because of the complexities of the causal nexus and the presence of so many variables that are difficult to evaluate or isolate, it is premature to assert that coca produces poor health or intellectual deterioration. Chronic coca consumption has not been correlated with any serious long-term disease in the way cigarette smoking has been connected with heart disease and cancer. According to the United Nations Commission Report, local physicians in Peru and Bolivia say that coca users have no more cardiovascular illness than the rest of the population. The differences on intelligence tests discovered by Negrete and Murphy, which are probably similar to differences correlated with alcohol consumption, do not represent an unambiguous causal relation. Without any suggestion of a mechanism that can be experimentally established, there is simply not enough evidence that the use of coca is a cause rather than an effect of the minor evils associated with it.*

Benefits were once claimed for the use of coca in the thin air and cold of high altitudes, especially by Carlos Monge and his colleagues at the Institute for Andean Biology. Joel M. Hanna has recently revived some of these claims in a modified form. He emphasizes especially the advantage of increased heat retention through peripheral vasoconstriction and asserts that coca is not used for work at low altitudes because there heat retention is a disadvantage. But reports from the Amazon basin belie his contention that coca is not used for work in hot regions. In any case, Gutiérrez-Noriega has given convincing reasons for believing that any benefits conferred by high-altitude life on coca and vice versa are not very significant. He points out that in the time of the Incas most of the population did not use coca and that even today Indian women, whites, and *mestizos* manage to adapt to the conditions of the Andes without it. He notes that coca is used in the Amazon but not in sections of Ecuador, Argentina, and Chile that are at the same altitudes as the coca-chewing regions of Peru and Bolivia; and also that Tibet, where the natives show powers of endurance equal to or greater than those of the South American Indians, has no equivalent stimulant. He denies that there is any consistent rela-

---

* For a profound study, at the epistemological level, of the difficulty in isolating variables for testing in retrospective research, where the phenomenon under examination has not been produced "to order" by the experimenter, see Paul Meehl, "Nuisance Variables and the Ex Post Facto Design," in *Analysis of Theories and Methods of Physics and Psychology*, ed. Michael Rodner and Steven Winokur, Minnesota Studies in the Philosophy of Science, vol. 4 (Minneapolis: University of Minnesota Press, 1970), pp. 373–402. Meehl points out that in any field where there is no accepted model or picture of the nature of cause and effect relations, it is impossible to decide rationally which variables to regard as potentially relevant to causal questions and which ones to factor out, whether by matching or by choice of sample population.

tionship between altitude and amount of coca consumed and points out that in some coastal areas settled by migrants from the mountains the coca habit remains common. What counts, he says (providing data), is not primarily altitude but the presence of coca plantations nearby—in other words, availability of the drug. Gutiérrez-Noriega also denies that coca is useful in the long run for heavy physical labor at high altitudes. Opposing the contention that the inhabitants of the Andes constitute a climatically determined racial variation with more resistance to the toxic effects of cocaine than other peoples, he asks why, in that case, Monge and others do not assume that the Indians' susceptibility to the useful stimulant effects of the drug is also reduced.[11]

Gutiérrez-Noriega admits that the respiratory stimulant and antifatigue action of coca may be useful on the heights in an emergency. In fact, amphetamines have been shown to counteract temporarily the effects of oxygen deficiency, and coca is a popular remedy for *soroche*, or altitude sickness. But, as we have noted, stimulants cannot produce energy but only regulate its employment and distribution. People who are physiologically adjusted to mountain conditions should have less rather than more need of stimulants for this kind of purpose. The adaptation of the respiratory and cardiovascular systems that acclimatizes them to low oxygen pressure should not require the complement of a drug; there is also no clear reason to believe that a stimulant harmful at sea level would be innocuous at an altitude of 10,000 feet.

Effects of Chronic Cocaine Use— Moderate Doses

If the evidence on the long-term effects of coca consumption is inconclusive, information on the effects of chronic cocaine sniffing or injection in small or moderate doses is even more sparse and entirely anecdotal. The works on cocainism * by Maier and by Joël and Fränkel concentrate on a condition parallel to alcoholism in which the drug has damaged its user's health and disrupted his life. No one has systematically studied the kind of cocaine use that is analogous to social drinking and is connected with serious abuse along the same kind of con-

* This is a term it might be useful to revive, along with morphinism, heroinism, barbituratism, caffeinism, nicotinism, and so on. By designating other forms of drug abuse with terms parallel to alcoholism, we would be emphasizing that they are neither necessarily worse nor essentially different.

tinuum that joins social drinking with alcoholism. We have already discussed the desired effects of "social snorting." The main undesirable effects, aside from financial depletion, are of several kinds: nervousness, irritability, and restlessness from overstimulation, sometimes extending to mild paranoia; physical exhaustion and mental confusion from insomnia; undesired weight loss; fatigue or lassitude in coming down; and various afflictions of the nasal mucous membranes and cartilage. Just as occasional headache, nausea, hangover, or embarrassing behavior is not enough to prevent people from drinking alcohol, so users of cocaine do not regard the evils we have just described as serious enough to outweigh its virtues. Although Maier designates the habit of using the drug in this intermittent fashion with the formidable appellation "periodic endogenous cocaine addiction" (*Kokainsucht*—*Sucht* means literally "craze" or "mania"), he states that it causes no lasting psychological or physical harm and is not a serious illness. Joël and Fränkel refer to a "symptomless, safe-appearing picture" in cocaine sniffing that may be misleading.[12]

There is very little to be said about this in any systematic way. As a matter of taste and temperament, some people do not enjoy cocaine, as some do not like alcohol. One man we interviewed said, "I didn't like the way it affected my friends, the way the people around me acted when they snorted it. . . . If I had snorted coke . . . I would think about what I had to do ahead of time, rather than let it come naturally. . . . It hardly seems as though anybody really happy . . . is involved with cocaine, that I know." Another said, "Most of the people who are heavily into coke, coke freaks, the coke scene, I don't like . . . heavy-duty macho craziness." There were complaints of a hollow feeling that demands more and more of the drug, jitteriness, being "strung out," and later, fatigue: "It's too jittery. If you do a whole lot of it, you get sort of zombied out and you tend to stare at things." (A "whole lot" meant about five grams sniffed.) Others dislike the effect on their nasal and oral membranes. The rock musician Paul Kantner has been quoted as saying (in 1972), "I stopped using coke a year and a half ago, when it was obvious it had become more dangerous than useful to me. . . . You can function and work relatively clearly on it, like for 12 or 15 hours straight. . . . But it's not controllable. . . . And when you're heavily into it, it makes you cold to people. Also it can get you physically fucked up [he is referring to the nasal passages]."[13] A woman we had previously interviewed wrote this account for us of the misgivings that eventually caused her to cut down her consumption of cocaine and then give it up except for rare occasions:

Once I started having my own coke around, it was easy to snort more and more of it. What had been a once-a-week treat became almost a daily necessity. Because coke made me feel better—more on top of things and able to pay strict attention to the longest work projects—I did not heed what I considered to be the negative aspects: the amount of money I was putting into a drug, the colds and dripping nose that inevitably followed a lot of use, the number of cigarettes I enjoyed smoking after using a lot of coke, and the frazzled efforts to come down and sleep after a day of work and drugs. I could tell I was abusing myself physically by doing all the coke I was doing. . . . My involvement with work and drugs allowed me to get by thinking about me, which meant thinking about being alone and depressed. It got to be that I didn't want to come down. . . .

I became worried about seeing shadows flitting by in the corner of my vision. It was harder to control my feelings of urgency about getting some work done or seeing to a request by someone at the station. I experienced a constant sense of something pressing on me, urging me to work faster, live faster. . . . I needed to use more depressants—alcohol, Librium, and sleeping pills—to cool myself out. Talking to other people when I was on coke became difficult. I didn't have time for it, or I found the effort to concentrate on what they were saying too difficult. I had noticed other people experiencing similar reactions. . . .

I wonder what role cocaine played in unearthing parts of me that had previously been inaccessible. Sometimes I see it as beneficial. And at other times I see it as having scattered parts of me all over the place. Parts of me that I must now work hard to make sense of and live with. So my feelings about cocaine are mixed. I fear it and am somewhat excited and attracted to it. I don't like the way it contributed to my refusing to see the consequences of the way I was living. I don't like the way it tempted me to ignore my physical and mental well-being. And I see this in other people who love coke.

Most, but not all, people who begin to feel the way this woman did apparently find it possible to do what she did.

More commonly, even people who had experienced some of the undesirable consequences of chronic use and were wary of cocaine, especially in large amounts, told us that they would probably use more if they could afford it. Some who are already using a large amount will occasionally just leave it alone for a while. The cocaine dealer "Jimmy" told Richard Woodley that if he put his finger in his nose and drew blood, or found himself becoming too hot-tempered, or his speech becoming slurred, he would stop using the drug for a day or two. Every once in a while he would stop for a longer time to "clean my system and heal my nose up." He also had to cope with fatigue from lack of sleep.[14] Some people find that these inconveniences make cocaine unappealing to them and others do not.

Aleister Crowley was a British poet, novelist, literary eccentric, cult leader, and drug and sex experimenter of the early twentieth century. In diaries, memoirs, novels, and pamphlets he left a record of the effects of cocaine and other drugs on him over a period of more than two decades. His writings are especially interesting because they illustrate the great variations in attitude one habitual user can feel. In a pamphlet he wrote in 1917 in opposition to the Harrison Act and the outlawing of cocaine, he described the drug as safe for a wise man but not for a fool. The happiness it produced, he wrote, was not passive and placid (he may have been contrasting it with opiates) but self-conscious. He also described, apparently not from his own experience, the hallucinations (including insects under the skin) and senseless craving that could accompany severe abuse. He opposed legislation against cocaine and suggested that the "drug fiend" who abused it would serve as a warning to his neighbors.[15]

At that time Crowley was certain that he and not cocaine was the master. Later he became less sure that he or anyone else could use it wisely. In his *Confessions* he recounted the case of a woman friend who took up to one-fourth ounce (about seven grams) of cocaine a day and had to switch to morphine and then alcohol in order not to destroy herself. He wrote, "I admire her . . . for her superb courage in curing herself." Despite his apprehensions, Crowley continued to use the drug: at one point in his life he was sniffing heroin continually and also indulging in three or four bouts of cocaine a week.[16]

Crowley's diaries of 1914 to 1920 show the ambivalence most clearly. He describes the cocaine effect: "The first dose produces a curiously keen delight, rather formless. . . . There's a memory-throb and a promise of new life. The next dose or two creates a curious nervousness. . . . It reminds one of the timidity of a boy before seduction. . . . This state is succeeded by a kind of anxiety and restlessness, not unlike that of a man who means to spend the evening in some kind of amusement, can't make up his mind what to do, and is irritable at his own indecision. . . . I flutter about, I toy with things. . . . The next stage is that I am aware of the master in the saddle. . . . We are off, a long, level, easy gallop, every muscle glowing with delight, the lungs intoxicated with deep draughts of pure sweet air, the heart strong and the brain clear. I am intensely happy, utterly calm, wholly concentrated." But "With big doses . . . the mind seems paralyzed; I am nailed to two or three thoughts, usually quite meaningless. . . . I notice how my mind's reaction to the experiment is always fear-laden," much more than with other drugs (apparently a mild paranoid reaction); and "the next day, even after a long night's rest, is likely to find me dull, bored, heavy," a weariness from which more cocaine provides a tempting release. At times, "I've had about a gramme and I feel nothing but a sort of nervousness." After a while, "I am now bored by the experiment . . . cocaine results are monotonous." Eventually he is writing that "cocaine's pleasure is not worth the candle." [17]

The alternation of praise and derogation in Crowley's writings is particularly interesting because he apparently had not experienced the more lurid symptoms of severe cocaine abuse that we have mentioned and will discuss in more detail. In 1922, shortly after the period when the diary entries we have quoted were written, he wrote and published an autobiographical novel, *Diary of a Drug Fiend*, in which his inner debate continues. By this time Crowley had begun to use heroin, prescribed by a physician for asthma, and at the start of 1922 he was abusing both drugs and suffered from itching, vomiting, insomnia, diarrhea, and restlessness (which sound like cocaine effects or heroin abstinence symptoms). Still he favored the free sale and use of psychoactive drugs and denied that a "drug habit" existed. [18]

In the novel itself cocaine sets off a romance, as it stimulates the hero sexually and makes him feel himself "any man's master." He is "like a choking man . . . released at the last moment, filling his lungs for the first time with oxygen." The effects on his beloved are similarly exhilarating, and marriage and a honeymoon with plenty of cocaine and extravagant descriptions of rapture follow. Physical pleasures are "etherealized"; the stimulation of cocaine is "calm and profound," not coarse like that of alcohol; it produces "ecstatic excitement and inextinguishable laughter." But disillusionment begins to set in. "Cocaine is merely Dutch courage." It destroys one's power of calculation and cannot be taken in moderation. Man and wife now start sniffing heroin "when the cocaine showed any signs of taking the bit in its teeth." The woman writes in her fictional diary: "I can't think about anything except getting H. I don't seem to mind so much about C. I never liked C much. It made me dizzy and ill." And later, "I wonder whether it's H or C or mixing the two that's messing up my mind." But when cocaine becomes available again, she writes, "And I thought I didn't like it! It's the finest stuff there is." But again, when the cocaine is gone, Crowley's hero calls it "no good without the H." He emphasizes the difference between the physical craving for heroin and the "moral" pull of cocaine. Each time he is able to get more cocaine he feels "recovered divinity" and self-confidence.

Eventually he becomes involved in a scheme to sell the drug, and his ambivalence reaches great intensity: "On the one side I was exuberantly delighted to find myself in possession of boundless supplies of cocaine; on the other I was enraged with mankind for having invented the substance that had ruined my life, and I wanted to take revenge on it by poisoning as many people as I could." His wife objects to his going into business with a "murderous villain" and calls his cocaine sniffing "a vice pure and simple." Her own craving for heroin she regards as a physical disease and not a moral failing. After some guidance from a guru figure modeled on Crowley himself, the hero ends up confident that he can use cocaine "as a fencing-master uses a rapier, as an expert, without danger of wounding himself." He concludes that psychoactive drugs are "potent and dangerous expedients for increasing your natural powers" and that "the taking of a drug should be a carefully thought out and purposefully religious act." [19] Crowley ultimately gave cocaine credit for helping his writing (including *Diary of a Drug Fiend*) and some of what he called his "magickal rituals," which included copulation. But he eventually transcended his desire for large amounts of the drug, while taking heroin on prescription for the last 15 years of his life.[20]

Rhinitis

This is the most common physiological problem experienced by steady users who do not take large overdoses. Some cocaine users think the sugar and other substances used to cut the drug in the illicit trade, or a residue of hydrochloric acid from insufficient washing in the last stage of production of the hydrochloride salt, cause this affliction. Joël and Fränkel suggest that it may be partly a mechanical effect of continual sniffing on an anesthetized mucous membrane.[21] But it seems to be mainly a consequence of sympathomimetic constriction of blood vessels in the nose by particles of cocaine that lodge there. This eventually produces necrosis of the nasal tissue from lack of blood and sometimes reactive hyperemia, or congestion. In extreme cases there may be perforation of the cartilaginous part of the septum, the wall between the nostrils; but this seems to be rare today. Bacterial infections also occur. The most common symptoms are runny, clogged, inflamed, swollen, or ulcerated noses which may be painfully sensitive and frequently bleed.

A Russian physician, Leon Natanson, examined 98 heavy cocaine users (1.5 to 10 grams a day) in a hospital for venereal disease and a clinic in Moscow in 1920. Ninety-four had nasal lesions, and 89 had perforated septa. He discounted syphilis as a cause, because the lesions appeared only in the cartilage and because only cocaine-using syphilitics had them. Reviewing the literature in 1936, Natanson noted that another physician had examined 32 upper-class cocaine users; 10 had perforations of the septum and 10 others had ulcerations. In an experimental study on animals with cocaine vapor administered through the nose, another researcher had noticed no nasal effects; but Natanson concluded that the doses he used were not high enough.[22]

The nasal syndrome was so common in Germany in the 1920s that cocaine users had a slang name for it: *Koksnase*. It is so common today that, according to one of our interview subjects, "People on the air who sound obviously stuffed up will say to the audience, 'Gee, I have this terrible cold'; it's a big in-joke." Several people we interviewed reported nosebleed, sores, runny noses, or sneezing fits that would go away if they stopped using cocaine for a while. Joël and Fränkel mention that in some of their patients the bridge of the nose was sunken in the cartilaginous part.[23] Because of its sympathomimetic effect, cocaine at first stimulates respiration and dries the nasal mucosa; so it was once used, like amphetamines, for hay fever, asthma, and colds (by Crowley, for example) in spite of warnings that the temporary relief would give way to rebound congestion and rhinitis. Today some cocaine users carry a container of nasal spray decongestant like phenylephrine with them to counter these effects, or rinse the nasal passage with warm water after sniffing. Others recommend taking care in sniffing to prevent degeneration of the cartilage, if nothing else, by allowing the drug to touch only the upper, bony part of the septum.

Effects of Chronic Cocaine Use— Large Doses

Literary descriptions based on observation convey some feeling of the consequences of a damaging cocaine habit. The condition of cocaine abusers among the Paris demimonde after World War I is described in several passages of the novel *Cocaine*, written by an Italian journalist, Dino Segré, under the pen name of Pitigrilli and first published in 1921. In spite of its decayed romanticism and melodramatic mannerisms, Segré's novel is useful because it shows evidence of firsthand familiarity with the situation it describes. In it a warning is delivered to a young man by his slightly older mistress: "You're still in time. . . . I know the workings of that dreadful and deadly powder. You have not yet reached the stage of frightful depressions, the period of brooding and destructive melancholy. You can still smile, though your blood be filled with venom. You are in the first stages yet, when one again becomes a mere boy." Restless craving is described: " 'My nights,' they will tell you, 'are agitated by dreadful shivering; insomnia tortures me; it is atrocious to be without the drug, but the thought of not knowing where to get it is even more atrocious.' " "The hands of cocaine-addicts," Segré writes, "seem always to be on the verge of some convulsion which is held in check with tremendous difficulty. . . . Their nostrils become monstrously dilated to catch imaginary particles of cocaine hovering in mid-air." At the last stage they plunge, "brain reeling in a veil of thick gloom, into utter degradation, down, down through the pit and toward final misery." Cocaine is an ambivalent symbol in the book: the hero addresses his dazzling mistress by the name of the drug, and it also serves as "a symbol of the death to which we all succumb . . . the fierce and subtle and sweet death—truly, a thing of black shadows, like some nameless cataclysm, which we inflict upon ourselves voluntarily." [24] A few of the phenomena described here, especially "dreadful shivering," sound more like results of opiate addiction, and might be discounted as a case of mistaken identity. But Segré elsewhere describes accurately the exhilarating immediate effect of cocaine and convincingly portrays the scenes where it is used. Even if only a small minority of cocaine users are ever reduced to this condition, it remains a possibility.

Vladimir Nabokov portrays a cocaine abuser in a short story written in 1924:

> Too-frequent sniffs of cocaine had ravaged his mind; the little sores on the inside of his nostrils were eating into the septum. . . . During the leisure hours when the crystal-bright waves of the drug beat at him, penetrating his thoughts with their radiance and transforming the least trifle into an ethereal miracle, he painstakingly noted on a sheet of paper all the various steps he intended to take in order to trace his wife. As he scribbled, with all those sensations still blissfully taut, his jottings seemed exceedingly important and correct to him. In the morning, however, when his head ached and his shirt felt clammy and sticky, he looked with bored disgust at the jerky, blurry lines. Recently, though, another idea had begun to occupy his thoughts . . . his life had wasted away to nothing and there was no use continuing it. . . .
>
> He kept licking his lips and sniffling . . . suddenly he could stand it no longer. . . . He locked himself in the toilet. Carefully calculating the jolts of the train, he poured a small mound of the powder on his thumbnail; greedily applied it to one nostril, then to the other; inhaled; with a flip of his tongue licked the sparkling dust off his nail; blinked hard a couple of times from the rubbery bitterness, and left the toilet, boozy and buoyant, his head filling with icy delicious air . . . he thought: how simple it would be to die right now! He

smiled. He had best wait till nightfall. It would be a pity to cut short the effect of the enchanting poison.[25]

Maier lists the symptoms of chronic cocaine abuse in more clinical language as nervous excitability, anxiety, hypersensitivity to noises, flight of ideas, graphomania (compulsive scribbling), memory disturbances, mood swings, feebleness, senile appearance, emaciation, heightened reflexes with muscular unrest, fast pulse, impotence, insomnia. There is also a syndrome he calls "cocaine insanity" that sometimes arises after chronic abuse: optical and auditory hallucinations, delusions of persecution and grandeur, jealousy, violent tendencies—often with clear consciousness and insight on the part of the abuser. In this condition minor frustrations may cause energetic suicide attempts. In the late stages of chronic cocaine intoxication, there may be spontaneous abortion, twitches, cold extremities, and even paralysis.[26]

Maier also describes in some detail the hallucinations or perceptual distortions induced by an excess of cocaine. They are usually related to increased perceptual sensitivity and derived from external stimuli that are objectively observable. They may be overpowering even when the victim recognizes them for what they are; one patient heard the voice of a friend in a distant city saying that she needed his help, and, knowing that it was a hallucination, still boarded a train and traveled to that city. Perceptual distortions include faraway voices sounding near at hand, inanimate objects taking on physiognomies, "electricity" flowing through the fingers, random street noises organized into the sound of marching feet, and so on. There may be distortions of the kinesthetic sense including feelings of lightness, flying, or lengthened extremities, and even a sensation of having left one's body. Sniffers, according to Maier, usually suffer only auditory and visual hallucinations. Injectors more often have tactile hallucinations, especially the sensation of small animals under the skin, also known as "coke bugs"; this may be associated with larger quantities of the drug or faster entry into the bloodstream. Maier points out, incidentally, that the victim remains temporally and spatially well oriented and that most cocaine abusers never have hallucinations at all.[27]

How to classify the distorted or anomalous perceptual states induced by stimulants is a diagnostic and terminological problem related to the general inadequacy of our categories for describing all such conditions. It is always difficult to define degrees of perceived reality and implied deleteriousness. Without going into these questions, we can note some comparisons between the hallucinations and delusions in cocaine abuse and those of alcoholism. With cocaine the ego attends to the ordinary environment along with the distortions and delusions more than with alcohol. The visual hallucinations are usually brighter, more colorful, and more closely related to the mind's normal feeling-complex. Motor disturbances (except punding) and giddiness are much rarer than in alcoholism. Joël and Fränkel point out that the cocaine abuser may want to go on taking the drug even though, with the typical clear consciousness, he *knows* that it is making him paranoid. The accompanying self-magnification is too enjoyable.[28]

In Maier's experience, the delusions that sometimes accompany cocaine abuse may lack the association disturbances, blockages, and seemingly senseless outbreaks of emotion and leaps of thought characteristic of schizophrenia, but otherwise they resemble paranoid states. Joël and Fränkel judge that the most common psychological effect of acute and chronic overdose is irrational fear of burglars, policemen, or other enemy strangers. A cocaine abuser may sit in front of his door holding a revolver for hours, lying in wait for an imaginary intruder. Lawrence Kolb mentions a man who "used a hatchet to attack a laundry bag in his bathroom because he believed it contained a policeman." Maier notes that delusions of grandeur may be related to occupation: the physician thinks he has found a cancer cure, the engineer believes he has invented a perpetual motion machine, the soldier in wartime imagines he has discovered a new weapon or an infallible military strategy. This kind of delusion is an exaggeration of the questionable self-confidence and sense of mastery induced by stimulants; Maier notes that the confidence with which cocaine abusers express their delusory ideas has great suggestive power, especially over wives and girl friends.

A common element in the delusions and hallucinations experienced by the cocaine abuser is the tendency toward an exaggerated sense of meaningfulness, especially of environmental details. It often results in a need to *order* things that first shows itself in phenomena like a compelling desire to pull a glass back from the edge of a table, senseless counting, picture hanging, or avoidance of sidewalk cracks, and a sudden anxious need to look for some object systematically even thought it is in sight. Joël and Fränkel assert that this last condition was so common that it had a slang name among German cocaine users: *Suchkokolores*.[29] The need to put everything in order may be related both to the obsessive repetitious stereotyped activity known as punding and to the paranoid delusions that objectify and organize intellectually the anxiety produced by an excess of cocaine. According to Joël and Fränkel, some of their patients regarded

grimaces and chewing movements as distinguishing signs of heavy cocaine use. Intravenous cocaine in dogs administered for a long time eventually produces the same kind of head weaving, chop licking, and jaw
snapping as amphetamines. Joël and Fränkel note a direct transition in
their patients from senseless writing, counting, teeth grinding, and pac-
ing to full-scale paranoia. The condition observed by Ellinwood in his
study of amphetamine psychosis seems to apply to cocaine as well: "emotional enhancement of certain forms of thinking, examining, attention to
details and significance, interacting with certain forms of hyperactive peripheral attention, suspiciousness, and fear . . . to produce paranoid delusional systems." [30]

As a further introduction to this topic, here is an esthetically elaborated
fictional case study by William Burroughs, whom we have already quoted
on "popping coke in the mainline":

> One morning you wake up and take a speedball [heroin or morphine and
> cocaine], and feel bugs under your skin. 1890 cops with black mustaches
> block the doors and lean in through the windows snarling their lips back from
> blue and gold embossed badges. Junkies march through the rooms singing the
> Moslem Funeral Song, bear the body of Bill Gains, stigmata of his needle
> wounds glow with a soft blue flame. Purposeful schizophrenic detectives sniff
> at your chamber pot. It's the coke horrors. . . . Sit back and play it cool and
> shoot in plenty of that GI M [morphine].

Burroughs also reports on coke horrors from sniffing: "I knew this cop in
Chicago sniff coke used to come in form of crystals, blue crystals. So he
go nuts and start screaming the Federals is after him and run down this
alley and stick his head in the garbage can. And I said, 'What you think
you are doing?' and he say, 'Get away or I shoot you. I got myself hid
good.' " [31] Making allowances for the exaggeration in Burroughs' comic-
terrifying literary manner, we can say that he hits off very well many of
the psychological themes in extreme stimulant abuse: tactile and other
hallucinations, paranoia, the use of depressant drugs to calm down.

The reported cases of psychosis and other ill effects are mostly from the
period 1885–1930, when cocaine abuse was more common than it is
today and cocaine abusers came to doctors' offices more often. Maier
claimed to have studied 100 cocaine psychoses in Zurich in the early
1920s, aside from his cases of "chronic cocainism." He includes in his
book 35 case histories—28 men and 7 women—of acute and (mainly)
chronic cocaine intoxication. Of these, 18 managed to achieve enduring
abstinence from the drug, 5 relapsed, 4 killed themselves, and the fate of
the rest was not known. Twenty-nine of the 35 were between 17 and 30

years old, but 4 of them were physicians in their 40s. Occupations and social status varied, but most were professionals or other members of the middle class.[32]

Five of the cases exemplify what Maier calls cocaine addiction (*Kokainsucht*). Their symptoms were troublesome but not severe, little more than exaggerations of the inconveniences most social users of cocaine tolerate. One man said that the only ill effect he noticed was a feeling of great fatigue on coming home from work that he knew could be relieved by more cocaine. He regarded this as a warning signal and simply stopped using the drug. A law student experienced perceptual sharpening and distortions while using cocaine. Another man, who had never sniffed more than 1.5 grams a day or on two successive days, had sexual fantasies of beautiful women, then homosexual fantasies in which he imagined himself a woman, and eventually sadistic and masochistic fantasies. A woman who sniffed 4 to 5 grams a day became sleepless and lost her appetite. Her weight fell by 10 kg and acquaintances noticed that she was alternately lively and anxiously dreamy. She, like the other cases of *Kokainsucht*, gave up the drug without much difficulty.[33]

Maier discusses his largest group of patients under the heading "chronic cocainism." Four of the twelve recovered, one committed suicide with barbiturates, four relapsed, and the fate of the others was unknown. In this condition alternating states of exaltation and depression were common, even when there was no psychosis or delirium. Loss of sexual feeling was universal. One man remembered hunting through a house at night for nonexistent mice with his friends and at another time asking a chambermaid why the hotel orchestra was playing until dawn. Another would go on a cocaine bender or run for two weeks and then lie in bed for three days, "sweating frightfully." In still another case, a man started by sniffing one gram of cocaine on Saturdays and found that it excited him sexually and did no apparent harm. As he raised the dose and took the drug more often, up to four grams in a night, depression and self-reproaches began whenever he abstained. He lost all sexual interest and began to feel spiders crawling on his skin. When he had taken a great deal of cocaine he felt an urge to drink alcohol and smoke tobacco. At the time he was examined he was pale and thin, and had difficulty getting up in the morning. He lacked energy and felt apathetic. He often thought that people on the street were staring at him or detectives following him.

In another case a 19-year-old student came to Maier because he was nervous and his physical and mental faculties were declining. He had been sniffing four to five grams a day and was stealing to get money to buy the drug. He would often fall asleep in class if he did not have it. His symptoms included tactile hallucinations, pounding of the heart, and a tendency to mechanical repetitive movements. He had no sexual feelings but saw images of beautiful women in the clouds and spoke to them. Once he started writing or talking he could not stop. Sleep was unpleasantly heavy and dreamless. During cocaine intoxication colors were brighter and more beautiful, small sounds seemed loud, he heard whole symphonies when clocks struck, he felt able to solve all problems but thought it unnecessary to do so. During abstinence, on the other hand, every prospect was displeasing and man too was vile. Eventually he stopped taking cocaine, but Maier is not sure whether he kept his resolution to "cure myself entirely." [34]

Maier also discusses something he calls subacute cocaine delirium, which tends to arise after cocainism is well established and lasts a few hours to a few days. The fictional cases described by Burroughs would probably fall into this category. Maier divides these cases, some of which also might be called cocaine psychoses, into three types: euphoric with delusions of grandeur; paranoid-anxious; and dreamy-passive with cinematographic images. Of course, there are also various combinations of the three. We have already discussed many of the symptoms. In one case a man thought his boss was hiring journalists to put allusions to him in the newspapers and paying actors to blow their noses conspicuously in public places in reference to his habit. There were other, less amusing kinds of paranoia. For example, a young man began to think that his father, from whom he had been stealing, was sending men to persecute him; he threatened these imaginary persecutors with weapons.[35]

We have already described the symptoms of what Maier calls cocaine insanity, a more prolonged and systematized version of subacute delirium. Maier discusses four cases, including two physicians who were what he calls morphiococainists, injecting both drugs intravenously. In one case the physician took morphine for headaches, then used cocaine to reduce the pain of the injections. The combination caused hyperexcitation, shortness of breath, and anxiety. His wife also took cocaine, and both of them then used scopolamine to calm down. The wife took an overdose of this drug and died. He began to have visual hallucinations, said he had to follow his wife, and believed electric currents were running through his body, that he was about to die, and that he could hear people plotting against him. Even before he reached this condition, as he took less and less morphine and finally cocaine alone, he had suffered hallucinations, jealous delusions, and torturing anxiety, and his career had fallen into ruins. Eventually he gave up cocaine and was cured.

In an even more serious case, a physician was given cocaine to increase his energy during an attempted withdrawal from morphine. He began to occupy himself with delusionary research projects, thinking that he had found new cures for astigmatism and cataracts, and that spectacles should be made out of protein because glass tended to concentrate the effects of an invisible destructive ray. He wandered about the house at night and heard imaginary burglars. At times he believed he had a cancer cure or a new theory of colors. Eventually his colleagues brought him to Maier's clinic because they thought he might harm a patient. Upon arrival his condition was described as euphoria with flight of ideas. He had microscopic hallucinations and expressed sexual fantasies. He did not speak willingly of his discoveries because he knew that others thought him mad. By that time he was injecting five or six grams of cocaine a day and about 300 mg of morphine. His morphine and cocaine were taken away, but eventually he went back to them and died of arteriosclerosis combined with an overdose of one or another or both of the drugs.[36] A similar case was reported in 1886 by D. R. Brower: a physician who injected up to a gram a day of cocaine intravenously and suffered from pallor, emaciation, and insomnia, became irritable, believed he had a mission to give everyone the drug, carried a pistol and threatened vengeance on doubters, neglected his practice and alienated his friends, and fell into poverty.[37]

The self-description of a physician who was a cocaine abuser, published in 1920, sums up many of the phenomena we have been discussing. "Dr. Schlwa" first took morphine in 1916. In October 1917 he began to use cocaine to aid withdrawal. As he raised the dose, signs of psychosis appeared and he quit both drugs. But in November 1918, at a moment of deep depression and nervous exhaustion, he began to take morphine and then cocaine again. He came to a clinic sleepless, talking cheerfully with an edge of anxiety, and complaining of "worms" in his body and "animals" in his room which he knew were hallucinations. He spoke of a fear of burglars which was beginning to turn into a systematic persecutory delusion. After injecting 50 to 200 mg of cocaine subcutaneously, he said, he felt euphoria, a fast pulse, and slow deep breathing, with a desire to talk and run. After a half hour there was less euphoria and more nervousness, and a desire for more cocaine—with morphine to reduce the anxiety.

After three weeks of one gram of cocaine a day, 100 to 200 mg subcutaneously produced painful, deep, slow breathing, fast light pulse, and cold extremities. The effect was now more paralytic than excitant. The hallucinations started to take control. Taking 400 mg of cocaine and 600 mg of morphine subcutaneously, he felt orgasmic euphoria. In the later stages of chronic intoxication the cocaine effects dominated those of morphine. A little of either drug would eliminate first his potency and then his sexual desire. He regarded the feeling of mental activity under the influence of cocaine as "bustling without depth" that left nothing of value behind. Sleep was dreamless. The stages of euphoria, psychosis, and paralysis were sharply separated.[38]

It is not clear why some cocaine users succumb to chronic abuse and psychosis when most do not. We can hardly expect to be able to answer this question adequately for cocaine when it has so far proved unanswerable in the much more numerous and better studied cases of alcohol, amphetamines, and opiates. There may be a personality type predisposed to the abuse of some particular drug or of drugs in general, but no one has been able to define that personality type for any practical purpose like diagnosis and prevention. The questionable research technique that finds "personality defects" and "pathology" *post factum* in people who have either actually harmed themselves or others with drugs or gotten into trouble with the law by using drugs (like marihuana) that have failed to receive respectable society's capriciously granted seal of approval is not only delusively easy to use by stretching definitions but often hypocritical and otherwise morally odious. (We will have more to say about this when we discuss the subject of drug dependence in Chapter 8). Maier does not presume to find any common pattern in the life histories of his cocaine abusers.

The most we can say is that in all probability whatever conditions are apt to produce amphetamine psychosis and amphetamine abuse are also apt to produce cocaine psychosis and cocaine abuse. The 50-year-old reports on symptoms of cocaine abuse bear an uncanny resemblance to more recent descriptions of amphetamine abuse. The Swedish physician Nils Bejerot has been impelled to write (his italics): "*the well-established amphetamine-Preludine-Ritaline-toxicomanias in all essentials resemble cocainism.*" [39] He makes a good case for his assertion. But cocaine, as it is habitually used today, does not produce these symptoms nearly so often as amphetamines, partly because of a difference in pharmacological properties and partly because the illegality and exorbitant price of cocaine make it hard to obtain large quantities.

Because of this resemblance, it is useful to consider what is known about amphetamine psychosis. In some cases amphetamine only precipitates schizophrenia in an ambulatory schizophrenic or preschizophrenic who has been using it because it seems to combat the incipient symptoms of the disease. But amphetamine abusers may also be wrongly diagnosed as preschizophrenic or schizophrenic on the basis of symptoms actually caused by the drug. O. J. Kalant, for example, believes that in 109 of 201 cases of psychotic reactions associated with amphetamines that she studied, the drug alone was responsible. Psychosis occurs most commonly when a chronic abuser takes a larger amount than usual in a short period of time. Many cases of amphetamine psychosis are probably unreported or misdiagnosed as toxic psychoses caused by other drugs or as paranoid schizophrenia. The condition is often distinguishable from the latter only by blood tests for drugs or its short duration—only a few days or perhaps weeks after the drug is withdrawn. It is sometimes asserted, however, that a predominance of visual hallucinations, relatively appropriate affect, and a setting of clear consciousness, correct spatiotemporal orientation, and hyperacute memory of the psychotic episode distinguish amphetamine psychosis from schizophrenia. In assessing the incidence of amphetamine psychosis it is important to note that amphetamine abusers may live among fellow "speed freaks" who tolerate their paranoid symptoms, or they may learn to discount the paranoia as an effect of the drug.[40] Although cocaine may not be so fertile as amphetamines in producing either chronic drug abuse or psychosis, most of the symptoms and conditions, including the high incidence of misdiagnosed and unreported cases, are probably similar.

It is doubtful whether cocaine ever causes a chronic psychosis, i.e., one that persists long after the drug is withdrawn. Heilbronner studied this question as early as 1913 and concluded that there was no typical chronic cocaine psychosis and that the apparent examples were only cases in which cocaine-induced symptoms "colored" an endogenous psychosis. Maier agreed with Heilbronner and stated that any paranoid condition remaining a few months after the cessation of cocaine abuse was probably schizophrenia. The analogy with amphetamines may again help us here. Although chronic amphetamine intoxication may have some role in producing a longer-lasting disturbance, it is not easy to distinguish this effect, if it exists, from endogenous or reactive schizophrenia.[41]

Gutiérrez-Noriega has produced an inhibition of the central nervous system resembling catatonia experimentally in dogs by the continued administration of large doses of cocaine. He injected 5 mg per kg per day intravenously into 13 dogs and 10 mg per kg per day into 2 others, for a period of one to three years (the equivalent of 350 mg and 700 mg in a 150-pound man). Five of the animals began to adopt cataleptic postures with waxy flexibility (preservation of abnormal limb positions) for several hours after injections, and showed either motor inhibition or alternating states of inhibition and excitation. They also exhibited stereotyped movements, sometimes so complex that they looked like a ritual dance. During the cataleptic states, symptoms of parasympathetic excitation like bradycardia (slow heart rate) and copious flow of saliva appeared. The change from an inhibited to an excited state was often marked by piloerection (hair standing on end), a sign of strong sympathetic nervous system arousal. In the cataleptic states, they were insensible to pain—an analgesia deeper than that produced by barbiturate narcosis or general anesthesia. Barbiturates suppressed the analgesia and all the catatonic manifestations, both excitatory and inhibitory, but actually intensified normal locomotor activity. Ephedrine and other sympathomimetic drugs intensified all the symptoms suppressed by barbiturates, including the bradycardia normally associated with parasympathetic action.[42]

Gutiérrez-Noriega and others believe that cocaine-induced inhibition of the central nervous system is an active process caused by selective arousal of inhibitory neurons (neurons that suppress further nerve activity when stimulated) and not a form of depression or narcosis. Excitatory and inhibitory neurons are closely associated in the body, and it is apparently not even clear whether epinephrine acts by arousing the former or blocking the latter. Larry Stein and G. David Wise suggest that the norepinephrine liberated by amphetamine serves largely to depress behaviorally suppressive cells in the brain, i.e., disinhibits rather than directly excites. Cocaine potentiates both excitatory and inhibitory responses to externally introduced norepinephrine in sympathetic nerves. Gutiérrez-Noriega believes that the reason why barbiturates suppress the cataleptic states induced by cocaine and produce a subsequent "noncatatonic" motor excitement may be that they have more depressant effect on inhibitory neurons than on excitatory neurons. It is as though chronic cocaine intoxication sometimes produced a condition in which inhibitory and excitatory neurons in the sympathetic system were in fierce competition for control over behavior.[43] The analogy between this condition and some of the manifestations of catatonic schizophrenia in human beings is striking; cocaine psychosis, like amphetamine psychosis, has been proposed as an experimental model for studying the functional psychoses that might help in determining their neurochemical mechanisms. We shall discuss this further in Chapter 7, which deals with the uses of cocaine in medicine and psychiatry.

Chronic Physiological Effects—Large Doses

The physiological effects of chronic administration of large doses of cocaine have never been studied systematically in human beings. Vervaeck refers to tachycardia, arrhythmia, syncope, and angina pectoris in chronic cocaine users. But Joël and Fränkel, as we mentioned, rarely found cardiovascular problems in cocaine sniffers. They point out that in cocaine abuse at its worst the immediate cocaine effects are less important than complications from the ensuing malnutrition, exhaustion, nervousness, and general debility, which can be very serious. The most common specific physiological symptom is the rhinitis we have discussed; bronchitis from constant irritation of the mucous membranes occurs in some heavy users; Joël and Fränkel also report a generalized loss of pain sensitivity similar to the effect produced in dogs.[44]

In animal experiments, 15 mg per kg daily subcutaneously (the equivalent of about a gram in a 150-pound man) in dogs for six weeks produced no physical ill effects. In an experiment by Gutiérrez-Noriega and Zapata Ortiz on rats, even 100 mg per kg orally produced no chronic effects, although at 200 mg per kg growth and reproduction rate were retarded, and 1.6 grams per kg caused death in a few days. The highest tolerable subcutaneous dose was 155 mg per kg daily. (Animals with more complex brains appear to be more susceptible to the drug's psychological effects and possibly to its physical effects too. If this is true, man would of course be the most susceptible species. Gutiérrez-Noriega and Zapata Ortiz estimate the resistance of a rat to cocaine to be 20 times that of a man.) In another early experiment with rats, 75 mg per kg subcutaneously every 24 to 72 hours (the equivalent of 5 grams in a 150-pound man) caused successively restlessness, muscular weakness, fast breathing, paralysis of the hind legs, twitching, convulsions, and death after 14.2 days in males and 34.8 days in females. There was an average weight loss of 7.8 percent. Although only five rats of each sex were used, the enormously greater resistance of the females can hardly have been a chance artifact. We have seen no attempt to duplicate this extraordinary (and implausible) result and no reference to it in the later literature.[45]

Other effects that have been noticed or tested are teratogenic potential (danger of producing deformed young) and liver damage. Cocaine has proved to be the least teratogenic of several drugs, including a number of opiates, in experiments on hamsters.[46] Liver abnormalities are more positively correlated with use of cocaine. Gutiérrez-Noriega, in an experiment on chronic cocaine intoxication in dogs, produced fatty degeneration, dilation of capillaries, and other forms of liver damage. He suggests that the transformation of cocaine to ecgonine releases methyl alcohol, which is known to be toxic to the liver. Vincent Marks and P. A. L. Chapple, in their study, "Hepatic Dysfunction in Heroin and Cocaine Users," found that of 89 patients at a British psychiatric hospital who took intravenous heroin and cocaine, 80 showed liver abnormalities. When they stopped using the drugs the liver returned to normal. Marks and Chapple state that liver damage is rare in amphetamine, barbiturate, and cannabis users. They postulate a direct toxic effect of heroin and cocaine in combination, although they also think contaminants in the injected solutions might be part of the problem.[47] Morphine and heroin administered to normal animals and men cause no liver damage even over a long period of time; so, unless an adulterant or hepatitis virus is the cause or there is a special effect of heroin and cocaine in combination, we must assume that cocaine is the toxic agent. The liver is presumably damaged, in a reversible process, by detoxifying large amounts of the drug in a short time.

After admitting that information on physical changes caused by chronic abuse of cocaine is sparse, Maier goes on to discuss a case cited by Eugenio Bravetta in 1922. The man in question had sniffed cocaine for a year and died of an overdose (one of the rare cases, if the account is correct, of death from sniffing). According to his wife, he had been using no other drug. His diary described the progression from euphoria to anxiety to hallucinations as his habit became worse. Autopsy showed ulceration of the nasal septum, swollen and hyperemic brain and lungs with infarctions, enlarged spleen, enlarged liver with a hard consistency, degeneration of the walls of blood vessels in the brain and thromboses and hemorrhages in the smallest of them, and fatty infiltration of nerve cells. Unfortunately, it was not clear how much of this was caused by acute poisoning and how much by chronic abuse. Maier notes that many of the symptoms, especially nerve cell degeneration, are also characteristic of acute morphine poisoning and some infectious diseases. But the hemorrhages and thromboses in the brain and lungs, he believes, were produced specifically by cocaine's constricting action on blood vessels.[48]

We can now sum up the scarce information on the physiological and psychological effects of chronic use of cocaine. There is no clear evidence linking the ordinary habitual use of coca in South America causally with any mental or physical deficiency or disease. Although *coqueros* suffer from malnutrition and its consequences more often than men who do not chew coca, the use of the drug seems more likely to be an effect than a cause of poor diet. But there is little reason to believe that if coca should be harmful at low altitudes it might be less so or even beneficial at higher ones. Sniffing (or drinking in small doses, as in Freud's case) usually produces no more serious psychological problems than irritability, nervousness, and insomnia, with occasional depression and fatigue on coming down. Physiologically, the most common problem is rhinitis. The symptoms of chronic overuse or abuse are very similar to those of amphetamine abuse, although they apparently occur more rarely with cocaine. This is apparently a pattern produced by stimulant drugs' neurophysiological mimicry of an emergency situation. In particular, the temporary psychoses in amphetamine and cocaine abuse seem to be nearly identical. Chronic intoxication may produce a paralytic or cata-tonialike condition. Physiologically, malnutrition, exhaustion, and general debilitation are the most common effects of severe abuse. Cardiovascular problems may or may not arise: the evidence is inconclusive. The teratogenic potential of cocaine appears to be low (possibly because it is so quickly detoxified and excreted), but there is evidence of reversible liver damage.

A Note on Cocaine and Opiates

This matter is worth discussing separately because chronic abuse of the two drugs at the same time has been common. Although otherwise dissimilar, they have a good part of their recent history in common. The law has yoked them together and doomed them to bear the same burden of public contempt and fear. Physicians and academic authorities have too often allowed this to influence them; for example, a book entitled *Drugs and Youth,* published in 1969, devotes one section to "Morphine, Heroin, and Cocaine," thus accepting legal categories rather than phar-macological ones as a basis for discussion.[49] Writers who know better perpetuate an unnecessary confusion in this way. If we want to begin to be reasonable about the real nature and dangers of psychoactive drugs, we must avoid the casual use of socially accepted misclassifications.

The association in use that causes such misclassifications, however, is (or once was) genuine. Cocaine, as we saw, began its career as a cure for morphine addiction in the 1880s, and soon the term "morphiococainism" entered the medical vocabulary. By 1913 A. Friedlander was observing that in Europe physicians rarely saw pure cocainists. Joël and Fränkel, writing in 1924, agree that until World War I most serious cocaine abusers were also morphine or heroin addicts. When both opiates and cocaine were forced underground, the association continued, although the social status of their users changed. William Burroughs wrote in 1956 that "I have never known a habitual cocaine user who was not also a morphine addict." The drug-taking habits of opiate addicts are relatively well known—at least for those addicts, probably a small minority, who come under the control of government and medical authorities. These authorities often monitor most of their activities and inspect their urine too. In a 1970 study of 422 male addicts at New York treatment facilities, John Langrod found that 66 percent of them had used cocaine, 47 percent more than six times. Proportionately more blacks were in the latter group: 54 percent, compared to 44 percent of the Puerto Ricans and 38 percent of the whites. Usually they injected it along with heroin. Ten percent had used cocaine before they took up heroin. David E. Smith and his colleagues at the Haight-Ashbury Free Medical Clinic interviewed 303 heroin addicts in 1971 and found that 10 percent of them had used cocaine moderately to heavily. In 1972 they interviewed 147 more and found that the proportion had risen to 20.7 percent. Of "new junkies" at the Haight clinic (those who began taking heroin after January 1967), only 0.5 percent used cocaine in early 1970, but 16.3 percent in the period May 1970 to July 1971. Among methadone maintenance patients 18.5 percent in a 1972 study at Philadelphia General Hospital had cocaine in their urine. At the National Institutes of Mental Health clinic at Lexington, a third of the methadone patients in a 1973 study had used cocaine at least once and about 20 percent used it often. In Great Britain 564 registered addicts were receiving cocaine in 1968 (a number that had risen from 25 in 1958 and 171 in 1963). The corresponding numbers were 2,240 for heroin and 198 for morphine. During 1967 and 1968 many doctors replaced cocaine with methamphetamine in prescriptions in the belief that it was less dangerous. Since 1968, when private practitioners were denied the right to prescribe opiates or cocaine to addicts and treatment centers were established, legal prescriptions for cocaine have stopped almost entirely.[50] The use of cocaine by opiate addicts is a very small corner of the contemporary scene anywhere in the world. The time is long past when anyone could say that he had never known a habitual cocaine user who did not also take opiates. Cocaine is now much more likely to be combined with alcohol, marihuana, or hashish for sedation.

Since at one time such a large proportion of cocaine abusers also used opiates, it has been suggested that cocaine is rarely dangerous when used alone. Freud, as we saw, insisted that cocaine had "claimed no victims of i⁺s own." In fact, many people who *inject* cocaine use opiates too, and these are the ones most likely to come to the attention of physicians. Most cocaine sniffers have little or no contact with opiates and, concomitantly, few symptoms of severe abuse. Opiates create a steady physiological need and a distinct abstinence sickness; although the desire or craving for cocaine can be very strong, it does not produce intense physiological symptoms on withdrawal and it can be more easily used intermittently. Therefore the real problem may appear to be the combination of opiates with cocaine rather than cocaine alone.

But pharmacologically it makes no sense to say that cocaine is relatively harmless unless combined with opiates. The drugs are usually complementary in their effects, and the psychological and physiological consequences of cocaine abuse can be worse than those of opiate addiction. In the 1880s Erlenmeyer said that the only thing as bad as cocaine abuse was alcoholism. (Today he would have to add at least barbiturate and amphetamine abuse.) Burroughs observes, "A morphine addict can live to be 90. . . . Their general health is excellent. . . . On the other hand, cocaine, methedrine, and all variations of the benzedrine formula are ruinous to health, even more so than alcohol." A 1914 article on "Narcotic Addiction" expresses the same opinion: "A pure cocaine addict very rarely complains of violent cramps and diarrhea upon withdrawal of the drug. On the other hand, cocaine is more destructive than the opium derivatives, and its effects upon the body appear very early compared to the physical changes observed in morphine fiends." Louis Lewin in *Phantastica* writes of the "picture of degradation worse than Hogarth" in cocaine abuse, and declares that unlike a morphine addict, the cocaine abuser cannot "mask his disorder." In an experiment performed in 1968, monkeys allowed to administer both morphine and cocaine by means of separate catheters implanted in their jugular veins used both drugs, cocaine dominating during the day and morphine at night until the animals became so disoriented that no pattern was discernible. The combination produced delirium, motor impairment, anorexia, emaciation, and death in two to four weeks. The effect was like that of cocaine alone; morphine alone in the same experiment was continued for over a year without any damage to health except infections. According to one writer, in Britain, "most heavy heroin users take grain-for-grain doses of cocaine. . . . It is possible that much of the harm attributed to heroin is in fact caused by cocaine hydrochloride." [51] The most severe symptoms of morphiococainism, except for the opiate abstinence syndrome, were always cocaine effects. Pure opiate addicts, for example, rarely suffer hallucinations, paranoia, and psychoses: as Burroughs remarks, they tend to be drearily sane.

Here is an account of his cocaine problem by a man who first sniffed heroin, then injected it intravenously, then began to use cocaine as well (legally, in Great Britain):

> I was on cocaine from then until 1964, a good solid five years. And it really ruined me; I lost job after job, and I couldn't work, and this and that, and we were on assistance, and, oh, it was great, a great performance. . . . Oh yeah, I

die for it every time I think of it. But it does me absolutely no good. I get terrible hallucinations, I get the horrors something awful, I'll fall asleep over my work, and I'll swear and all this. . . . I get as paranoid as you could possibly get, or at least as possible for me to get. . . . I see secret tunnels opening in the walls and everything else, you know, and it's really too much. And yet, if, by some mischance, they started to give it to me here I wouldn't refuse it. . . . Oh, I would try to rationalize it, I suppose, that I would manage it better this time or something. But I wouldn't.

At one time this man was taking 500 mg of heroin and 500 mg of cocaine a day intravenously. He withdrew several times from heroin, and found it harder when he was also taking cocaine. He was addicted to heroin at the time of the interview, in the early 1970s. According to the authors of the book in which his case is reported, he never had any trouble earning a living for his family except during the cocaine period.[52]

In fact, one function of opiates in the combined abuse of the two drugs has been to prevent the cocaine from ruining the user's mental or physical health. It is possible (and for the addict sometimes necessary) to maintain oneself on opiates, but not on stimulants, at a high level intravenously. Cocaine abusers often have to turn to some sedative drug, and morphine and heroin are among the most effective ones available. We have already mentioned the cases cited by Aleister Crowley, including his own (although he never injected but only sniffed the drugs). William Halsted is an even more significant example because he had such a distinguished professional career after he rid himself of the cocaine habit and became a morphine addict. Lawrence Kolb, writing in 1962, cites the case of a physician who, after using cocaine for 6 years and suffering 12 "fits"—apparently delirium or psychoses—turned to morphine to counteract what he called the horrible effects of the cocaine and continued to use morphine alone, without pleasure, for 21 years. Burroughs sums up: "The nervousness and depression resulting from cocaine use are not alleviated by more cocaine. They are effectively relieved by morphine. The use of cocaine by a morphine addict always leads to larger and more frequent injections of morphine." [53]

Since opiates were a pharmacological answer to the condition of anxiety produced by cocaine, morphine addicts "cured" with the help of cocaine often went back to morphine to cure themselves of the cocaine habit. This was one of the main sources of "morphiococainism." We have already mentioned some cases recorded by Maier and Mayer-Gross' "Dr. Schlwa." Magnan and Saury in 1889 described several others. In one of them a man who used cocaine in an attempt to rid himself of a morphine habit that was doing him little harm except for abstinence symptoms found himself taking both drugs. After a month he was injecting one gram of cocaine a day and after two months began to have hallucinations and became hyperexcitable. He stopped using cocaine and continued with morphine for six months; these symptoms disappeared. Then he began taking cocaine again, this time up to two grams a day, and suffered tactile hallucinations, tremors, and convulsions.[54]

The acute effect of the heroin-cocaine combination taken intravenously—a speedball—is particularly interesting. It has been described this way: "Coke hit my head, a pleasant dizziness and tension, while the morphine spread through my body in relaxing waves." In an experiment on mice, cocaine but not procaine or imipramine (an antidepressant) heightened morphine analgesia.[55] Obviously, opiates and cocaine complement each other instead of canceling each other's effects: "I felt them as distinctly different highs. It wasn't as if I felt a mixture of the two," said a man we interviewed. We have mentioned that cocaine may actually increase the lethal power of heroin in the doses used by addicts. So the acute as well as the chronic effects of a combination of cocaine and heroin may be both desired and dangerous.

We must emphasize here, although we make the point more fully in another context, that nothing is unique about this combination except its peculiar social status: unusually heavy legal penalties and public disapprobation for the use of either drug. Methamphetamine and other amphetamines have often been substituted for cocaine in the speedball and may be more dangerous. Cocaine abusers sometimes use barbiturates to calm down. Although these drugs antagonize cocaine's acute toxic effects, they are themselves dangerous and addictive. Depressants that heighten each other's lethal effect, like alcohol and barbiturates or alcohol and heroin, may be the worst combination of all.

The combination of cocaine and heroin is simply the least socially respectable modification of a common American and European cultural pattern: the use of "uppers" and "downers," alternately or together, to change one's mood chemically at will. A prescription mixture of morphine and cocaine under the name of Trivalin was used in Europe in the 1920s, in much the same way that amphetamine-barbiturate combinations with names like Dexamyl and Obedrin were used in the United States in the 1950s and 1960s. In Great Britain a mixture of morphine, heroin, and cocaine with gin or brandy, known as Brompton's cocktail, is used to assuage the pain of dying cancer patients. "Its value in terminal care is unsurpassed," according to the author of a recent British book, *Intractable Pain*.[56] Opiates and cocaine do not necessarily constitute the most dangerous of the stimulant-sedative combinations, especially if the drugs are inhaled but possibly even if they are injected. Amphetamines, barbiturates, and alcohol kill more people and ruin more lives than either cocaine or heroin. Of course, that is largely because they are so much more freely available, but it is not obvious that freely available heroin and cocaine would be any worse. In fact, heroin was sold in Sweden until 1956, under its chemical name of diacetylmorphine, in several popular cough syrups, apparently without doing appreciable harm.[57] Cocaine and heroin have come to their present status by a historical anomaly that we will discuss in Chapter 10. Clearing up the confusion about these drugs requires more than an analysis, necessary as that is, of their pharmacological properties.

7

COCAINE IN MEDICINE
AND PSYCHIATRY

TODAY the range of recognized medical uses for cocaine, outside of South American folk medicine, is very narrow. But the alkaloid was once used, and coca still is used, for a much wider variety of medicinal purposes. Like opiates, alcohol, amphetamines, and other drugs affecting the central nervous system, cocaine seemed to relieve the symptoms of many otherwise very different illnesses and functional disturbances. The use of the word *panacea* has long implied scorn, but before the full development of modern medicine, with its goal of influencing the underlying causes of clearly, usually etiologically, defined diseases instead of or in addition to providing symptomatic relief, the centrally acting cure-alls were among the most important items in the pharmacopoeia. Even 40 or 50 years after the heyday of the opiate- and cocaine-containing patent medicines, the amphetamines could be introduced to physicians and the public with a proposed range of applications very similar to those of cocaine in the 1880s and permitted to run the same gradual course of disillusionment in the medical community. Of the old panaceas only morphine retains something of its former range of application, and even it is used much more sparingly and under much stricter medical supervision than it once was. But new synthetic substances like the barbiturates and tranquilizers have been found to serve some of the same purposes. Medicine continues to have its panaceas, although it is more shamefaced about them. Nevertheless, there has been a tendency over the last century to expel the use of such drugs from the category of medicine and transfer it to the categories of superstition, religion, fun, or even disease and crime. We will say more about this historical development, which has been fateful for attitudes toward cocaine, in Chapter 10.

Because a drug like cocaine can make one feel better in so many different circumstances, it is not easy to classify the therapeutic applications. An unconventional but revealing approach is to consider how South American Indians classify the uses of coca. They have been taking the drug for over a thousand years, and their tradition contains more empirical knowledge about its effects than could possibly have been accumulated in the single generation devoted to its investigation by Western medicine in the nineteenth century. Horacio Fabrega, Jr., and Peter K. Manning, in a 1973 study entitled "Health Maintenance Among Peruvian Peasants," analyzed the functions of coca and other herbal drugs in the village of Huarocondo, Peru, population 6,000, located 25 miles north of Cuzco at the altitude of 12,500 feet. Most of the people chew coca. Fabrega and Manning listed the herbs in common use and the conditions socially defined in the village as amenable to herbal treatment. Then they interviewed a sample of 40 adult males, asking whether a given herb was useful for treating a given condition. If more than 70 percent of the subjects considered an herb helpful for a given condition, it was regarded as a standard remedy. Coca in one form or another was a standard remedy for more problems than any other herb—8 out of a total of 18 listed. Not surprisingly, it was the accepted treatment for hunger (100 percent) and cold (98 percent). It was used to improve spirits (93 percent) and provide physical strength (93 percent). It was also used for two folk illnesses: *el Soka*, a condition of weakness, fatigue, and general malaise apparently resembling neurasthenia as defined by Freud and Mortimer (73 percent); and *el Fiero*, a chronic wasting illness (73 percent). Coca was also the remedy of choice for stomach upset and pains (100 percent) and for colic, or severe gastrointestinal disturbances including diarrhea, cramps, and nausea (73 percent). Other herbs were preferred for acute high fever, headache, respiratory infections, severe mental disturbances (*la Locura*), and gastrointestinal disturbances believed to be caused by emotional difficulties (*la Colerina*).[1]

Coca is also reported to be used in the form of leaf powder or tea for stomach ulcers, rheumatism, asthma, and even malaria. Coca tea is a remedy for the nausea, dizziness, and headache of *soroche* or altitude sickness, and it is routinely served to tourists arriving at hotels and inns in the high Andes. The juice from the chewed leaf may be applied to the eye to soothe irritation, or gargled for hoarseness and sore throat. Coca also contains vitamin C and some B vitamins, and it is sometimes said to be an important source of these nutrients in the Andean diet.

The therapeutic applications favored by nineteenth-century physicians for coca and cocaine were sometimes defined differently from those common in Peru. Freud's list includes "nervous stomach disorders," asthma, diseases of the vocal cords, sexual disinterest, and convalescence from typhoid fever; Mortimer's includes uremia, vomiting in pregnancy, convalescence from yellow fever, skin conditions, stimulation of uterine contractions in childbirth, and appeasing thirst in diabetes. But except for morphine addiction, alcoholism, and surgical anesthesia, nineteenth-century medicine covers largely the same ground as the Peruvian Indians. In an appendix to his book on coca, Mortimer lists the responses to letters he sent to "a selected set" of over 5,000 physicians in 1897 asking for their observations on coca. Of the 1,206 replying, 369 said that they had used coca in their own practices. (By that time doctors were suspicious of coca because they were familiar with the dangers of cocaine abuse.) Common observations were that coca increased appetite (113 of 369 responding), raised blood pressure (88), stimulated circulation (107), strengthened the heart (117), improved digestion (104), stimulated the mind (109) and the muscles (89), improved respiration (40), and served as an aphrodisiac (60). Forty-four of the 369 claimed failure to get any results. The most popular therapeutic applications were debility, exhaustion, neurasthenia, and overwork. Smaller numbers recommended coca for anemia, melancholia, bronchitis, angina pectoris, and other conditions. Only 21 physicians thought there was a dangerous tendency to form a "coca habit." [2]

To use the CNS-stimulating effect of cocaine or coca for symptoms described as neurasthenia or exhaustion often meant little more than taking the drug to forget one's troubles or make them easier to bear—functions also served at that time by opiates and later by amphetamines, barbiturates, and tranquilizers. In fact, the distinction between the use of a substance as a medicine and its use as a drug taken for fun, like the distinction between health and pleasure, is not nearly so well defined as we usually prefer to believe and often dependent mainly on social factors like the authority under which the drug is taken. In cultures where medicine has not freed itself from religion, and in cultures like nineteenth-century Europe and the United States where the categories of medicine and pleasure overlapped more obviously than they do in our society today, it was even easier to regard the euphoriant effect of centrally acting substances as medicinal—"a harmless remedy for the blues," as the

Louisville Medical News called cocaine in its editorial in 1880. The use of this particular drug for this kind of purpose—making the user feel good when he feels bad or better when he feels only not so bad—is now indulged only on the various illicit drug scenes and not under medical authority or even with the approval of physicians.

The following account, written for us by a young woman we had previously interviewed, shows how an illicit drug can be used not for carefree pleasure but as a prescription for respite from psychological pain and misery:

Cocaine—a quarter of an ounce of good coke is going for $475 or more right now. That's expensive happiness for sure, but a price I'm still willing to pay. I've been snorting cocaine for three years and she [Lady Cocaine] has been a super friend to me during this time, especially the last two years when I've been going through hard times. . .times that society labels "mental illness." The last two years have included four psychiatric hospitalizations for me and private therapy twice a week. The Lady has pulled me through some rough moments better than anything else could work for me at the time. Today she's with me again. . .helping me share some intimate thoughts on paper for people I don't know.

I'm 28 years old. A lot has happened in the last two years to totally change the life I was living before. Before. . .I was able to have some close friends and to function in good control. I went to college after high school, graduated and went on to get a master's degree in nine months. After completing my master's I started working in public television and in two years I had reached the status of producer/director. I was happy with my work and doing good programs I felt proud of. I was also living with a man I was very much in love with.

Things started down for me when things started down for public television. . .when Nixon cut the funding for PTV in 1972 and many of us found ourselves without jobs. I moved to Boston to be with the man I loved, still hoping to get another job in my field. The job never happened and I started to lose confidence in myself. A few months later the man I was living with left, and the pain of that loss was great, adding to my insecurities. Things with my parents were not good. . .our communication was poor or nonexistent. I felt that they really didn't understand why I had to escape all I was raised with and leave home. I guess I really didn't understand it all either, but I did feel the pain of the separation from them and our differences. Then my grandfather died, someone I felt very close to, and that loss contributed to my fears. I started to withdraw, spending most all the rest of 1973 in my room, afraid. . .very afraid. Oh, I had gotten a job in a radio station, but it wasn't public television and my work was not in production. I had a mid-management paperwork job with a super amount of pressure. I was lonely and unhappy in a new place without friends. I didn't feel very good about myself and consequently didn't believe anyone could feel very good about me either. I didn't trust people who were friendly to me. . .wondered what they wanted from me. . .what they were after. I wasn't comfortable out of the house and I left the house only for work or brief stabs at trying to get close to people again. Things deteriorated. Each relationship I tried with a man ended in pain, and the relationships became shorter and shorter. During this time one of the men I was seeing turned me on to cocaine.

COCAINE. . .without it I would have stayed in my room all the time. It got me through work during the day and out of the house at night. I was afraid of people. . .of what they would do to me. Coke made me not so afraid. The men I saw during the rest of 1973, I saw thanks to the help of coke. Coke made me confident again. I didn't know what was happening to me except that I was scared. But with the help of The Lady I could function in the outside world without letting anyone know something was wrong.

I really feel that without coke I would have freaked out a lot sooner than I did. Oh, there were other drugs too. . .grass was a regular and had been for some time, and I enjoyed acid a few times a year too. But with the exception of brief encounters with other drugs (speed, downs, etc.) cocaine was the important drug for me. I certainly couldn't count on acid to help me function better in the real world. . .it took me deeper into my fantasies in a beautiful way, but I couldn't ask acid to help me feel sane. Grass was a good escape, but I couldn't function well at work under it and it didn't make me feel I could handle things I was afraid of. It just helped me escape the feelings of fear, hiding in a constantly stoned world, beyond reality. With coke the reality was still there, still waiting to be dealt with, but *I* was different. . .I could handle it. I remember so many days at work that year when I would end up in the bathroom hysterical, feeling I was losing control. But I had coke with me then and after snorting a few lines I was back in control. I felt good and able to go back to my desk and deal with everything that was overwhelming me before. I'm not saying that cocaine was the solution to my emotional problems, but it did put me back in control and allowed me to stay on at work, to see and be with people, to continue to function when I don't think I would have been able to without it.

I wish coke had been the total solution. But by the beginning of 1974 even cocaine wasn't enough. I was doing about a half a gram a day and it was getting very expensive and there was never enough. Things were getting very out of proportion for me and I was having trouble telling what was real from what was not. Some people were telling me that my problem was drugs, but I knew without drugs I couldn't hold on. By March even drugs were not enough. My dog was my only friend and I was starting to get confused over what was me and what was my dog. Sometimes I felt that he was getting sick, was going to die, and other times I felt it was me. . .and someone was trying to kill me with fear. A friend of mine from college who's a psychiatric social worker told me she thought I needed some help. On March 7, 1974, I started seeing a psychiatrist. Less than a month later I was in a mental institution.

The first time inside was very scary. The reality of being in a mental institution was something I couldn't face. WHAT WAS HAPPENING TO ME? WHY WAS I HERE? Everything had always been all right before, hadn't it? Look at all I had accomplished up until the last year. I didn't remember ever being afraid before. . .in fact I was always the person who could handle things, the person others turned to when they needed help. I just didn't know what was going on and I was scared to death to find out. I asked some friends to sneak in some cocaine for me, which they did. And I did the coke. . .before group therapy, before team meetings, before, during, and after all the confronting, heavy times. I just couldn't stand what was going on with me and around me there without her help. Cocaine allowed me to stay there and begin to work on some things. Without The Lady's help I would have killed myself to escape the pain.

After a month in the institution I ran away. None of the therapists there thought it was right to run, neither did my psychiatrist, but I just felt I couldn't take it anymore. The problem was I couldn't take it on the outside either. I continued to see my psychiatrist during the month I was out, and things continued to get more and more confused for me. My arm looked a mess. . .long razor blade cuts were all over it. . .ugly scars that looked like I was feeling. I suppose I did the work, ripping the shit out of my arm, out of myself. . .but I never remembered picking up the blade. I was only working a few days a week and many of those days I couldn't get through even with cocaine. I started hallucinating, I became even more scared of everyone and everything. I couldn't sleep because I was afraid of the horrible nightmares. I couldn't eat because everything tasted so bad. Finally I couldn't even leave my room, and I still didn't know what was going on with me. I decided to give up trying. Luckily my psychiatrist helped me make another decision. . .to return to the mental institution for safety and to stay this time until I felt I could handle the outside world again. So after a month outside, trying to make it on my own, I went back again. . .inside. . .and stayed this time for three months until I could get a medical discharge.

The second time I was institutionalized they knew about the cocaine. Of course I was searched when I got there, but they went a step farther. . .they did not allow me to have visitors for some weeks, and then even when I could have visitors they watched me. . .watched to make sure no cocaine was brought in. They wanted me to get in touch with all the fears. . .to feel all the pain. They gave me their kind of medication to ease the hard times. . .Thorazine, Stellazine, Mellaril, Haldol, and I hated it. It made me feel like a zombie. . .very drugged up and unable to function well. At least on cocaine I could function, it didn't impair my ability to think or to feel. In many ways cocaine gave me the courage to face things, where the antipsychotics kept me in a dull stupor. Oh, I hated a lot of things about being institutionalized, but I also knew there was good in it too. Where cocaine left off, the institution took over. . .surrounding me with warmth and support. It was quite a few weeks before I did any cocaine again, and I survived. But I survived with 24-hour intensive support. Once I was discharged from the institution that 24-hour intensive support was gone. Seeing my psychiatrist twice a week wasn't enough and even though I made an attempt to do without The Lady I couldn't make it very long. I was still very afraid. In fact I found it a lot harder being out of the institution than it had been being in. And the main reason was the lack of total support. I turned to cocaine for that support again.

The past year since I was discharged I've continued to work with my psychiatrist twice a week, I've been able to hold down jobs for months at a time, and I've continued to count on Lady Cocaine for support. A few months ago I was institutionalized again two more times, but both were short-term crisis situations.

From my experience, and from others', it seems like it takes a long time to work through what the last few years have been like for me. I'm still working, still trying to get to a place where I'm not so afraid of people anymore. Drugs are playing an important role in helping me maintain control when things get rough. . .giving me the support I need to not face being institutionalized again. Haldol has been a help, but if I could afford to buy enough cocaine so I wouldn't need the Haldol, I'd do it. I still have trouble being outside of the house, still have trouble feeling comfortable with people, and I can't do these things without some drug in my body. Haldol works on a day-to-day level. . .it's in my blood cooling me out, but sometimes it's not enough. And for those times I do cocaine. . .it makes me feel strong and confident, it lets me unfreeze my lips and feel comfortable enough to share some feelings with people, it helps me to continue at times that I don't want to live.

There are negative effects from cocaine too. The main one I see is money. It costs so much and it's terribly hard to get good cocaine that's not cut with a lot of speed. And then there are the cocaine jitters. . .when I've been on a coke binge and done a lot more than my body can physically handle. Another problem is the sneakiness that seems to go with doing coke. I don't tell people I'm on coke because they are either very against it and put me down for using it, or they are very for it and it's too valuable to me to share. Making coke deals has been a drag too. . .waiting on street corners in bad sections of town for my connection to show, not being able to make a connection when I desperately need to. Cocaine is a desperate drug for me. Sometimes I feel like I need cocaine to survive and like I would do almost anything to get it. Those memories are not good ones. But I would much rather deal with those bad times than be defenseless without cocaine and have to face the real world alone.

The hope is that through therapy I can grow to find the supports I need in people, and I won't need any drugs anymore. But now. . .for those times when the world is crashing around me, give me cocaine to pull me through.

We are in no position to evaluate this account. It is impossible to say whether this woman has been alleviating or exacerbating her problems by relying so heavily on a drug that can itself cause serious mental disturbances. What is clear is that she regards cocaine as a kind of therapeutic agent, like the Thorazine (chlorpromazine) and Haldol (haloperidol) that doctors have prescribed for her.

Other medical applications of coca and cocaine that we have described are presumably related to their effects on peripheral nerves; most of them can be classified as digestive, respiratory, or local anesthetic. But even here it is hard to separate the central stimulant from the peripheral effects. The singer or actor who drank Mariani's wine could hardly know how much of the improvement he thought he noticed was caused by local anesthesia or constriction of blood vessels in the throat and how much by euphoria and a feeling of mastery. As for stomach and intestinal upset, the gastrointestinal tract is probably the most common site of psychosomatic symptoms. The use of coca or cocaine in convalescence from long-lasting, debilitating diseases, or for relief of misery in situations like General Grant's last illness, represents a similar combination of central and peripheral effects.

It might seem that the book has long been closed on the question of using cocaine for these purposes; the medical community has decided that the dangers and the availability of alternatives make any possible benefits unimportant. But the decision has not been entirely unequivocal; for example, Joël and Fränkel report that cocaine was still often prescribed in the 1920s for many of the conditions mentioned by Freud and Mortimer. Even today cocaine is occasionally prescribed for nonsurgical purposes; for example, in the 1970 volume of the *Journal of the American Medical Association*, there is a letter from a New York doctor who has been using 50 mg of cocaine in an eyedrop solution and asks if it might induce drug dependence. The editors assure him that it will not.[3] Some physicians and others have expressed interest recently in reintroducing coca or oral cocaine for stomach ache and as a respiratory stimulant and tonic for the vocal cords. In effect, this would be a return to the days of preparations like Mariani's wine and the original Coca-Cola, which were never actually shown to be harmful and fell into disrepute for reasons related only indirectly to possible medical dangers.

The use of cocaine as a surgical anesthetic, which first made the drug famous in the 1880s, remains almost the only one medically recognized today. The property that makes it possible is unrelated to the central and sympathomimetic stimulant effects and has been duplicated in a series of synthetic local anesthetics that do not produce these effects and are otherwise structurally dissimilar to cocaine. Local anesthetics block the conduction of impulses within nerve cells to which they are applied directly. They may act on any part of the nervous system and any kind of nerve fiber. They affect sensory fibers before motor fibers and the smaller sensory fibers before the larger ones, so that pain is the first sensation to be paralyzed, followed by cold, heat, touch, and pressure. The local anesthetics act on the nerve cell membrane by interfering with the large transient increase in its permeability to sodium ions that occurs in normal nervous system functioning. This increased permeability is produced by a slight depolarization of the membrane that permits the operation of the sodium pump, a process essential to transmission of the nerve impulse. The effect can be nullified simply by raising the external sodium concentration. It is not known exactly how the mechanism works, but it appears to have something to do with the calcium that controls membrane permeability. The most commonly accepted theory is that local anesthetics compete with calcium for some receptor sites that control sodium and potassium conductance, occupying the crucial positions and preventing the calcium from changing the permeability.[4]

There is a difference of opinion about whether any of the local anesthetics except cocaine are properly described as CNS stimulants. It is often said that any of them may stimulate the brain but that only cocaine affects the cerebral cortex. An apparently more plausible view is that the synthetic local anesthetics produce only sedation and disorientation, sometimes accompanied by a restlessness caused by blockade of inhibitory neurons and mistaken for stimulation.[5] In any case, cocaine is the only local anesthetic that affects catecholamine transmission. Any of these drugs in large enough doses, however, may cause convulsions, unconsciousness, and death by respiratory paralysis. Vasoconstriction, which localizes the drug at the nerve, prevents quick absorption, causing the anesthesia to last longer and also reducing systemic toxicity by allowing destruction of the drug to keep pace with absorption. Among local anesthetics only cocaine is a vasoconstrictor.

Ideally a local anesthetic should be water-soluble, low in toxicity, applicable to the mucous membranes as well as by injection, and not irritating to body tissues. The onset of anesthesia should be fast and its duration neither too long nor too short, and the drug should do no permanent damage to nerves. Procaine, the substance first synthesized in 1899 by Einhorn (its hydrochloride salt is commonly known as Novocain), once seemed to have the best available combination of these properties for most purposes. It was the most commonly used local anesthetic until the 1950s, when a new synthetic, lidocaine, began to replace it. The main drawback of cocaine is its potentially high and, more important, unpredictably variable toxicity. For this reason it is no longer used in infiltration anesthesia (subcutaneous or submucous injection), nerve block or conduction anesthesia (injection into or near a nerve trunk), or spinal anesthesia. It is still used in 1 to 4 percent solutions topically—that is, applied to mucous membranes—in the ear, nose, and throat, and more rarely the eye.

Within a restricted range, cocaine was once used quite commonly. A study published in 1954 of practices at ten large American city hospitals from 1948 to 1952 shows substantial and not diminishing use of the drug as a topical anesthetic or a topical supplement to other anesthetics, with a very low proportion of deaths. Of 600,000 reported applications of anesthesia, 46,200 or 6.6 percent involved topical anesthesia. Cocaine was used as a primary anesthetic in 20,300 (43.7 percent) of these cases and as a supplement in another 7700 (16.6 percent). It was the most popular topical anesthetic.[6] Since then it has apparently been replaced in most instances by lidocaine and other agents, but its use in minor ear, nose, and throat surgery, where its vasoconstrictor action also helps to prevent bleeding, continues; for example, a 1965 article in the *British Journal of Plastic Surgery* recommends it for the correction of simple fracture of the nose.[7] The major medical use of cocaine today is to prevent retching and pain in instrumental examination of the mucous membranes of the upper parts of the digestive and respiratory tracts: the procedures called laryngoscopy, pharyngoscopy, bronchoscopy, and esophagoscopy. In oral surgery, cocaine is occasionally used to shrink the nasal mucosa or as a topical anesthetic in operations on oral-antral fistulae (pathological passages from nose to mouth). There are also still a few ophthalmological applications. It may serve as an adjunct to infiltration anesthesia with other agents in cataract surgery, and it is also sometimes injected under the conjunctiva (the mucous membrane of the front surface of the eyeball) along with epinephrine to break adhesions of the iris to the lens capsule and produce mydriasis (dilation of the pupil). But these uses of cocaine are now rare. Surgeons avoid cocaine in glaucoma operations today, although they once used it in cases like that of Freud's father, because here they want contraction of the pupil rather than dilation.

Most of the physicians we consulted rarely used cocaine. But in a 1975 article in the *Annals of Otology, Rhinology, and Laryngology,* Nicholas L. Schenck recommends it as the anesthetic of choice in otolaryngology. Procaine is too weak to be used for mucous membrane anesthesia, lidocaine is often too transient in its effect (15 minutes, as opposed to one hour for cocaine), and other local anesthetics are too highly toxic. Schenck emphasizes the usefulness of cocaine in nasal surgery, where its vasoconstricting powers and its shrinkage of mucous membranes are important, and in diagnosis and treatment of tracheobronchial lesions— especially when fiber optic or tubal examination is necessary. He was able to find only three deaths from topical application of cocaine in the medical literature. Although he suggests using a drug of low toxicity like lidocaine for laryngoscopy and other procedures that take only a short time and do not require mucous membrane constriction or hemostasis (arrest of bleeding), he prefers cocaine for most purposes. Schenck's article may indicate a trend toward more use of the drug in ear, nose, and throat surgery.[8]

In veterinary medicine cocaine is still used for epidural anesthesia (anesthesia of the dura mater, the outer envelope of the brain), as a nerve block in diagnosing lameness of horses, and even as a central stimulant in shock and collapse. It is also used orally to prevent vomiting in cats and dogs. The usual dose for horses is 300 to 600 mg; for dogs as a gastric sedative it is 4 to 8 mg.[9]

The long list of past medical uses of cocaine and the short list of present ones do not entirely exhaust its possibilities. Recently interest in cocaine as a diagnostic tool in certain mental disturbances has revived. This interest has appeared sporadically among psychiatrists for nearly a century. In the days when cocaine was being prescribed for many minor functional disturbances, physicians also experimented with it in more serious conditions like severe depression. Freud comments cautiously on this in one of his papers. He points out that psychiatry has many drugs to "subdue overstimulated nervous activity" but few to "heighten the performance of the depressed nervous system." He mentions that cocaine has been used to combat hysteria and hypochondria, and then refers to experiments on melancholics in which a slight improvement was claimed. He concludes, "On the whole one must admit that the usefulness of cocaine in psychiatric practice has yet to be proved; however, its possibilities do seem to justify a careful inquiry."[10] But soon afterward the danger of cocaine abuse came to be regarded as so high that this inquiry was not worth conducting. Not until the introduction of amphetamine as a medicine more than 40 years later did psychiatrists test extensively and systematically the effect of a powerful CNS stimulant on severely depressed patients—and find it wanting. Today the tricyclic antidepressants are securely established as the main pharmacological treatment for severe depression, and no one supposes that the use of either cocaine or amphetamines for this purpose will be revived.

The short-term diagnostic use of cocaine has also been occasionally recommended at least since 1901, when Hans Berger (later the inventor of the electroencephalogram) noticed that a subcutaneous injection of the drug temporarily dissolved a months-long catatonic stupor. Reporting in 1921 on his observations in this and other cases, Berger concluded that in the early stages of catatonic schizophrenia, if there has been no organic brain damage, cocaine eliminates the cortical inhibition that, he believed, causes the condition. Writing in 1924, Ulrich Fleck found similar effects, although he rejected Berger's explanation. He injected 100 to 200 mg of cocaine subcutaneously into 34 patients, 18 of them described as catatonic schizophrenics in a condition of stupor. There were no ill effects and in some of them the drug allowed access to their minds. One catatonic was able to reflect critically on his previous condition: "I didn't know who I was and what I was and how I was." The effect on severely depressed patients was similar. Euphoria occurred in only 2 of the 34 patients, and 10 of them actually seemed more depressed. The character-istic effect was greater liveliness and ability to communicate, without change in mood or alleviation of schizophrenic symptoms. Fleck con-cluded that cocaine was better than ethyl alcohol for bringing catatonics out of stupor, and thought it might have some diagnostic usefulness. But he advised against using it in therapy because of its transient effect and the danger of habituation.[11]

Inspired by Fleck's work, Gustav Bychowski made further studies on catatonics in 1925, also by injecting 100 to 200 mg subcutaneously. His results were similar. The drug produced heightened communica-tiveness, greater suggestibility, and more emotional rapport. The expres-sion was dissociated and structureless, whether it took the form of stream-of-consciousness talk or movement. Bychowski suggested that in view of what was now known about the brain stem, Berger had been wrong in thinking that the effect of cocaine on schizophrenics was mainly cortical. He concluded that the only immediately practical use for the drug in these cases was diagnostic, although the sudden and short-term psychological contact might also be put to some therapeutic use.[12]

In a controlled experiment in 1926 Arno Offerman administered 50 to 100 mg of cocaine subcutaneously to a number of psychiatric patients. In one case described as a total stupor of catatonic or hysterical origin, with catalepsy and muteness, an injection of 70 mg gave the patient's mask-like face a normal expression and even made him smile. He answered questions clearly, said that he felt good, and attributed his condition to a disappointment in love. There was no trace of schizophrenic incoherence in his speech. After the experiment, according to Offerman, he continued to be more accessible than he had been before. In three other cases of catatonic stupor, cocaine had no effect, and in a fourth there was some incoherent liveliness. Among postencephalitic patients, some with Par-kinsonism (muscular rigidity with slow rhythmic tremors in the extremi-ties), cocaine tended to produce more articulate and coherent speech, more movement, and a better mood. Offerman believed that this sug-gested a brain stem ganglia effect. He noted that some of the patients had already begun to beg for more injections of the drug, and he interpreted this, in line with accepted medical attitudes toward cocaine, as requiring rejection of any use in therapy because of the danger of abuse.[13]

In 1927 August Jacobi published still another paper on this subject. After describing the results obtained by Berger, Fleck, Bychowski, and Offerman, as well as other researchers who had been more disappointed in the effects, Jacobi reported his own experiment on 24 mentally ill pa-tients, 18 of them schizophrenics. In 16 of the 24 there was an increase in expressiveness of speech and in grimaces and gesticulations. In 7 cases he established some rapport with the patient for the first time, and in 2 instances he felt able to speak of "a direct glimpse into the life of the soul." Euphoria was not common, or at least rarely obvious from behavior; a desire or even a compelling drive to talk was the most significant effect. Jacobi concluded that the use of cocaine was of interest for the diagnosis of intelligent and articulate patients who otherwise could not express their thoughts. He too rejected any therapeutic application.[14]

In 1934 Erich Lindemann and William Malamud examined the effect of sodium amobarbital, hashish, mescaline, and cocaine on two patients described as neurotic and four described as schizophrenic. The dose of cocaine was 30 to 50 mg subcutaneously. A schizophrenic woman who had been hallucinating and acting in a dreamy and childishly playful manner began to cry and speak fast in interrupted sentences as though talking to herself. She took the right side of her body, representing God, to be reproaching the left side, representing Satan. The effect lasted two hours. A woman who had been depressed and later paranoid, refusing to eat or be touched, became restless and antagonistic when given 15 mg of cocaine. After another dose of 20 mg half an hour later she had auditory hallucinations and was combative and excited for an hour. A man described as seclusive, impulsive, grimacing, and silly became restless and hilarious and produced a series of memories in stream-of-consciousness form. Later he became evasive and uncommunicative. A man described as neurotic who was depressed and tended to bark rather than speak also became restless and talked to himself. The other patient described as neurotic, a theatrically excited man who was trying to convert everyone to a new religion he had invented, became irritable, demanded discharge from the hospital, and talked in a stereotyped singsong fashion about his religious ideas.

Lindemann and Malamud, possibly because their subjects were not catatonics, interpreted their results less favorably than the earlier German researchers. Cocaine seemed to make the patients more self-assured, demanding, aggressive, and distant or critical in attitude to the hospital and the examiner. One of them (the paranoid woman) openly expressed cannibalistic desires. Both cocaine and sodium amobarbital made the subjects more communicative, but the barbiturate tended to bring them emotionally closer to the physicians while cocaine made them indifferent, arrogant, inclined to "stand up for their own level of regression." Oddly, in a later experiment conducted by Malamud and others on 14 male schizophrenics, subcutaneous doses of less than 50 mg of cocaine had no influence clinically, although 50 mg subcutaneously produced relaxation and warmer affect. Amphetamine (20 mg intravenously) caused only anxiety and tension.[15]

After a hiatus of over 30 years, Robert M. Post and his colleagues have recently revived experimentation with cocaine on depressed patients. In one controlled, double-blind study of 16 depressed patients, intravenous cocaine (infused over a 90-second period) starting at 2.5 to 8 mg and working up to 5 to 25 mg had marked, highly correlated effects on mood and vital signs like temperature, pulse, blood pressure, and respiration. Individual variability was great: the dose it took to raise pulse rate and blood pressure significantly varied from 2 mg to 16 mg. The most common reactions to the drug were profound emotional release and tearfulness; patients appeared distraught but reported that they felt much better, presumably because of the emotional catharsis. Post and his colleagues conclude that infusions of cocaine producing moderate physiological arousal create a sense of well-being, while higher doses mobilize intense mixed affect, a combination of elation and uneasiness about uncontrolled release of feelings. In another experiment, cocaine administered orally starting at 35 mg to 60 mg a day and gradually increased to 65 to 200 mg twice a day at 9:00 A.M. and 10:00 A.M. produced no consistent mood changes or physiological effects but did reduce the total amount of sleep and therefore the amount of rapid eye movement (REM or dreaming) sleep.[16] Post's work is continuing and may produce results useful for psychiatry. Intravenous amphetamines in doses of 15 to 30 mg have been used for diagnostic purposes; an advantage of cocaine for work with depressed patients would be its shorter period of activity, which makes the effect more controllable.

But the most interesting proposed psychiatric application for cocaine is not therapeutic or diagnostic; it is the use of cocaine effects in theoretical models of psychosis. We have noted that cocaine psychosis closely resembles amphetamine psychosis, which is in turn often indistinguishable from paranoid schizophrenia. Recently investigators have been making use of this resemblance to study stimulant-induced psychoses, treating them as analogues of the functional (natural—reactive or endogenous) psychoses that may have a related neurochemical basis. This work may contribute to a reclassification of the schizophrenias and other forms of psychosis and suggest new ideas about their origin.

Stimulants are clearly preferable to other drugs for such heuristic models. In other drug psychoses acute reactions rarely develop into complex delusional belief systems as they do in stimulant intoxication and schizophrenia. Also, most drug psychoses are "toxic," with clouded senses and disorientation, unlike amphetamine or cocaine psychosis and most schizophrenia. Evidence from both pharmacology and the clinical observations of blind raters in controlled experiments shows the effects of so-called psychedelic or hallucinogenic drugs like LSD and mescaline to be quite unlike schizophrenia and other functional psychoses. Some of the feelings in incipient schizophrenia—a sense of having broken out of the conventional ways of seeing things, intense joy, dread, or awe—resemble psychedelic states, but in schizophrenia these feelings are soon integrated or encapsulated in restricted patterns of autism or thought disorder. The mimicry of functional psychoses by stimulants is more accurate than that of any other known chemical substances.[17]

It is becoming clear that various kinds of mental disturbance may often be stages of a common development of variations on a single pattern with a common origin. For example, the hypomanic character, whose condition is an attenuated form of the manic phase of a manic-depressive cycle, is described as self-assured, aggressive, affable, self-satisfied, uninhibited, glib, boastful, ambitious, free-spending, and boundlessly energetic. But if he is crossed or thwarted, his emotional exaltation may be replaced by anger, irritability, resentment, and even paranoid delusions. The hypomanic differs from the paranoid or paranoid schizophrenic because he suffers no solitary and asocial tendencies or personality disintegration and because his hopes and fears produce empathy rather than merely puzzled discomfort in observers. Nevertheless, there is a connection. Paranoia and paranoid schizophrenia are associated with high aspirations and inability to accept defeat. Delusions of persecution are projections of the victim's own aggressive inclinations, and the compensatory exaggerated self-assurance makes for associated delusions of grandeur. Catatonic schizophrenia is superficially quite different from paranoid conditions but apparently related. In the most common phase, catatonic stupor, it is marked by grimaces, clenched muscles, catalepsy (the patient maintains body positions impressed on him and initiates no spontaneous movement), muteness, and apparent unawareness of the external environment. But the term "stupor" is somewhat misleading, since the immobile patient has not actually lost his powers of thought, fantasy, and observation. The symptoms of the other phase of catatonic schizophrenia (a given patient may experience both or one of them) are disorganized hyperactivity, impulsive behavior, unpredictable violent tendencies, disregard of cleanliness, terrifying or ecstatic hallucinations, hostility, and resentment. Catatonic stupor may resemble the blank impassive response sometimes experienced in severe depression or bereavement, or the instinctive immobility reaction of some animals in the face of danger.

The excited catatonic state often seems the same as acute homosexual panic, in which voices accuse the patient of homosexual practices.

There is a clear relationship between paranoid schizophrenia and stimulant-induced psychoses. Ellinwood has named three symptoms as defining characteristics of amphetamine psychoses: visual hallucinations; auditory hallucinations that include voices talking directly to the victim; and delusions of persecution or gross paranoid reactions. Elsewhere he mentions hyperalertness, overreactions to slight movements in the periphery of vision, distorted perception of faces, compulsion to find meaning in details, a feeling that the body is transparent, and intense terror of losing control. The unique effectiveness of the antischizophrenic phenothiazine drugs in treating amphetamine psychosis heightens its resemblance to paranoid schizophrenia.[18]

Conditions resembling hypomania and catatonic schizophrenia can also be produced by stimulant drugs. The thesaurus of adjectives applied to the hypomanic patient might also serve for someone who had taken a large dose of cocaine or amphetamines. In many reported cases of cocaine intoxication there is the same tendency to pass from cheerful confidence and expansiveness to delusions of persecution. On the other hand, some phenomena in the late stages of chronic stimulant intoxication, especially the cataleptic postures produced by cocaine and amphetamine in rats and dogs, distinctly resemble the symptoms of catatonic stupor. Maier describes "cocaine paralysis" in human beings as a state of silly imbecility with senseless, unsystematized delusions of grandeur. Joël and Fränkel distinguish three stages of chronic cocaine intoxication. The first consists of psychomotor excitation, heightened powers of attention, sensory acuteness, fast thought associations, and cheerfulness. Then some delusions, hallucinations, anxiety, attention disturbances, and compulsive psychomotor activity; and finally psychomotor inhibition, sensory paralysis, slow difficult thought processes, and feelings of indifference or misery.[19] These three stages in many ways resemble the psychiatric classifications of hypomania, paranoid schizophrenia, and catatonic schizophrenia, respectively.

Ellinwood, along with several colleagues, has been examining the amphetamine model of psychosis in a series of recent papers based largely on animal experiments. He and Abraham Sudilovsky, in a 1972 paper, note that the repetitious stereotyped behavior called "punding" is a characteristic feature of chronic amphetamine intoxication in both man and animals. The common denominator is a pattern of searching and examining. In man an initially pleasurable suspiciousness or tendency to look for meanings in environmental details often turns into paranoid delusions. In animals incapable of producing delusional thought systems, generalized investigatory stereotypies like sniffing that are only slight exaggerations of normal curiosity tend to degenerate into more restricted repetitious behavior. In the late stages of chronic amphetamine intoxication, experimental animals also assume bizarre dystonic postures resembling catalepsy. These are probably produced not by a failure of muscular coordination but by loss of initiative in changing attitudes or postural movement sets. Autonomous, dysynchronous movements of body parts occur; for example, a cat will go about its ordinary activities with one of its hind legs in an awkward raised position. Ellinwood and Sudilovsky conclude that both paranoid and catatonic schizophrenia have in common with chronic amphetamine intoxication a restriction of attitude and perception, which in the latter two conditions is related to stereotyped postures and movement. The mechanisms that regulate the smooth flow of behavior from one "set" to another fail, and actions and attitudes become fixed, autonomously organized complexes. The victim loses his willingness to change his emotional and physical attitudes in accordance with the demands of the situation. Ellinwood and Sudilovsky summarize: "A paranoid catatonic continuum is proposed as the clinical expression of chronic amphetamine intoxication in lower animals, and the need for earlier observation of the process in humans is stressed. We postulate that perseveration and distortions of postural-motor attitudinal sets are common to both possible pathways of the psychotic process." [20]

In an earlier paper Ellinwood had noted that in animals predisposing "personality" and environment affect the reaction to amphetamines; for example, relatively wild farm cats become aggressive and hiss while tame city cats show curiosity and sniff. Later Ellinwood and M. Marlyne Kilbey studied the influence of environment and "prepotent behavioral patterns" on the form of amphetamine stereotypy. They noted that not only simple motor reflexes but more complex attitudes like following behavior in a dog, and even thoughts, can fall into stereotyped patterns. We may be reminded of Crowley's comment that under the influence of a large amount of cocaine his mind was "nailed to one or two thoughts." Ellinwood and Kilbey suggest that in the early stages stereotypy is generalized enough to incorporate any prepotent behavior induced by the environment and disposition of the individual animal. Later species-specific motor activities become dominant. That would presumably include the grooming or picking at the skin with the hands in primates (men and monkeys), which human beings rationalize by perceiving insects or worms. [21]

The most useful feature of the stimulant model is its suggestion of neurochemical mechanisms for the various stages or classes of psychosis. The changes in behavior seem to depend on whether dopamine or norepinephrine effects predominate and whether the neurotransmitter is being potentiated (by release, reuptake blockade, or heightened receptor sensitivity) or depleted. In the early hypomanic stage of stimulant intoxication the dominant influence is release of norepinephrine. Injections of amphetamine or cocaine increase the rate at which animals stimulate electrodes placed in the medial forebrain bundle or reward center of the hypothalamus. Disulfiram (Antabuse), a drug that prevents the body from converting dopamine to norepinephrine, inhibits this effect, and an injection of norepinephrine nullifies the inhibition. Norepinephrine is the main neurotransmitter in this region and probably accounts for the euphoriant and aphrodisiac effects of stimulants. The enhancement of alertness, motor coordination, and attention is thought to be mediated by the influence of the reticular activating system on cerebral arousal mechanisms. The nerve fibers that carry this influence are apparently also norepinephrine fibers.[22]

The more intense forms of stereotyped behavior and bizarre postures, on the other hand, are related to dopamine action in the corpus striatum. In experiments on cats with large daily doses of methamphetamine (Methedrine), Ellinwood and Sudilovsky discovered that the milder forms of stereotyped behavior were most common after 90 minutes on the first day. Before that, they concluded, norepinephrine release was masking or blocking the dopamine effect, which became obvious only when norepinephrine was depleted. Species-specific motor stereotypies and bizarre postures resembling those of human catatonics became more common later in the experiment, when dopamine may have been acting in the almost complete absence of norepinephrine. When disulfiram was administered along with methamphetamine the stereotyped behavior was more intense and began sooner; this tends to confirm the hypothesis that the presence of norepinephrine somehow prevents it. The changes in the form of stereotyped behavior and its eventual degeneration into a catatonialike condition suggest that methamphetamine continues to release and deplete dopamine and stimulate receptor sites for it while synthesis of dopamine fails to keep pace. Eventually, according to Ellinwood and Kilbey, a recycling movement in the dopamine system without sensory feedback causes motor disorders. They suggest that a similar progression sometimes occurs in schizophrenia.[23]

Solomon Snyder has made some related observations on norepinephrine and dopamine. Dextroamphetamine blocks the reuptake of dopamine and therefore reduces, for a while, the muscular rigidity and tremors of Parkinsonism, which are caused by inadequate dopamine synthesis. The butyrophenones and phenothiazines, drugs effective in treating both schizophrenia and amphetamine psychosis, also have effects that resemble Parkinsonism. They apparently produce these effects by blocking dopamine receptor sites in the corpus striatum. The more potent a butyrophenone or phenothiazine is in combating schizophrenic symptoms, in general, the more such effects it produces. Snyder believes that the dopamine receptor blocking action of these drugs is related to their therapeutic effect. Phenothiazines also block norepinephrine receptor sites, and pretreatment with phenothiazines increases the rate of intravenous self-injection of cocaine and amphetamines in laboratory animals. Apparently phenothiazines antagonize the reinforcing property of these stimulants, since they have the same effect as a decrease in dosage per injection. Other depressants like morphine and pentobarbital do not have this effect, which provides further evidence that the pleasure-giving powers of stimulants are related to release of norepinephrine.[24]

Snyder suggests that the alerting effect produced by norepinephrine may generate the search for significance that turns the undifferentiated schizophrenialike dopamine effect into a paranoid delusional system. As an alternative, he proposes that in an amphetamine-induced paranoid state that is not schizophrenic, the paranoid hypervigilance produced by norepinephrine protects the patient from more flagrant dopamine-mediated schizophrenic symptoms.[25] This hypothesis accords with Ellinwood's observation that the presence of norepinephrine prevents the appearance of stereotyped behavior and cataleptic postures in cats given methamphetamine.

David L. Garver and his colleagues describe an experiment that epitomizes and confirms some of the conclusions of Ellinwood, Snyder, and others. They administered 2 mg per kg of dextroamphetamine by gastric tubing to four selected monkeys in a laboratory colony every 12 hours for 3 to 20 days. The monkeys were periodically rated for visual checking (changes in eye direction, a measure of hypervigilance), motor activity, repetitious stereotyped behavior, and physical distance from other animals in the colony. The dextroamphetamine increased all these. Other animals avoided the intoxicated monkeys, who gradually became socially isolated. Possibly they were giving fragmented and unintelligible behavioral cues analogous to the confused schizophrenic thought, language, and behavior that produce puzzlement and discomfort in normal people. The stereotyped behavior became more solitary as time went on, in a process analogous to autistic withdrawal in human schizophrenics. These solitary stereotypies were associated with a reduction in hypervigilant visual checking. AMPT (the drug that prevents the synthesis of both dopamine and norepinephrine) and chlorpromazine (a phenothiazine) blocked all the dextroamphetamine effects. Pimozide and haloperidol, which act primarily by blocking dopamine receptors, prevented visual checking but not the other symptoms. Garver and his colleagues conclude that the hypervigilance is mediated by norepinephrine and the motor activity and behavior that resembles psychosis by dopamine. They believe that psychosis in human beings may be caused by excessive release of dopamine, by supersensitivity of dopamine receptor sites, or possibly by a change in the critical balance between acetylcholine and dopamine neurotransmitter systems operating in a single brain region. Antipsychotic drugs that block dopamine receptor sites could then be remedying a condition of acetylcholine deficiency as well as one of dopamine excess.[26]

Stimulant-induced psychoses resemble another symptom complex besides paranoid schizophrenia: the form of epilepsy that originates in the temporal lobe of the brain. Epilepsy is a condition in which periodic seizures are caused by a disturbance in the electrophysiochemical action of brain cells. Temporal lobe epilepsy, as opposed to grand and petit mal seizures, consists of trancelike attacks and confusional episodes lasting usually for a few minutes, but sometimes for hours or days. Some of the symptoms that may appear are delusions of persecution, a sense of being influenced by mysterious malevolent forces, distorted perception of faces, visual hallucinations, depersonalization, simple or complex stereotyped behavior—all resembling acute paranoid schizophrenia and cocaine or amphetamine psychosis. Ellinwood notes that electrical stimulation of certain areas in the temporal lobe, especially the amygdala (a region connected to both the hypothalamus and the cerebral cortex), produces first attention or orienting behavior, then frightened cringing or aggression in animals. Repeated administration of amphetamine induces a characteristic pattern of electrical activity in the amygdala of the cat. Ellinwood believes that "amphetamine stimulation of the limbic system and the associated temporal lobe can heighten as well as distort the emotional interpretation of experience and memory." He mentions the unusually intense sense of reality and the emotionally charged memories produced by human beings in both amphetamine psychosis and temporal lobe epilepsy and also under electrical stimulation of the temporal lobe. There is some work suggesting resemblances between cocaine-induced seizures and temporal lobe epilepsy.[27]

Since there is no obvious connection between dopamine excesses and the abnormal neuronal responsiveness of temporal lobe epilepsy, Ellin-wood suggests that stimulant psychoses, functional psychoses, and tem-poral lobe epilepsy have a common source in chronic hyperactivity of the reticular activating system (RAS) arousal mechanisms. This, he believes, causes a change in the neuronal substrate of cognitive processes that in-duces an intense sense of significance, often called apophany, and eventually delusional belief systems. Ellinwood also believes that chronic RAS hyperarousal can cause associative circuits in cortical and subcor-tical regions to become autonomous, producing the kinds of isolated dy-synchronous behavioral configurations observed in the late stages of chronic amphetamine intoxication. The stimulation of internal arousal mechanisms may also cause the so-called paradoxical calming effect of amphetamines on hyperactive children. One commonly accepted hypoth-esis is that these children are at the mercy of every irrelevant stimulus in the environment because they are deficient in internal mechanisms for attention, concentration, and retention of the emotional significance of stimuli. Amphetamines are said to reduce this excessive susceptibility to distraction, presumably by their effect on the RAS.[28]

Almost all the work on stimulant-induced model psychoses so far has been done with amphetamines. References to cocaine have been con-fined to a few *obiter dicta* about the near-identity of cocaine psychosis with amphetamine psychosis. But cocaine should be particularly useful in this kind of research because it is chemically quite different from the whole range of amphetamine congeners, all of which resemble the cat-echolamines in molecular structure. It would also be important if cocaine affected the neurotransmission process primarily, if not exclusively, at a different point from the amphetamines, influencing receptor sensitivity more and release or reuptake of catecholamines less. It is particularly sig-nificant that, as we pointed out in Chapter 3, cocaine seems to have proportionately less effect than amphetamines on dopamine, as opposed to norepinephrine, transmission. This may account not only for the ex-perimentally demonstrated fact that amphetamine causes stereotyped be-havior at much lower doses but also for the (admittedly less well-es-tablished) clinical observation that cocaine produces fewer psychoses. Snyder observes: "By developing drugs which readily differentiate the roles of particular neurotransmitters in regulating various behaviors, one might then administer such drugs to humans and thereby determine the neurochemical substrata of those behaviors, such as psychosis, which seem to be peculiar to humans." [29] Cocaine and possible related drugs are promising for this purpose.

Robert M. Post has recently begun the systematic study of cocaine model psychoses. After noting that the biological changes produced by cocaine mimic those associated with stress, he points out that acute stress raises catecholamine levels in the brain and chronic stress lowers them (by depletion). He suggests that the disconnection of cortical-sub-cortical integration by cocaine could affect cognition and affect in the same way as functional psychoses, and he proposes the idea of a continuum of manic, depressive, and schizophrenic functional psychoses analogous to the progressive effects of chronic cocaine intoxication. Unfortunately, so little information on cocaine is available that Post continually has to appeal to the amphetamine analogy.[30]

As these observations suggest, many uncertainties and loose ends remain. Nevertheless, the stimulant model of psychosis seems to hold some promise. The basic process that schizophrenia, amphetamine psychosis, and cocaine psychosis have in common may be the neurochemical reproduction of a state of chronic emergency. Disturbances in the brain tracts that regulate the impact of sensory stimuli on feeling and thought may be the cause of the unmanageability of the schizophrenic's world and the source of the resemblance between functional psychoses and stimulant effects. Experiments using cocaine to elaborate and test the model might concentrate on its differences from amphetamines in the site of action and the degree and kind of effect on dopamine and norepinephrine. Variations on the chemical formula of cocaine that preserved some of its central stimulant power might also be useful as drugs of the kind Snyder calls for to differentiate the roles of various neurohormones. Whatever the prospects may be, certainly there are few more important applications for a drug than helping to determine the origins of schizophrenia.[31]

8

COCAINE AND
DRUG DEPENDENCE

A MONSTROUS TANGLE of social, psychological, and pharmacological issues surrounds the concepts of drug addiction, habituation, and dependence. There is practically no agreement on how to determine when someone is addicted to or dependent on a drug or even on what these terms mean. To some, this implies simply that psychopharmacology is an embryonic rather than a fully developed science. They associate the prevailing confusion with the absence of a suitable model or paradigm (to use Thomas Kuhn's now familiar term) that would permit scientific research to proceed along clearly defined lines. But the emotion with which the issue is debated suggests that something more is involved. In fact, the ideas put forward on this subject resemble the tentative, groping theories of the early stages of a science less than the polemical stances and moral certainties of a conflict of political ideologies. Phrases like *drug addict, drug habit,* and *drug dependence* are not only emotionally fraught but morally and socially loaded, especially when we substitute the names of particular illicit substances for "drug." The ideas involved seem to be what W. B. Gallie has called "essentially contested concepts": like such words as *freedom, democracy, tyranny,* and so on, they make no sense apart from their use by conflicting social groups that necessarily differ on their interpretation. To clarify the issues in this discussion is less like evaluating scientific theories of a more or less known phenomenon than like moderating a political debate.

Concepts of Addiction

The loaded character of the concepts we are talking about is indicated by the way ordinary words, by having "drug" prefixed to them, become captives of the medical vocabulary. An addiction once meant a strong habitual inclination to do something. Shakespeare had a character in *Othello* say, "Each man to what sport and revel his addiction leads him." Even in the middle of the nineteenth century, Gladstone could refer to "addiction to agricultural pursuits." It is true that the word always suggested devotion, binding, being given over to something. But to most of the public today it means nothing more than a mysterious and utterly debasing enslavement to a dread chemical called heroin. The old sense of the word, as in expressions like addiction to television, detective stories, or sex, is felt to be metaphorical if not jocular: addiction per se means heroin use. This forecloses discussion of the otherwise not entirely implausible notion that a strong habitual inclination to take heroin is no worse than a strong habitual inclination to watch television. (If Halsted had taken to some equivalent of television instead of to his opiate, would his surgical career have been so successful?) But the original meaning is probably unreclaimable. The word has become too intimately incorporated into medical and pseudomedical jargon.

The pseudomedical definitions of addiction once popular among physicians and drug researchers are characteristic of the confusion of moral, social, and pharmacological categories that has been so baneful in this field. Until the 1960s the main contrasting term was *habituation*. The influential World Health Organization (WHO) definition of 1950, reformulated in 1952 and given its final form in 1957, described drug addiction as a state of periodic or chronic intoxication produced by a drug, which also involved a compulsion to continue taking the drug, a tendency to increase the dose, a psychological and "generally" a physical dependence on its effects, and a detrimental effect on the individual and society. "Habituation" implied no "intoxication," involved a "desire" rather than a "compulsion" to take the drug, produced some psychological but no physical dependence and therefore no abstinence syndrome, and was detrimental, if at all, primarily to the individual.[1] What this amounted to was an attempt to give a medical justification for the received opinion that the use of some drugs, mainly opiates, was very bad, and the use of other drugs not so bad—a habit rather than an addiction.

There is no need to point out the vagueness and logical errors in these definitions, for they are no longer used. Soon it became apparent that abuse of "habituating" drugs like amphetamines and barbiturates could be just as serious as abuse of opiates and just as detrimental to society. The proliferation of new drugs made it difficult to determine whether any particular case was "addiction" or "habituation." "Habituation to addiction-producing drugs" and "addiction to habituation-producing drugs" came to seem necessary as qualifications to the official terminology. Finally this terminology was abandoned as hopelessly confused and unwieldy, and the words *addiction* and *habituation* were dropped as general descriptions of chronic drug use and abuse.

Medical researchers still use the word *addiction* in several ways. The most precise and limited of these (although not the only reasonable one), a definition clearly medical and not pseudomedical, is the one we have adopted in this book. It is as follows: drug addiction is a condition induced in higher mammals by chronic administration of certain central nervous system depressants like opiates, alcohol, and barbiturates; a gradual adaptation of the nervous system produces a latent hyperexcitability that becomes manifest when the drug is withdrawn and includes physiological symptoms that are interpreted as a physical need for the drug. This definition implies no moral or political attitude. It makes no pretense to solving the problem of free will by distinguishing "compulsion" from "desire" and does not presume to judge what drugs are detrimental to society. The condition it describes is no doubt unfortunate and unpleasant, but no more so than the condition of a diabetic who needs insulin to function normally. In fact, it is distinctly less so, since the diabetic will die without insulin but the addict suffers only an abstinence syndrome that is temporary and, except in the case of barbiturates, rarely fatal. In this way of using the word *addiction* we do not prematurely fix blame on the drug or the drug user. We insinuate no moral overtones and intimate nothing implausibly horrible and debasing. The abstinence syndrome, the main evidence of addiction, is simply a fairly well-defined set of physiological and psychological symptoms, different for different depressant drugs. It is neither the main cause nor the most important consequence of drug abuse, and its presence or absence is not a proper criterion for moral judgment.

Withdrawal Reactions

Cocaine produces no abstinence syndrome in the sense implied by our definition of addiction, since it is not a depressant. It may, however, cause the same kind of uncomfortable but physiologically unspecific feeling of need that is associated with nicotine, caffeine, and amphetamines. An interview subject who had taken both cocaine and heroin intravenously said, "Cocaine is mostly just an urge or yearn—it doesn't take your body out of proportion like heroin. . . . Really you can have a great urge for it, but as far as physical, no." Aleister Crowley, as we noted, remarked on the difference between the "moral pull" of cocaine and the feeling of a physical need for heroin. Burroughs writes, "There's no withdrawal syndrome with C. It is a need of the brain alone—a need without body and without feeling. Earthbound ghost need"; also, "I have talked with people who used cocaine for years, and then were cut off from their supply. None of them experienced any withdrawal symptoms. . . . Addiction seems to be a monopoly of sedatives." Gutiérrez-Noriega admits that *coqueros* have little trouble abandoning the habit when coca is not available; one of his studies with Zapata Ortiz showed that there were no significant physiological changes after withdrawal of coca.[2] Another comment from one of our cocaine-using subjects: "It's not physically addicting, you don't feel sick or anything. You just feel you *need* some cocaine, and your whole direction is to get some. It's like quitting smoking or something. You're strung out." Or again, "I'd wake up feeling angry and nasty and mean till I got going again on cocaine for that day."

The other major undesirable aftereffect of cocaine abuse can be described as a withdrawal reaction only in a very limited sense. The lethargy, lassitude, or irritable depression that may succeed overstimulation as the body recovers are indirect effects of the drug itself rather than its absence; they resemble a milder version of the amphetamine "crash." If the stimulation is moderate and the letdown that follows correspondingly slight, more of the drug may provide tempting relief; possibly that is how all nicotine dependence and some amphetamine and cocaine dependence are produced. But in its more severe forms the stimulant letdown does not cause a desire for the drug. In the terminology of laboratory psychology, the symptoms do not have secondary reinforcing value like opiate abstinence symptoms.[3] In other words, an amphetamine or cocaine abuser feeling the effects of overindulgence does not immediately seek more of the drug to relieve his misery. On the contrary, he may have to wait out and sleep off the effects of his body's reaction to the overstimulation originally produced by the drug before he can take an interest in its euphoriant powers again. The pattern of "runs" and "crashes" observed in stimulant use by laboratory animals as well as on the street is quite unlike the gradually developing steady need that arises with opiates. The stimulant abuser stops taking the drug when he can no longer tolerate its effect (although it is often already too late), and starts taking it again when he considers himself well enough to tolerate it; the opiate addict may stop not because he can no longer tolerate the effect but, on the contrary, because his tolerance has become so high that it is impossibly expensive to continue; he may start again at the height of the withdrawal crisis because he can no longer tolerate the *absence* of the drug's effect. Although the more severe symptoms that follow amphetamine or cocaine withdrawal are of course most likely to arise in heavy chronic abuse, they are basically an acute reaction to overstimulation—not the consequence of a slow adaptation of the nervous system that makes the drug necessary for normal body functioning, as in the case of opiates. (This seems clearer for cocaine than for amphetamines, which produce tolerance and also withdrawal symptoms that resemble depressant abstinence syndromes in some ways.) Incidentally, it should be obvious without saying, but apparently is not, that opiates and stimulants as well as alcohol can be used in such a way that they do not produce any of these consequences.

It is not certain that any distinct physiological reaction to the withdrawal of psychoactive drugs exists apart from the depressant abstinence syndromes, but the research on sleep patterns initiated by I. Oswald and V. R. Thacore may provide some leads. It has been discovered that certain drugs reduce the amount of REM (dreaming) sleep and cause a rebound overcompensation on withdrawal. That is, one dreams less while taking the drug and more for a while afterward, as though to make up for the loss. Dreaming normally begins about an hour after we fall asleep and occupies about one-fifth of the night. It seems to be psychologically necessary in some way, since people who are experimentally deprived of it by wakening during the night start to compensate by falling into REM sleep sooner than usual. Cocaine is one of the drugs that produce this same deprivation and overcompensation, along with amphetamines, alcohol, opiates, barbiturates, and meprobamate (Miltown). Cocaine and the amphetamines, in particular, reduce REM sleep simply by reducing total sleep; they do not change the ratio. But even in people so habituated to amphetamines that their sleep is normal, withdrawal may cause compensatory dreaming to the point where it occupies half of a now much longer night's sleep. Although some antidepressants not implicated in drug abuse, like imipramine, also cause REM sleep rebound, it has not been shown that any drug which produces a withdrawal reaction does not also produce both REM sleep suppression and rebound overcompensation.[4] All the drugs in our list above may be providing an escape from the tensions and disappointments of waking reality that acts as a partial substitute for the wish-fulfillment process of dreaming. At any rate, this suggestion is compatible with vom Scheidt's conclusions about Freud's cocaine episode (see Chapter 2). If the idea proves to have any value— and this must remain speculative—it would be the first one effectively connecting a definition of drug withdrawal reactions with a distinct kind of behavior or activity (rapid eye movements in sleep and the accompanying reported dreams) and a distinct neurophysiological condition (the electroencephalogram patterns in REM sleep, which resemble those of the waking state rather than other kinds of sleep). Incidentally, Oswald has proposed a mechanism for the REM sleep suppression and rebound. He suggests that the brain normally synthesizes protein during REM sleep and that the rebound after drug withdrawal indicates a process of brain cell repair.[5]

Tolerance

Pharmacological tolerance is usually described as an adaptation of the nervous system to the effects of a given amount of a drug which makes it necessary to keep taking more of the drug to get the same effects. With depressants, it is a sign that an abstinence syndrome may appear on withdrawal, and is therefore evidence for addiction in the restricted medical sense we have adopted. In fact, CNS tolerance to depressants, as opposed to metabolic tolerance produced by acceleration of breakdown in the liver, is almost interchangeable with "latent hyperexcitability," since the reason a person comes to need more and more of a sedative drug is that the nervous system becomes more and more sensitive to the excitatory stimuli blocked by the drug. In this sense, at least, tolerance seems to be a reasonably clear concept.[6]

Unfortunately, beyond this area of clarity lies a region of conceptual caprice, anarchy, and gloom—especially where stimulant drugs are concerned. What exactly are the same effects? How does one distinguish between simply wanting more of the same and becoming tolerant to it? And is the phenomenon in any way specific to drugs? After all, everyone knows that pleasures tend to pall. How is the call for stronger wine dif-

ferent from the call for madder music? There are no conclusive answers
to these questions. Amphetamines, for example, are usually said to produce tolerance, since some users increase the dose to levels that might
have been fatal at the beginning. Tolerance to the appetite-reducing effect and sometimes to the euphoriant effect develops rather quickly, and
the EEG and sleep patterns may also return to normal after a while. On
the other hand, physicians use amphetamines in the long-term treatment
of the disease called narcolepsy, and in the treatment of hyperactive
children, without having to raise the dose. Laboratory animals injecting
the drugs voluntarily do not steadily increase the dose. Nils Bejerot, who
has examined over 4,000 intravenous abusers of Preludin (phenmetrazine), an amphetamine congener, states that most of them increased
the dose rapidly at the beginning and adjusted to a high level, but went
no further after that.[7] Is this a case of a pharmacological tolerance reaching its limits or a matter of titration to a desired dose? Individual variability is important: some people, finding that their amphetamine pill no
longer seems to work, will stop taking it; others will take more pills; still
others will feel an intense need for their daily dose but no desire to increase it.

Whatever the case may be with amphetamines, it is generally agreed
that there is no consistent need or tendency to increase the dose of cocaine. A common opinion is one expressed by a man we interviewed: "I
would notice myself getting off from it quite a bit after a few lines, and
then after ten lines I wouldn't notice so much. . . . But I don't think you
develop tolerance over time. If you do it once a week for a year, you're
gonna get off on it as much in the last week of the year as you will in the
first week." In other words, the pleasure may pall with constant use, but
there is no specific pharmacological tolerance. Joël and Fränkel state that
users of cocaine become neither tolerant nor sensitized (i.e., requiring
less for the same effect); they contend that an overdose has the same effect on a habitual cocaine abuser as on anyone else, and they find that
even after a six to eight months' pause, the same dose acts in just the
same way. Maier, admitting that he disagrees with most other observers,
states that some cocaine users do keep raising the dose; he concludes
that the issue remains in doubt. Rats continually given large doses apparently develop no tolerance. Bejerot suggests that cocaine users may
increase the dose at the beginning until they reach a desired amount, and
then hold their intake constant, as Preludin users do.[8] The question of
cross-tolerance with amphetamines has not been studied. (There is clinical evidence of cross-tolerance among the amphetamines.) The apparent

absence of tolerance to cocaine may be related to the speed with which it is eliminated from the body.

There is evidence that continual use of high doses of cocaine, without periods of abstinence, actually causes sensitization to some of its more toxic effects. A. L. Tatum and M. H. Seevers, after experimenting on rabbits, dogs, and monkeys, concluded that all these animals became hypersensitive to other stimulants like ephedrine as well as to cocaine itself after chronic use. They administered 3 mg per kg daily subcutaneously to a dog (the equivalent of 200 mg in a 150-pound man) and found that the stimulant effect became greater and greater until, after 16 months, this dose, which had originally produced few observable effects, caused violent bodily motion. After 25 months, the animal was near convulsions after each dose. Tatum and Seevers believe that this phenomenon and the exaggerated reaction to other stimulants suggest a fundamental functional change in the nervous system. Robert M. Post and Richard T. Kopanda attribute the effect to a "pharmacological kindling" analogous to the generalized convulsions that occur after continual electrical stimulation of certain regions of the brain, especially the amygdala, at levels that initially fail to produce seizures. They believe that the kindling process may also explain why patients with psychomotor epileptic seizures sometimes eventually develop psychotic symptoms, and they suggest that a change in the brain after repeated overstimulation may account for some aspects of the naturally occurring psychoses. We have seen in Chapter 7 that Ellinwood makes a similar suggestion about chronic overstimulation of the reticular activating system. Gutiérrez-Noriega found that a dose of cocaine that was harmless at the start would kill some experimental dogs after a year or more of chronic intoxication. He also found that intoxication with carbon tetrachloride (which causes fatty degeneration of the liver) made animals much more sensitive to cocaine. He suggests that cocaine becomes more lethal after chronic use because it damages the liver.[9] This sensitization resembles some effects of chronic amphetamine intoxication and seems to be quite different from the failure to "read" the effects properly that may make the first few experiences with cocaine disappointing.

H. O. J. Collier has invented an interesting speculative neurophysiological explanation of psychoactive drug tolerance that suggests a reason why it is not induced by cocaine. His reasoning is based on an assumption that the cause of drug dependence is a drug-induced change in the number or efficiency of receptor sites on nerve cells. Some drugs increase the number or sensitivity of receptors for endogenous neurotransmitters, while displacing the transmitters from neuronal storage pools to supply these receptors. In this process they create a need which they also serve, but inadequately. That is, they cause the receptors to demand more of the neurotransmitter than the body can synthesize, and answer the demand by releasing the transmitter from a limited neuronal storage pool. Eventually more and more of the drug has to be used to displace less and less of the neurotransmitter to serve the heightened needs of more and more receptor sites. This is what we call tolerance: a need to keep increasing the dose to obtain the same effect. The withdrawal reaction is caused by the excess of oversensitive receptor sites demanding transmitters the body cannot supply.

But instead of raising sensitivity to the body's own neurotransmitters, a drug might transform the receptor sites so that they become more sensitive to it (the drug) than to the endogenous transmittters; or it might create extra receptor sites that are sensitive to it rather than to the body's own chemicals. In that case the drug would fully satisfy the need it created. It would not have to deplete an endogenous substance, becoming more ineffectual as that substance became less available. There would be no need to use more and more of the drug to obtain its effect; in other words, no tolerance. But a withdrawal reaction would occur, because the additional or transformed receptor sites sensitive to the drug alone would be demanding it in its absence. Collier considers morphine a prototype of the first kind of drug and cocaine a prototype of the second kind. Obviously, various combinations of these mechanisms and different effects on different neurotransmitters are possible. As we have noted, amphetamines seem to work primarily by releasing catecholamines. This would make them drugs of the first kind. Cocaine, as we have observed, is now sometimes said to affect mainly receptor sensitivity. But each of these drugs probably also has some actions of the other kind, and that might account for ambiguities and variability in the evidence.[10]

Dependence

Today, the word *dependence* has replaced the word *addiction* in most general descriptions of chronic drug abuse. In 1969 the WHO Expert Committee on Drug Dependence defined it as follows: "A state, psychic and sometimes also physical, resulting from the interaction between a living organism and a drug, characterized by behavioral and other responses that always include a compulsion to take the drug on a continuous or periodic basis in order to experience its psychic effects, and sometimes to avoid the discomfort of its absence. Tolerance may or may not be present. A person may be dependent on more than one drug." [11] Incidentally, to indicate the cultural bias associated with this definition, we should point out that the WHO list of substances under international narcotics control, presumably because they produce dependence most easily, includes cannabis and the psychedelics (LSD, psilocybin, mescaline, etc.) but not alcohol or nicotine.[12]

The WHO attempts to confer some diagnostic value on its very broad definition of "dependence" by qualifying it according to the kind of drug involved. "Drug dependence of cocaine type" is defined as follows:

> . . . a state arising from repeated administration of cocaine or an agent with cocaine-like properties, on a periodic or continuous basis. Its characteristics include:
>
> (1) An overpowering desire or need to continue taking the drug and to obtain it by any means;
>
> (2) absence of tolerance to the effects of the drug during continued administration; in the more frequent periodic use, the drug may be taken at short intervals, resulting in the build-up of an intense toxic reaction;
>
> (3) a psychic dependence on the effects of the drug related to a subjective and individual appreciation of these effects; and
>
> (4) absence of physical dependence and hence absence of an abstinence syndrome on abrupt withdrawal; withdrawal is attended by a psychic disturbance manifested by craving for the drug.

The definition of "dependence of amphetamine type" is similar but refers to a tendency to take increasing amounts, accompanied by some tolerance; also, it mentions no "psychic disturbance" accompanying withdrawal but rather states simply that there is no characteristic abstinence syndrome.[13]

The change from the older terminology to the rather vague word *dependence* had several virtues. One could speak of a continuum of degrees of dependence rather than two contrasting conditions (habituation and addiction) of basically different nature. The differences between drugs could be incorporated into the definition by means of qualifying phrases, so that the misconception of drug addiction as a unitary phenomenon or disease entity would be avoided. Another point about this definition is that it identifies withdrawal reactions and tolerance as secondary issues by qualifying all references to them with "sometimes" and "may or may not." The very vagueness and apparent neutrality of the word *dependence* seem attractive. Questions about psychoactive drugs involve all the complexity and indefiniteness of relations between mind and body and between individual psychology and social setting. A term that leaves the field open, without making premature ideological or pseudoscientific commitments, may be helpful.

Unfortunately, words used in this context tend to become charged with cultural tensions. The innocent word *addiction*, once it was captured by the medical, psychiatric, and police vocabulary, never really escaped. Even *habit* has had its troubles. One of our interview subjects, who had injected cocaine several times a week for some years, said it was not habit-forming like heroin. The man who says something like, "I've been smoking two packs of cigarettes every day for 20 years, but I can take it or leave it; it's not a habit," means, first, that he considers himself a free man and not a slave, and second, that tobacco is not an exotic drug used by incomprehensible and contemptible social outcasts. *Dependence* seems to be safe from this kind of dissolution of meaning, because people in general use it for too many purposes to allow it to be captured by the word *drug* and incorporated into a mythology. But this view may be too optimistic. "Drug dependence" has not yet become current in popular speech, but when it does, it may go the way of the old "drug addiction."

Meanwhile, there is much dissatisfaction with the term even in its present sense. Some authorities, who might be called humanists, do not like the ring of the word *dependence*. To them it suggests a mitigated form of enslavement and falsely implies that people who use drugs are not doing what they have chosen to do. They point out that in the vague sense in which the WHO uses the word, one can be dependent on a parent, a religious ritual, or trousers. Words like *compulsion, craving,* and *overpowering need* that are used to explicate *dependence* in the WHO definitions apply just as often to love of chocolate cake, or for that matter to love of another human being, as to desire for a drug; or else they are merely scare rhetoric to incite punitive campaigns. To like and want a drug, and to feel some disappointment when it is not available, is no different from liking and wanting and being unhappy in the absence of anything else. One of the strongest statements of this position is Thomas Szasz's remark that a euphoriant that does not produce dependence (he refers to addiction, but the principle is the same) is like a nonflammable substance that ignites.[14] And of course, anything that makes people feel good qualifies as a euphoriant.

Writers who take this attitude often consider themselves defenders of the right to pleasure against life-denying, restrictive bourgeois morality, or defenders of individual liberty against the encroachments of state authority and medical technocracy, or both. Where the emphasis lies makes a difference. For example, Norman Zinberg and John Robertson, in

Drugs and the Public, state disapprovingly that the idea of drug dependence has "puritanical" overtones.[15] They are presumably aligning themselves with the hedonists. Szasz, on the other hand, proudly identifies his attitude with the morality of Puritans (and Black Muslims). For him no drug has a power to enslave or produce dependence from which people must be protected against their will, for their own good, by physicians or the government. A person is neither more nor less responsible for misuse of psychoactive drugs than for any other bad habit or vice. The concept of drug dependence encourages the idea that some chemical compounds have a mysterious power over the will that only coercive authority can cope with; it is a denial of individual freedom and moral autonomy.

On the other side are researchers trying to develop a more scientific model of drug dependence. They often, although not necessarily, have a more censorious attitude toward the use of psychoactive drugs than the humanists and show fewer signs of social rebelliousness. They hope to demonstrate and measure drug dependence in experimental animals in terms of reinforcement schedules, operant conditioning, and other conceptual paraphernalia of behavioristic psychology—and maybe, some day, in purely neurophysiological terms. Unfortunately, in the view of these scientists, the WHO definition and all definitions like it are so imprecise that they permit debaters' points about dependence on harmless objects and dependence that has nothing to do with drugs. The WHO definition, they believe, is too shaggy and needs to be shaved down, made operational, for laboratory purposes. The disagreement is more than a debate within the medical profession; it is a philosophical conflict, almost a war between the notorious two cultures.

Craving and Compulsion

In the WHO definition of drug dependence, the only condition said to be *always* present is "a compulsion to take the drug on a periodic or continuous basis in order to experience its psychic effects." Stripped of jargon and the implication of enslavement, this says little more than that when someone takes the drug, he or she often enjoys it and wants to go on using it. The question is whether, when all references to abstinence symptoms are removed, any meaning is left in the word *compulsion* (or its near-synonyms, *craving* and *overpowering need*) that is not merely rhetorical and prejudicial.

There is no doubt that people often want to keep taking cocaine, and the desire can be an intense one. An extreme case is the patient who told Maier, "It is a horrible torture when one has no more cocaine. . . .I knew that I could kill someone in cold blood to get cocaine." Bruce Jay Friedman's character Harry Towns has a less sanguinary approach; he only wants cocaine so badly that at 4 A.M. he contemplates digging up the grave of a gangster whose coffin, so he has heard, contains a large amount of the drug that was stashed there and forgotten.[16] Here are some less hyperbolic comments from our interviews: "After shooting some I would immediately want to shoot some more, and so whenever I had some I would shoot up the whole amount, I never could save any. But I never had more than about four shots on hand at one time. So I guess in that way I was dependent on it, just because I was always wanting to repeat the rush." "When you're doing it, you're tied to it a lot, on the same level that you would be tied to cigarettes. You're just tied to it on a psychological level. A little bit on the physical, but very minor." "It's very drawing; it grabs you right away."

Another observation commonly made (and also confirmed by animal experiments) is that desire for cocaine, in spite of its intensity, is not necessarily persistent in the absence of the drug. Burroughs writes in *Naked Lunch*:

> Your head shatters in white explosions. Ten minutes later you want another shot. . .you will walk across town for another shot. But if you can't score for C you eat, sleep, and forget about it. . . .The craving for C lasts only a few hours, as long as the C channels are stimulated.

And in an article on drugs:

> The desire for cocaine can be intense. I have spent whole days walking from one drug store to another to fill a cocaine prescription. You may want cocaine intensely, but you don't have any metabolic need for it.

A writer in *Playboy* describes sniffing cocaine and feeling a mild paranoid reaction (to either the cocaine, or the hashish he smokes with it, or the situation): he suspects his companions of being police agents. Now "the coke might bury the enormous load of anxiety I'm suddenly carrying. . .the worst part of the entire experience, whatever the experience has been, was the fact that it took me ten long minutes to get beyond the feeling that the coke would set me free." [17] A man we interviewed said, "I do have a problem with just not doing it for a period of time while I have it. I don't mind being without it in between times. . .but I wouldn't like having it and not doing it. Therefore I would just plain not have it sometimes."

   This peculiar attractive power was the most disturbing thing we encountered in our research and interviews on cocaine. Andrew Weil
writes, "Whenever I have been around groups of people with access to
large amounts of good cocaine I have noticed how difficult it is to leave
that substance alone. One expression of this difficulty is that any amount
of cocaine set out for use gets used: it is unheard of for any to be left over.
And it seems very easy to get into a pattern of taking cocaine all the time
if a lot of it is around." [18] Several of our interview subjects said that desire
for cocaine could make people act out of character; for example, "Dealing
cocaine is so unpleasant because people will lie, people will be sneaky,
they're very suspicious about how you're cutting it, they treat people in a
way that they wouldn't treat them otherwise. It's different from grass, it
becomes very necessary to people, they get an attachment that's unique
and makes them act in surprising ways. And it's *not* just the paranoia of
its being illegal. . . . I found myself acting in ways that I'd never done
before around marihuana or any other drugs that I'd been involved with."
We would add that this person ultimately had less trouble giving up
cocaine than tobacco, but the expense and danger were strong incentives
to abstinence. Subjects sometimes said that, like Burroughs, they would
spend whole days making arrangements to buy cocaine; a few said they
felt able to lie, cheat, and steal to obtain it in a way they would not do for
anything else, even though they denied that it did them any harm. This
phenomenon may not be universal or even common, but it exists. Laboratory experiments on animals, as we shall see, confirm the anecdotal evidence of cocaine's attractive power.
   Understandably, the WHO speaks of "a psychic disturbance manifested by craving" in withdrawal of cocaine. (It inexplicably omits this in
reference to withdrawal from amphetamines.) But even the WHO must
admit that the "psychic dependence" on the drug is related to "a subjective and individual appreciation of [its] effects." In other words, not everyone even wants to keep on using it, much less desperately craves it.
William Hammond (whom we quoted at length earlier on the effect of the
drug) reported that he gave cocaine to many patients and found that they
had less trouble giving it up than giving up coffee, much less tobacco or
alcohol. A patient told Joël and Fränkel, "So long as I don't hear about
coke, I have no desire for it." [19] Even a habitual cocaine user who loves
the drug told us, "It's not something that I sit at home and go into morbid
states about or go into depressions because of. You still have a vehicle,
you still have a way to get around, but you know you could get around
much better with *that*. . . .it's not a thing that you *demand* of yourself."
There are also those who, like Freud, feel a mysterious aversion to co-

caine, while approving its effects—as though it were something like cod liver oil. Again we must face the fact that vague talk about cravings and compulsions does not differentiate one drug from another or drugs in general from other objects of human desire.

In the 1940s Alfred Lindesmith developed a theory of opiate addiction that made "craving" in a more precise sense the central concept. He defined craving as the condition that arises when the addict *interprets* his abstinence symptoms and properly attributes them to the absence of the drug. The hook in opiates, he believed, was not positive pleasure but only relief of the physiological symptoms of abstinence. Unfortunately, it appears that most addicts take much more than they need to prevent or overcome withdrawal distress. Besides, according to I. R. Chein and his colleagues, craving as observed clinically is not correlated with addiction in the physiological sense. People may desire opiates with an intensity that can be called craving before they are addicted, and in fact it is this desire that causes addiction rather than the other way around. If we consider stimulants (nicotine, caffeine, amphetamines, cocaine), it becomes even more obvious that craving for a drug can exist without physiological addiction. But so can craving for other things. Chein suggests a common-sense definition of craving: a desire that is abnormally intense or abnormally unamenable to modification or produces an abnormally intense reaction when it is not fulfilled.[20] The only trouble with this, again, is that nothing about it is specific to drugs. One of the cocaine users we interviewed put it well by saying: "The psychological addiction part is confused in my mind with just liking it a lot and looking forward to having it again. . . . So the issue is to make a distinction between a great love for the substance and a need for it. If you love somebody, you might want to spend a lot of time with them, but you don't *crave* them. . . .maybe you do, after a while."

Evidence from Laboratory Psychology

What we know so far is only that sometimes some people intensely desire to consume certain substances called psychoactive drugs. The WHO Expert Committee and others have unsuccessfully tried to account for this desire with a quasi-medical concept of "drug dependence." Obviously a different way of looking at the problem is necessary. The approach of the "scientists" is to examine in detail the qualities of the substances and reproduce their effects in experimental animals. The purpose is to identify and measure drug dependence in observed behavior or physiology, without reference to psychological variability or social context. "Humanists," on the other hand, take the euphoriant potential of the substances themselves more or less for granted and seek other reasons for the desire or its intensity—and for moral approval or condemnation of it—in biography, environment, and culture. The "scientist" is interested in whether laboratory monkeys ingest cocaine more enthusiastically than alcohol or marihuana. The "humanist" is more concerned about why cocaine and marihuana are on the international list of "narcotics" and alcohol is not. Both sides have something worth saying. The question is whether, when they have done their work, any value is left in drug dependence as an analytical category.

The program of the behaviorist or scientific psychologists is summarized in this definition of drug dependence by M. H. Seevers: "Repeated use of psychoactive drugs leading to a conditioned pattern of drug-seeking behavior. A characteristic, predictable and reproducible syndrome is associated with each drug (or drug type)." Notice that this cleaned-up and shaved-down terminology contains no references to vague and unmeasurable cravings, compulsions, and desires. All the concepts are meant to be "cashed" in observational and quantitative currency. Whether this language manages to preserve the phenomena it tries to account for is a highly controversial question. To some of those whom we have called humanists, "conditioned pattern" and "predictable syndrome" are fighting words, weapons in a cultural war. At any rate, experiments conducted under the dominion of these concepts have caused Seevers to conclude, "Stimulants represent . . . the *most* psychologically reinforcing drugs and have the greatest abuse potential of all psychoactive substances." And in 1974 the chief pharmacologist of the Drug Enforcement Administration called cocaine "the most powerful of reinforcing drugs." [21]

Animals may inject stimulants in spite of seizures, and human beings may inject them in spite of psychoses. They can also produce a conditioned response: in animals habituated to amphetamines, an injection of salt water can set off a behavior pattern like the one produced by the drug. And metamphetamine administered once to a formerly habituated animal after a lapse of 12 months has produced the bizarre postures and movements of the late stages of the earlier chronic intoxication. [22]

Laboratory animals attached to machinery for the intravenous self-injection of drugs (a lever press attached to a catheter in the jugular vein) administer cocaine in the same way they administer amphetamines. In an experiment performed in 1968 by Gerald Deneau, Tomoji Yanagita, and Seevers, monkeys in harnesses were allowed to inject various psychoactive drugs at different times. They voluntarily took morphine, codeine, cocaine, a morphine-cocaine mixture, dextroamphetamine, pentobarbital, ethyl alcohol and, more sporadically, caffeine; they showed no interest in mescaline, chlorpromazine (a phenothiazine), or salt water. Two monkeys began administering cocaine at a dose of 0.25 mg per kg, and three others started at a dose of 1.0 mg per kg. The course they followed was a rapid but erratic increase in consumption ending in convulsions and death within 30 days. The pattern was self-administration around the clock for 2 to 5 days, followed by exhaustion and abstinence for periods of 12 hours to 5 days. The largest intake in any 24-hour period was 180 mg per kg. All the recognized symptoms of cocaine abuse appeared, including apparent tactile hallucinations and weight loss. The toxic symptoms quickly disappeared when cocaine was discontinued (none of the animals discontinued it voluntarily). The patterns of administration for dextroamphetamine were similar but the toxic effects somewhat milder. Grand mal convulsions and death did not occur. The monkeys began to take dextroamphetamine at a lower dose (0.1 mg per kg), but the maximum daily dose reached was only 9 mg per kg instead of 180 mg per kg. In a later experiment by R. Pickens and T. Thompson on rats, cocaine again proved to be an effective reinforcer, although when the dose per injection was set too high or low responding ceased. In this experiment, as opposed to Deneau's, no weight loss, ataxia, or convulsions occurred. In still another experiment, Pickens found weight loss and the development of tolerance with dextroamphetamine and methamphetamine but not with cocaine. The mean hourly intake of the rats in his experiment was .54, .78, and 6.1 mg per kg of dextroamphetamine, methamphetamine, and cocaine respectively, and this remained constant for all values of dose-per-injection that were neither too high nor too low to make the animals respond.[23]

Thompson and Pickens have also compared patterns of stimulant self-administration with those of opiate self-administration in rats. The animals increased their total consumption of opiates gradually over a 45- to 90-day period and then held it constant. The interval between infusions was variable. The dosage pattern for stimulants, on the other hand, was much more regular and established almost immediately: a cycle of rapid administration of large amounts followed by self-imposed abstinence. At the same dose per injection, rats administered six times as much cocaine as methamphetamine. With a fixed-ratio reinforcement schedule of five lever presses by the animal for each infusion of drugs, cocaine at 1.0 mg per kg per injection produced about 30 responses (6 infusions) an hour, and methamphetamine at the same dose produced 5 responses (1 infusion) an hour. Obviously the short duration of the effect of cocaine made more frequent doses necessary. When the machinery was disconnected from the drug supply, animals kept trying to obtain opiates (after a complete cessation at the height of the withdrawal crisis) at a low rate for a long time—weeks and sometimes months. They tried to obtain stimulants at a very high rate—as much as 2,000 to 4,000 responses an hour—for a short time, stopping suddenly after two to six hours. One interpretation of this difference suggested by Thompson and Pickens is that the desire for stimulants is controlled only by the reinforcing effects of the drug, while desire for opiates is also influenced by the negative reinforcement of withdrawal symptoms.[24]

Yanagita has made an extensive study of voluntary self-administration of psychoactive drugs by laboratory animals. Following Seevers, he describes the basic factor in psychological drug dependence as a "drug-seeking behavior-reinforcing effect." Secondary psychological dependence is said to arise because of withdrawal symptoms and the consequent withdrawal aversion, with or without "physical dependence" (addiction in our sense). In the experiment monkeys under restraint were trained in "practice" sessions to administer a standard dose of a standard drug (.1 mg per kg body weight of SPA, or 1-1, 2-diphenyl-1-dimethylaminoethane-HCl, a stimulant) by pressing a lever 100 times for each dose. Then the standard drug was replaced by salt water until the number of daily injections dropped 50 percent, and the test drug was substituted. Twenty-four hours later the experimenters began to double the ratio of lever presses to injections—to 200, then 400, and so on—at an interval of a fixed number of injections. In the case of stimulants, when the time between injections became more than 24 hours the self-administration behavior was said to be extinguished. The final ratio of lever presses to injections was used to measure drug-seeking behavior-reinforcing effect as a factor in dependence liability. In other words, the dependence liability of a drug was determined in part by how long and how hard an increasingly frustrated monkey would work to obtain it.

To determine the range of doses at which each drug had reinforcing power and could be used experimentally in studying dependence, Yanagita performed preliminary experiments designed to discover minimal effective dose, lethal dose, duration of effect, behavioral toxicity, and systemic effects as indicated by behavioral manifestations, autonomic signs, motor functions, and body temperature. In a special comparative cross self-administration test on stimulants, he allowed monkeys under restraint to self-administer intravenously salt water, the standard drug, and the test drug in turn for four hours daily, three days each. In this preliminary test cocaine produced a higher rate of injections at .025 mg per kg per injection (120 percent of the rate for .1 mg per kg of SPA) than phenmetrazine at the same dose (110 percent) or methamphetamine at .006 mg per kg (80 percent). With the necessary arithmetical adjustments, this test was used to help determine the relative amount per infusion of each drug that an animal would choose to take.

The dose chosen for cocaine in the final test was .11 mg per kg per injection, for methamphetamine .02 mg, and for racemic amphetamine (Benzedrine) .03 mg. At these doses the lever press-to-injection ratio that the monkeys reached before giving up averaged 5 times as great for cocaine as for methamphetamine and 9 times as great for cocaine as for Benzedrine. In other tests cocaine at .12 mg per kg produced a final lever press-to-injection ratio 7.5 times the ratio at .03 mg per kg; at .48 mg per kg it gave 12 times the ratio at .03 mg per kg. The highest ratio attained with cocaine at .12 mg per kg was 6,400; at 0.48 mg per kg, two monkeys reached 12,800. In nonaddicted monkeys, the highest ratio for morphine was also 6,400. (Heroin was not tested.) In addicted monkeys the highest ratio was 12,800.

Yanagita rated cocaine "very remarkable" for total reinforcement power, along with pentobarbital and the opiates morphine, meperidine, and codeine—higher than amphetamines, alcohol, and nicotine. He also judged it "very remarkable" for behavioral toxicity (not further defined) at doses the animals took voluntarily, along with alcohol, pentobarbital, and amphetamines—much higher than opiates, which were judged only "positive," and nicotine, which was rated "negative." The reinforcement power of cocaine was not greater than that of pentobarbital and opiates only because it did not produce addiction with its negative secondary reinforcing effect. One monkey attained a maximum lever press ratio of 200 for morphine before addiction and a ratio of 6,400 afterward; but monkeys did not reach a higher ratio for cocaine after they were habituated to it than before. Incidentally, the animals would not voluntarily take hallucinogens or tetrahydrocannabinol (administered orally by hand, since it is insoluble in water) at all.[25]

The results of these interesting experiments, unfortunately, are too ambiguous to serve as a measure of "dependence liability" in a conceptually precise sense. What Yanagita and others have measured with some precision is the reinforcing effects of particular doses of particular drugs in a particular confined situation. Cocaine is said to have greater reinforcing power than amphetamines because, although it takes more cocaine than methamphetamine (and about the same amount of racemic amphet-amine) in a single dose to arouse desire for the drug, the animal will eventually ingest much more cocaine in a given time, or work much harder for an equal number of doses. Larger single doses of amphet-amines may be unpleasant, because pleasure-producing norepinephrine is depleted, and the animal has to go back for cocaine more often because of the short duration of its effect. But reinforcement strength in this sense is not Yanagita's sole measure of dependence liability, even for drugs without negatively reinforcing withdrawal symptoms. He believes that the psychotoxic or behaviorally toxic effect of the doses animals vol-untarily administer is also relevant; for that reason he does not regard caf-feine or nicotine as serious drug dependence problems. But psychotoxic effects are far more indeterminate than measurable reinforcement strength. They are also of doubtful relevance in any strict definition of dependence as distinct from abuse. Whether animals prefer stimulants to other drugs, and cocaine to other stimulants, is one question, and whether the doses people take are likely to harm or kill them is another.

So, although this laboratory work provides impressive testimony to the attractive power and potential toxicity of cocaine, it cannot answer either conceptual questions about drug dependence or the question of abuse po-tential in human beings in ordinary social situations. Commenting on such experiments, a journalist in *Rolling Stone* remarks, "With an attrac-tion like that, cocaine doesn't need to be addictive in the same sense that heroin is. You can get strung out on it on some ragged semblance of your own free will." [26] But is the freedom only a "semblance"? Certainly an ex-periment of this kind cannot answer that question, since its conceptual categories exclude the idea of free will from the start. Aside from meta-physical issues, there is the question whether the conditions of the exper-iment in any way resemble the situations encountered by animals in the wild and people in society. A characteristic antiexperimentalist position on this issue appears in the remark by Zinberg and Robertson that the more carefully controlled an experiment is, the less it is like the use of drugs in the real world.[27] In Yanagita's experiment the animals were caged and under restraint. They had few sources of satisfaction except pressing the lever for drugs. Would they act the same way if the situation offered a variety of dangers and opportunities? Is the conditioned pattern more than a laboratory artifact? Does the earthbound ghost need for cocaine express itself even nearly as predictably in a normal human environment?

Dissatisfaction with behavioristic psychology is expressed by those who consider it not scientific enough as well as by those who consider it too "scientific." M. Fink, writing in a large collaborative volume entitled *Chemical and Biological Aspects of Drug Dependence,* objects to definitions of drug dependence "based on social and behavioral criteria, not on any characteristic biologic, biochemical, or neurophysiologic aspects of their use." He is impatient for the day when the concept of psychological dependence, which he calls "a weak interpretation of the available data," will be replaced by a hypothesis based on physical tissue responses. This, he believes, "would do much to relieve the endless arguments and sophistry, as to drugs to be proscribed and those to be allowed medical use and study, so characteristic of governmental and intergovernmental committees today." [28] Unfortunately, there are few substantial leads in this direction. Even the observations on REM sleep and its associated EEG patterns can hardly be said to achieve a reduction to physical tissue responses.

The difficulty in providing a purely neurophysiological explanation of drug dependence is illustrated by H. O. J. Collier's ingenious ideas, which we have already discussed in connection with tolerance. His attempt to sketch a unified theory of dependence fixes, quite naturally, on the hypothalamic reward (pleasure) and punishment (pain) systems. Stimulants are said to sensitize or multiply receptor sites on excitatory neurons in the reward system, and depressants are said to sensitize or multiply receptor sites on inhibitory neurons in the punishment system. This distinction cuts across the one between drugs that deplete endogenous neurohormones, producing tolerance, and drugs that do not. Collier's theory suggests a plausible framework for studying at least some aspects of what is usually called drug dependence. But in concentrating so hard on the hypothalamic reward and punishment areas, Collier ignores the fact that our perceptions of pleasure and pain, reward and punishment, are often mediated through the cerebral cortex. Thinking of the old WHO definition of addiction, or one like it, he writes, "Dependence is a pharmacological concept and its laws are those of nature; addiction is a sociological concept and its laws are those of man." [29] But perhaps dependence—on drugs or anything else—is produced in the hypothalamus only when cortical impulses react upon it. The cerebral cortex, with its functions of memory, fantasy, association of ideas, and concept formation, is in a way society's representative within the brain as well as the brain's representative in society. Maybe, then, "dependence" can no more be saved from sociology than the old term "addiction." For if it is primarily what we *think* that makes us want to use a certain drug in a certain way, then the culture we are educated into and the company we keep are overwhelmingly important. In the realm of psychoactive drugs, we are simply not in a position to distinguish between the laws of nature and the laws of man.

The question is how and how much anything identifiable as drug dependence can be discovered by examining universal mammalian or species-specific psychology as opposed to personal histories and social conditions. It is interesting that even neurophysiological speculations and experiments by advocates of methodological behaviorism tend to suggest the importance of these factors. Separating biological from social aspects of dependence is difficult. Suggestions of "social" conditions appear even in experiments designed to exclude them. Caged monkeys pressing levers for drugs are in an artificial environment free of the dangers and exigencies of ordinary animal life and also free of competing interests or temptations. They are dependent on drugs in the sense of having few other resources. Can it not be said that they are "culturally conditioned" to crave the drugs or that the situation they are in makes them "psychologically disturbed"? We may recall the report that the reaction of tame city cats to heavy doses of amphetamines is less aggressive than the reaction of wilder farm cats. So even animals unable to develop the complex individuality of human beings show drug effects dependent on upbringing and "personality."

Now consider from another point of view the plausible neurophysiological theory that the hypothalamic pleasure and pain centers in some way determine drug dependence. The trouble here, of course, is that they determine so much else as well. What is unique about some drugs? Henry Brill suggests an answer: they are "substances which enable [man] to *short-circuit* the environmental adjustment since they act *directly* on the pleasure-pain system. . . . Man shares with lower forms the ability to react with pleasure to such drugs, but only man has had the ingenuity to identify these pleasure-producing substances by his own initiative, and what is more important, to pass on this information as a part of his cultural heritage" (our italics).[30] But in what sense do substances usually thought of as pleasure drugs, and not others, "short-circuit" some processes that are clearly normal? Isn't taking aspirin a short-cut to disposing of a psychogenic headache? For that matter, don't people evade the environmental adjustments of food gathering by satisfying their hypothalamic hunger mechanisms with a restaurant meal? Could Freud have felt aversion for cocaine along with his intellectual approval of its effects if it had acted in some peculiarly unmediated, almost irresistible way? Or should we say that Freud was abnormal? Why are some kinds of short circuits (if that is what they are) acceptable and others not?

We can take a hint for an answer to these only partly rhetorical questions from the attitude of Aleister Crowley, who was certainly no scientist. For Crowley, an attraction of certain psychoactive drugs was that they permitted what he regarded as an unnatural evolutionary leap, a biological transformation that made the drug user something more than or different from an ordinary member of the human species. The drug fiend, he thought (and he used the phrase with relish), was a unique kind of explorer and adventurer, a man apart. This view of psychoactive drug use as a deliciously or impressively unnatural act suited Crowley's version of diabolism, with its "magickal rituals," very well. Brill's term "short-circuit" actually implies an interpretation similar to Crowley's, although with a negative or at best neutral valuation. A person who habitually cuts neurophysiological corners with the help of these substances is abnormal—at worst a dangerous "drug fiend" (of course, Brill is too sober a scientist to use this term) and at best drug-dependent—immature or unfree in some way. By comparing Crowley's attitude with Brill's, we can see that Brill is insinuating a culturally conditioned attitude toward drug *use* into his attempted neurophysiological explanation of drug *dependence*. This does not mean that he is especially censorious about drugs. He concedes that psychoactive drug dependence may be "evidence of an overwhelming need" of humanity in the face of "problems which grow out of social pressures," and therefore justifiable.[31] But the fact remains that just as Crowley's evolutionary leap is not biological but psychological and cultural—he is transported to exotic and above all *forbidden* realms— so Brill's short circuit is really a social rather than a physiological phenomenon.

The inadequacy of all notions about psychoactive drugs based on what experimental animals do, and the implicit social content of all dependence concepts, is amusingly illustrated by the fact that laboratory animals show no interest in tetrahydrocannabinol, while the WHO classifies it as a dangerous "narcotic" with dependence-producing powers that have to be coercively controlled. Not that animal experiments and neurophysiology are irrelevant: the neurophysiological effect of a drug's chemical action is obviously necessary to produce interest in it and the habitual inclination to use it. But this effect is far from sufficient to sustain a coherent conception of dependence. Nils Bejerot asserts (using *addiction* in a broad sense that resembles the common use of *dependence*): "The nervous disturbances and social grievances which may have contributed to the inception of addiction have *in principle* nothing at all to do with the actual process of becoming addicted. *Any man and any animal will develop an addiction if certain substances are introduced into the body in sufficiently large doses for a sufficient length of time*" (his italics).[32] He is certainly making a necessary distinction. But a person who is force-fed long enough will become grotesquely overweight, and yet Bejerot would hardly contend that the problem of obesity as it is encountered in the real world has in principle nothing to do with nervous disturbances and social grievances. The question is why people *want* to eat so much and not what an excess of food does to them. In fact, the effects of eating too much are so much less variable and equivocal than the effects of using psychoactive drugs—the "syndrome" is so much more "predict-able"—that a neurophysiological concept of excess-food dependence (based on, for example, a maladjustment of the appetite-regulating mechanism) may have *more* substance than a similar concept of drug dependence.

Individual Psychology of Dependence

Social and idiosyncratic psychological conditions, then, are as impor-tant in the development of drug dependence, if that is what it is properly called, as in the development of any other kind of dependence. But, in this field, most of the terms used imply moral and even political judg-ments, and we must not burden these necessarily loaded words further with unreasonable connotations. How this may happen is illustrated by another implication of the idea that psychoactive drugs are a short circuit to pleasure. The implied conclusion that the drug user is cutting corners, like an immature and impatient automobile driver, can be elaborated psy-chologically: the driver is a reckless (read "drug-dependence-prone") per-son. Or it can be explicated sociologically: our culture promotes haste (read "drug abuse") or our driver education programs (read "drug educa-tion programs") are insufficient. But in either case the analogy assumes that taking certain psychoactive drugs resembles *reckless* driving rather than simply using an automobile, which is after all a short way to get where you are going compared with riding a donkey or walking. Un-doubtedly some psychoactive drug use resembles reckless driving, but the terminology of "shortcuts" is only superficially neutral; it suggests a psychological or social problem where there may not always be one.

If we pay attention primarily to the individual rather than to social con-ditions, we come upon the questionable concept of the addictive or dependence-prone personality. This label has been used in a confused fashion both to complement and to contradict the notion of a dependence-producing drug. It is said that only an inadequate personality is inclined to take a bad drug (especially opiates), or that deficient persons make bad use of a good drug. In the first of these judgments, evasion of issues and premature fixing of blame on the chemical substance are obvious. The second judgment may also be misapplied: compare the half-truth propagated by the National Rifle Association that "Guns don't kill people; people kill people." Still, it is not absurd to speak of personalities, as well as drugs, associated with drug dependence. Some people seem to need the influence of a drug to gain respite from their troubles or surcease from pain. But we should not assume that *only* and *all* persons who are in some unique way "dependence prone" will use or misuse cocaine or any other psychoactive drug.

In other words, dependence-proneness is not a useful diagnostic category. It has always proved impossible to determine in advance who might take to the habitual use of which drug. The notorious difficulty of establishing causal relationships by means of retrospective research, which we have discussed in connection with the effects of chewing coca, contributes to the problem. It is too hard to arrange control groups or to compensate for the effects of differential availability. Besides, as we emphasized in discussing drug-induced psychoses and again in discussing the question of motivation, imputations of anterior psychological defect or deficiency in connection with drug use too easily turn into attempts to justify a socially accepted prejudice by scapegoating. As Thomas Szasz complains, the use of illicit drugs may be considered a symptom of some illness, and this illness may then in effect be circularly defined as drug dependence, i.e., the use of illicit drugs. Szasz also points out the absurdity of Ernest Jones' regarding the fact that Freud gave up cocaine as evidence that he was not an addictive personality, while treating his far more serious dependence on cigars far more casually. And he sarcastically suggests that today a William Halsted would not be admitted to the faculty of Johns Hopkins but only to the psychiatric division of its hospital, with the diagnosis of "personality disorder: morphine addiction." [33] Burroughs succinctly comments (he is writing in 1956):

> In Persia where opium is sold without control in opium shops, 70% of the adult population is addicted. So should we psychoanalyze several million Persians to find out what deep conflicts and anxieties have driven them to the use of opium? I think not. According to my experience most addicts are not neurotic and do not need psychotherapy.[34]

But attributing character disorder or defective personality type to people who use certain drugs can have a self-justifying effect if social policy is based on the attribution. As Bejerot observes, "The correlation between character disorders, social disturbances, and addiction seems to be very marked when an epidemic of addiction is under constant strong pressure from society." [35] He is commenting on Chein's contention that in the heroin addicts he studied, personality weakness rather than the drug was the reason for addiction. Since Burroughs' sensible observations about the Persians, the Iranian government has instituted a program of modernization and westernization that includes violently discouraging opium smoking and substituting the use of tobacco and alcohol. People have been executed by firing squads for selling opium. In these circumstances it has become reasonable to say that anyone who still wants to smoke opium in Iran, like anyone who still wants to use heroin in the United States, ought to have his head examined. The question is whether he should be called weak rather than courageous to the point of foolhardiness. In spite of the irony, our argument is a serious one. The private transaction between an aberrant personality and an overpowering chemical compound is a myth. If the drug user is not permitted to lead a normal life, he will obviously appear to be inadequate. The social conditions in which the drug is available determine the kind of person who uses it and the attitude taken toward his or her personality.

One fact relevant to the psychological analysis of drug users is that some of them (especially heroin users or addicts) become subjects of study only after they have been arrested or have agreed to confinement. This not only biases the sample (restricting it to those who are caught) but creates a situation in which judgments about their personalities reflect their newly created institutional position more than anything that could account for their original drug use. In an interesting book on American opiate policies, Troy Duster recounts some interviews with people who had known heroin addicts. When he asked them how they found out that their acquaintance was an addict, their answer was always something incidental like reading of his arrest in the newspaper or seeing needle marks on his arm. From his behavior, general appearance, and personality they would never have guessed. Yet *after* they knew that he was an addict, they started reading in his actions, appearance, and attitudes accepted signs of heroin addiction: "puffed up and nervous"; "glassy and starry eyes." Duster also points out that in 1890 the average opiate addict in the United States was a member of the upper or upper-middle class, and in 1920 the average addict was poor and often unemployed or a criminal. He asks rhetorically, "Is this to be explained by some dramatic transformation in the personalities of the privileged classes. . .and the lower classes?" Of course, the answer is that the social meaning of addiction and the circumstances in which it was detected changed, not the psyches of whole economic classes.[36]

An observation by Szasz provides further evidence of how misleading psychiatric diagnoses of drug users can be. He notes that legal, socially accepted psychoactive drugs are used symbolically in rites of passage to maturity, while illicit drug use is regarded as a sign of immaturity even when it is not seen as evil. Adults may laugh encouragingly as the adolescent chokes on his first tobacco cigarette or retches while learning to "hold his liquor like a man." Marihuana, on the other hand, is (or was until recently) identified with youthful rebellion, that is, with immaturity. One of the worries of parents who discovered their children smoking the weed was that they would never "grow up" in the socially approved manner. Growing up in any society involves a series of rituals, and the status of being a grown-up is based less on a mysterious psychological condition known as maturity than on having learned the proper rites, passwords, and symbols—among them use of the proper drugs. One who takes the wrong drugs, following a slightly different ritual and constructing a slightly different social reality, is then considered to be "fleeing" the only reality that exists. He is emotionally immature, weak, unable to face problems.[37] Of course, some "right" drugs may have such grossly obvious psychotoxic effects when abused that the abuser—say, an alcoholic—is said to be immature or fleeing reality. Other drugs like coffee and tobacco, no matter how overused or abused, practically never carry any implication of psychological deficiency or aberrancy. The power of Szasz's argument to undermine many claims about the personality defects of drug users should not be underestimated. It is especially interesting in relation to theories of drug dependence, since dependence may suggest childishness and therefore contain the hidden social symbolism Szasz has uncovered.

Duster says of the opiate problem that "its essence does not lie within the structure of the individual's psyche, but in the prevailing social interpretation and social meaning of narcotics use."[38] We agree that the essence of psychoactive drug problems is not in the structure of the user's psyche, and we would add that it is not in the chemical structure of the drug. Yet the almost illimitably vague word "compulsion" at the heart of the WHO definition of drug dependence is obviously meant to suggest precisely two such external forces overpowering the human will: a pharmacological mechanism and a drive coming out of what Freud called our inner alien territory. If we refuse to accept either of these forces as explanatory and insist on the importance of social context, what remains of the idea of drug dependence? If the pharmacological mechanism is not decisive, the drug is not all-important. If psychological weakness is not decisive, dependence is no longer an essential term. Carl N. Edwards describes the case of a patient "addicted" to a placebo, with dose increase, craving, and abstinence symptoms.[39] This classic example of iatrogenic drug dependence without a drug illustrates in parody the social nature of many of the issues involved, for here it was solely the institutionalized doctor-patient relationship that produced and sustained the dependence. Of course, in most cases the pharmacological qualities of the drugs and the psychological idiosyncrasies of the users are important, but they cannot sustain a unitary coherent conception of drug dependence, and it is not clear what can. In a way, this brings our position close to that of the humanists who are suspicious of the unitary concept on political or metaphysical grounds.

Social and Cultural Definitions of Dependence

We have written of the social definition of the nature of a drug and given some examples in connection with opiates, marihuana, alcohol, and other drugs, including cocaine. In fact, social rules and customs may define the consequences of using a drug in the most literal sense of the word *definition*. The old word *addiction* too often meant little more than "the consequences of opiate use," most of them produced by punitive legislation making it necessary to embark on a criminal career to obtain the drugs. If someone needed a barbiturate or whiskey to fall asleep every night, his habit was not described as addiction mainly because it was not as expensive as injecting weak solutions of heroin or as likely to lead to prison, and therefore did not disrupt the drug user's life or society's routines so much. If we impose a moral judgment on a social situation (especially one created by the law) in the guise of a medical or psychiatric diagnosis, the terminology we use, whether it is the language of addiction in the old sense or drug dependence in the new sense, will tend to lose its analytical value and even all meaning that is not merely rhetorical.

It is not surprising that some people regard the concept of drug dependence as little more than a device for finding a way to disapprove of certain drugs when it is hard to show what harm they do. It has certainly been used that way in attacks on marihuana. On this intellectual battlefield Szasz has laid down a particularly heavy barrage of his usual deliberately provocative analogies. We have mentioned his comparing a nonaddictive euphoriant to a nonflammable substance that ignites. (Obviously, one can read "dependence" wherever he refers to addiction.) Another of his polemical analogies is this: "Addictive drugs stand in the same sort of relation to ordinary or non-addictive drugs as holy water stands in relation to ordinary or non-holy water. . .trying to understand drug addiction by studying drugs makes about as much sense as trying to understand holy water by studying water." An addictive drug, in other words, is not a unique kind of chemical substance but something that has been named by an approved ritual. The content of the word is social. Szasz also refers to the foreigner's "addiction" to speaking English with an accent—a habit that cannot be broken by a mere act of will, and in fact may never be broken, but not something that has a mysterious ungovernable power over its "victim." He even calls abstaining from a drug (or dieting instead of eating heavily) a substitute habit or addiction.[40]

The gravamen of Szasz's charges is that by using terms like *drug addiction* and *drug dependence* we stage the drama of temptation and restraint wrongly. We center it on a chemical substance, treating men as organisms acted upon (by biological and psychological forces) rather than as agents responsible for their acts.[41] It is not necessary to accept Szasz's radical moral individualism and the associated classical liberal conception of the functions of the state and medical profession to agree with some of what he says. The consumption of any substance may be either a virtue or a vice—either a good or a bad habit—depending on what the consequences are, and people are normally asked to take responsibility for the consequences of their habits. But the term *drug dependence* places the chemical substance itself and the mere existence of a habit at the center of interest and treats them both as external forces operating in opposition to the individual will (or, in the case of strict behaviorism, excludes the concept of free will entirely). Naturally, this makes us more inclined to use external countervailing force, and to apply it to the drug itself and the users of the drug rather than to other sources of misery in the environment. We avoid making difficult and complicated decisions about whether this person's use of this particular drug in this particular way is a good or a bad habit; we do not permit the users to decide, either, for they are dependent, immature, unable to make choices for themselves. Szasz's puristic liberal solution for psychoactive drug problems is to permit any adult to take any amount of any drug he wants until his subsequent actions begin to harm others (he has no objection to attempts at dissuasion that respect the drug user's freedom) and then to hold him morally responsible only for those acts, not for the drug use itself. Without accepting this answer, we can admit that Szasz has helped to clarify an issue by pointing out how concepts like drug dependence may serve to prejudge social situations and restrict the choice of policies to deal with them.

Suppose we reject drug dependence as an explanatory hypothesis for the use of psychoactive substances, while admitting that people can be dependent (in some sense that does not negate their freedom) on drugs as well as other things. If we really want to create a typology of dependencies, categories like "dependence of cocaine type" and "dependence of morphine type" may be less illuminating than ones like "dependence of middle-class type," "dependence of working-class type," "dependence of American type," and "dependence of twentieth-century type." We can say that, for example, taking opiates was a middle-class form of dependence in 1890 and a lower-class form in 1930; the differences between the two practices were more significant than the similarities.

It happens that in our culture, energetic, active striving is considered more desirable than resignation or passive enjoyment. People who suffer from this dependence of American type on speed, activity, and novelty will often prize stimulant drugs, and society may often be tolerant of their desire to use them. That happened for a while with cocaine in the nineteenth century, and it has recently happened again with amphetamines. We may find stimulants especially congenial because their alerting and invigorating effect enables us to get on with what is regarded as the business of society. The reasons for enjoying a stimulant and the reasons for finding it useful easily become indistinguishable, and so the transition from the need to be alert and active to the need to use the drug is easy for the conscience. One is feeling good and doing something right at the same time.

Of course, cocaine is not yet used so often or so casually for business, except by entertainers, as caffeine or (until recently, at any rate) amphetamines. It is simply too scarce and expensive. But obviously if it were more easily available it would be used more. In this connection there is one conception of the dependence liability of a drug that implies no moral judgments or pharmacological theories and suggests no unique enslaving power. It is simply the answer to this question: In a given society at a given time, what is the chance that if this drug is freely available, many people will use it and some people will use a great deal of it? There is no precise scientific answer to this question any more than to other questions about the nature of men in society. The advantage of this formulation is that it puts the fact of drug dependence, or, better, the habit of using a drug, in its proper subordinate place. For our primary interest, after all, is not in how many people are ingesting how much of some chemical substance but in what the individual and social consequences are.

In this sense, cocaine here and now appears to be high in dependence liability, i.e., the likelihood that people who try it will form the habit of using it if they can obtain it. Several remarks from our subjects provide evidence of this:

> My one very bad habit is coffee—I'm still very addicted to some kind of stimulant. . . .I can tell you that a whole lot of the reason I'm not doing it [cocaine] now is financial. I can't afford to blow the rest of my inheritance. And the easiest way to blow it is to buy blow.

From another interview:

INTERVIEWER: Do most of your friends take cocaine too?
SUBJECT: Pretty much the same level I do. I know some people who have more money, who use it more regularly.
INTERVIEWER: So your reason for not using it more often is mainly financial?
SUBJECT: That's the only one, that I've thought of. . . .if that one were removed, then I would think about it again.

And from two other interviews:

> More people who were addicted to alcohol would also be addicted to cocaine; people who were addicted to coffee would be addicted to cocaine. If it were free and easy, I would be addicted to cocaine, probably.
> I've experimented a lot and found that when you can get it, coke is best.

Even many people who did not think that "coke is best" and had sometimes had unhappy experiences with it said that their main reason for not using more was high cost.

Our main point in this discussion is that dependence, stripped to its essentials, is a matter of who is likely to use the drug and how; any more elaborate definition tends to dissolve under analysis. Too many statements about cocaine dependence appear to have a pharmacological or psychiatric basis but in fact are of dubious relevance to either the pharmacological nature of the drug or the needs of the drug user and society. We also want to insist on the importance of cultural context in determining any drug habit. The answer to the question of dependence as we have finally formulated it may give some idea of how much cocaine is likely to be used, and how often a person will want to use it heavily if he uses it at all. It tells us nothing about the overwhelmingly important variable of availability. And it also implies nothing about the relation between moderate use and *excessive* use, misuse, or abuse. It is important to avoid using the word *dependence* in a way that prejudges questions of abuse and excess. These questions are the essential ones where social policy is concerned, and they are so complex and difficult in themselves that attempts to answer them cannot bear the further weight of confusion with secondary issues.

PART III

9

ABUSE POTENTIAL OF COCA AND COCAINE: THE POLICY DEBATE

THE HISTORY of attempts to define and explain drug addiction and dependence suggests that these terms have long been recognized as labyrinthine semantic traps. The notions of abuse and excess in the use of drugs have always seemed simpler and clearer; we all have some intuitive sense of what they are. Unfortunately, when we examine these intuitions more closely, they turn out to be just as individually variable, unreliable, confused, and culture-bound as most ideas about dependence and addiction. Practically everyone would agree that an alcoholic drinking himself to death is abusing a drug; yet a cup of coffee in the morning, an aspirin tablet for a headache, or, for an increasingly large number of people, an occasional marihuana cigarette are considered moderate and beneficial uses of drugs. But those who do not consider alcohol a drug would disagree even about the alcoholic, while Mormons, Christian Scientists, and food purists would disagree about coffee and aspirin. As for the range of quantities and substances lying between these extremes, there are almost as many opinions as there are individuals.

The law has introduced further confusion into the situation by in effect defining *any* use of certain drugs, like cannabis and heroin, and almost any use of other drugs, like cocaine, as excessive or abuse, while placing only minor restrictions on severely abused drugs like alcohol and tobacco. These apparently arbitrary classifications have their own obscure cultural roots, but certainly they are not based on any principle that has even a superficial appearance of abstract rationality, and practically no one any longer pretends that they are. In weighing the potential of various drugs for abuse, it is necessary to ignore these punitive statutes and their social consequences, like deaths from hepatitis and quinine, the opiate addict's need to steal to pay for his drug, or the attractiveness of the high-profit, high-risk trade in illegal commodities for men willing to take chances with their own lives and threaten the lives of others.

In searching for definitions of drug abuse useful in determining social policies on coca and cocaine, we might first turn to the practice of physicians. The World Health Organization's (WHO) 1969 official definition of drug abuse does so: "Persistent or sporadic excessive use inconsistent with or unrelated to acceptable medical practice." [1] But defining abuse by reference to excessive use tells us nothing at all, and certainly *any* excessive use of anything must be inconsistent with acceptable medical practice. The purpose of the WHO is to grant the medical profession an important role in determining what drug abuse is, but this tautologous and irrelevant definition gives no more idea of what that role should be than of the abuse it professes to describe. It would make sense to use accept*ed* medical practice as a guide if physicians were not so inconsistent in their treatment of psychoactive drugs. They bar the use of some (cannabis, heroin, mescaline, LSD) entirely, prohibit the use of others (morphine, cocaine) by laymen, dispense still others freely on prescription (many tranquilizers and, at least until recently, amphetamines and barbiturates), and exercise almost no control over the use of a fourth group (alcohol, caffeine, nicotine). Like the legal classifications with which they are linked in a circle of mutual justification, these categories have been established by historical accident and cultural tradition without any serious attempt at systematic rational justification. They are not very useful as an objective guide (if such there be) to problems of drug abuse.

The Disease-Crime Model

One of the apparent advantages of the concept of abuse or excess over the concept of addiction or dependence as a framework for discussing drug problems is that when we speak of abuse or excess it is more difficult to disguise a moral or political judgment as a medical or psychiatric diagnosis—more difficult, but not impossible. The trouble is that drug abuse has two senses. In the first sense it implies drug use that makes the user himself ill or miserable, so that he recognizes it as harmful or even as a cause of disease. In the second sense it implies drug use condemned by society, but not by the drug consumer. The law in its majesty or the culture in its traditional wisdom has determined, without the drug user's approbation, that he should not keep taking his drug because it harms other people or himself. In the first sense, drug abuse is a recognized bad habit—one that the user himself wants to eliminate because it causes discomfort or disease. In the second sense, it is a source of bad actions or itself a bad action and in effect associated with moral evil or crime. But the disease concept casts its shadow here too, and the safeguards of individual freedom ordinarily implied in the criminal law and in our moral judgments—for example, the demand for a clear specific showing of harm to a specific victim—are ignored. The accepted definitions of *harm* become loose, arbitrary, and tyrannical. After all, what has treating a disease to do with questions of individual liberty? In the name of curing illness we may allow vengeful envy, hypocrisy, and the passion for scapegoating free rein. So the concept of drug abuse comes to imply an objectionable mixture of unexamined and often stupid moral judgments with reasonable medical concern about the health of the drug user as he himself interprets it.

The unhappy practice of treating drug use of certain kinds as both a disease and a crime in varying degrees, together with the almost complete lack of rational principle in our choice of what kinds of drug use to treat that way, has made the very idea of drug abuse suspect to some morally sensitive people. This disease-crime model leaves the greatest possible room for arbitrary and irrational action and the least possible room for conceptual clarity. Not only are disease notions surreptitiously introduced into criminal proceedings or moral judgments, but willed, chosen, and unregretted acts become part of the definition of a disease. As Szasz points out, the idea of dis-ease properly implies discomfort on the part of the sufferer, who comes to a physician because he *wants* to be cured. The disease of drug abuse (or drug dependence) often involves nothing of the kind; instead, punishment for certain behavior in the guise of forced treatment is imposed on someone who does not admit that he or she is unhealthy. Nils Bejerot is almost brutally explicit about this: "the addict generally does not suffer from his disease, he enjoys it. . .the patients. . .must be. . .kept free from drugs for a long period, with or against their will." [2] They have abdicated the right to decide whether they want treatment for the "illness" simply by virtue of having contracted it.

An even more dangerous consequence of justifying coercive action against disapproved drug use as medical treatment is that it need not even be ostensibly for the user's own good. If we try to find ways of extending the disease analogy (it is no more than that), we are likely to conclude that drug abuse is an infectious, epidemic disease and drug users properly thought of as carriers, like rats in bubonic plague. Both Brill and Bejerot propose a public health contagious-disease model for psychoactive drug abuse. This naturally suggests preventive measures like quarantine that are imposed not for the sick person's own good but for that of other potential victims. Brill and Bejerot miss the morally central point that drug use does not spread independently of anyone's will, like a true infectious disease, but rather by persuasion and example. The kind of public policy they propose to base on their extended analogy, like most proposals for preventive detention in criminal law, is a clear and present danger to free speech and action.

As Szasz points out, it is consistent to treat the drug user as a helpless, sick victim or as a free man in control of himself, but not both. Yet authorities have usually acted in just such an inconsistent way whenever they have concerned themselves with psychoactive drugs.[3] This has been especially obvious in the case of American public policy on opiates. The Harrison Act, by outlawing all use of opiates except for "legitimate medical purposes," in effect introduced notions of good and evil into medical problems about the physiology of addiction. A series of Supreme Court decisions laid down the rule that maintenance—that is, preventing or putting an end to abstinence symptoms—was not a legitimate medical purpose. Addiction had in effect become a moral evil rather than a disease. A physician writing in the *Journal of the American Medical Association* in 1921 referred to "the shallow pretense that drug addiction is a disease."[4] He was at least being consistent: the law punished the addict as a willful criminal and society should regard him as one. Today the received professional wisdom is that "addiction" (in some vague sense) is a disease, yet addicts are still subject to arrest and criminal punishment; by the most abysmal of ironies, they are rarely permitted treatment for the only *actual* clearly recognizable disease (in a strict sense) directly caused by opiate use: the abstinence syndrome, which is cured by administering opiates. In effect, we deny medical treatment to a disease because of moral disapproval of the way it was contracted, then treat the actions that led to the disease as themselves signs of illness, usually some mental or psychological weakness, *without* retracting our moral disapproval or even removing the criminal penalties.

By surreptitiously and even unconsciously switching from the disease concept to the crime concept and back, authorities can evade both the institutional restraints on coercion built into procedures for dealing with criminals and the quite different ones built into procedures for dealing with sick people. This happens in various ways. As Isidor Chein observes of opiate addiction programs:

> That the basic concern of compulsory therapy plans is suppressive rather than therapeutic and that ideologies of suppression and of therapy do not easily mix is most evident in the qualifications that are generally introduced as to which addicts are to be eligible for compulsory therapy as an alternative to jail sentences. . .Thus, there is generally a limit set on the number of times a patient may avail himself of the therapeutic alternative, and a self-committed addict who seeks discharge before his time is up renders himself ineligible forever. *It is as if one were to declare that an easily cured patient is sick, but a hard-to-cure patient is a scoundrel* (our italics).[5]

Troy Duster points out that it is not anything they do that convicts opiate addicts under supervision of backsliding, but simply the presence of heroin in the blood or urine. He contrasts this situation with that of the drunken driver, who is tested chemically for alcohol only after his behavior has shown dangerous signs of intoxication.[6] Heroin in the blood is by itself evidence of the disapproved behavior of drug abuse, but there is no need to show that this behavior does any harm or has any serious consequences, because it is somehow at the same time a disease and ipso facto harmful—as though it were like hypertension. The use of alcohol, on the other hand, is by itself neither disapproved behavior nor disease, so something more than a chemical in the blood is needed to convict the drunk of drug abuse.

The most important restraint on our treatment of sick people is the requirement that the patient feel ill and want to be cured; the most important restraint on our treatment of criminals is the requirement that they have committed some specific clearly harmful act; by regarding the status of addiction to certain drugs or the habit of using them as somehow both an illness and a crime, we dispense with both restraints. There is no need to show evidence of a harmful act, because the drug use is after all in itself a disease; but there is no need for the consent to treatment ordinarily required with disease, because it is a crime, or at best a morally wrong, immature, irresponsible habit that must be forcibly suppressed. Finally, by treating certain habits as diseases and crimes without demanding the usual evidence of disease or crime, we produce pernicious consequences which we then falsely attribute to the habits themselves and treat as further evidence that they should be regarded as diseases and crimes.

This criticism of the disease-crime model of psychoactive drug abuse does not necessarily imply a liberal or permissive attitude toward the drugs. We need not wait until the alcoholic actually hurts someone in an accident while driving intoxicated or the amphetamine abuser assaults a companion while in a paranoid state before imposing coercive controls. We might treat some drugs as dangerous instruments, like firearms, and prohibit or restrict their sale and possession. But even where restrictions on the trade in firearms exist, the use or misuse of a gun is treated as a free act, not a disease. Besides, even the strictest prohibition of guns is aimed only at the possibility of misuse. It implies not that *any* use of a gun is abuse, but only that the potential dangers of misuse by a few people outweigh the probable benefits of proper use by most people. We should recognize, first, that drug abuse is not a disease in any sense (although it may cause disease), and, second, that there are no bad drugs but only bad uses of drugs. Then we might be able to reexamine our drug policies and put them on some more or less rational basis, whether permissive or restrictive. Of course, if we did this, we might discover that there was a better case for prohibiting tobacco and alcohol than for prohibiting marihuana, cocaine, or opiates. The clouds of confusion generated by the disease-crime model of drug abuse and the disease and crime it produces serve to prevent any such disconcerting and anxiety-producing dissolution and reconstruction of social attitudes.

A Framework for Analyzing Abuse Potential

In spite of the inconsistency and moral inadequacy of the disease-crime model, it at least gives some ritual recognition to the distinction between harm to oneself (disease) and harm to others (crime). Any social policy aimed at both preserving individual freedom and protecting the innocent should make this distinction central. (Our present policies certainly do not.) Yet it is difficult to determine not only how far the authorities should be permitted to go in saving people from themselves but where the line should be drawn between harm to oneself and harm to others. Should we let anyone who chooses to do so poison himself and ruin his life with drugs, as the liberalism of Thomas Szasz proposes? If some people are indirectly destroying the lives of members of their families or have become an economic and medical burden on the community because of drug abuse, at what point do we decide that they are harming others as well as themselves and institute coercive measures? These questions point toward the heights or depths of legal and political theory and cannot be answered here. But any drug policy that values individual freedom at all must at least recognize that it is more justifiable to impose coercive controls on a drug if its abuser harms others—even in some admittedly vague intuitive sense—than if he harms mainly himself.

Harm to self includes: psychological effects of acute intoxication, like psychosis, delirium, and acute withdrawal reaction; physical effects of acute intoxication, like overdose poisoning, exhaustion, pain, and loss of motor control leading to accidents—we do not include poisoning by contaminants or infection, since these are not produced by the drug itself; psychological effects of chronic use—mental, moral, or emotional deterioration of some kind; physical effects of chronic use—minor, serious, or fatal chronic illness and organic pathology. Harm to others includes: crime and violence; loss of psychomotor control leading to accidents involving other people; and finally, becoming a general economic and social burden or invalid. This type of harm is the vaguest and most dubious of all: first, because it is really an element in the harm produced by the other categories, and second, because whether or not the drug user becomes a social burden depends overwhelmingly on factors other than the pharmacological effects of the drug—especially the attitude society takes toward the drug, its social definition. We have included this factor mainly because it is nowadays one of the reasons most commonly given for strict punitive controls on psychoactive drugs. The overlapping vague category of "psychological effects," often conveniently identified with drug dependence to give the latter an undertone of evil, is another justification too often used for drug prohibition. It is like the last category except for its emphasis on the need to save the individual from himself rather than the need to protect society from the individual.

The Debate on Coca

Because the cultural and political contexts of the two practices are so different, we must distinguish chewing the coca leaf from taking cocaine. The most obvious difference is that with coca the question is whether to outlaw or restrict a practice that is legal and more or less socially accepted, while with cocaine the question is whether to liberalize our laws and attitudes about a practice that is illegal and condemned. The inertial force of established custom, that first law of politics, favors coca, while it opposes cocaine. Nevertheless, a great deal of passion has been generated on both sides in the debate about coca. In fact, it is often hard for someone with no emotional and cultural investment in the argument to see what the excitement is about. We shall examine here the statements made by advocates and opponents of coca and try to judge their motives and social context.

This is an exceptionally clear example of a situation in which the pharmacological effects of a drug are less important than the symbolism that surrounds the habit of using it. With a few minor exceptions, no one contends that coca causes any significant crime, violence, loss of psychomotor control, acute illness, or severe withdrawal reactions. The harm imputed to it almost always falls into the categories of chronic psychological or physical deterioration and general social deficit. As we have shown in Chapter 6, the evidence that coca actually does any damage of this kind is inconclusive: in the case of chronic physical illness, because of the difficulties of retrospective research, and in the other two cases for the same reason and also because of the descriptive inadequacy of the categories themselves. There does seem to be some connection between the use of coca and diseases associated with malnutrition, but the order of cause and consequence is opaque. Gutiérrez-Noriega declares, "At the start they use coca because they are not eating well. Then they do not eat because they are using coca." [7] But it is hard to determine how to arrange experiments to prove this, and what evidence there is suggests that lack of food leads to coca chewing, and not the other way around. The evidence for other chronic defects and deficiencies attributed to coca is even less conclusive.

Nevertheless, condemnations of the drug are often intensely passionate. The United Nations Commission report that declared coca to have "genuinely harmful economic and social effects" was very mild compared to some of the other judgments about it. Mario A. Puga writes, "It is an elaborate and monstrous form of genocide being committed against the people." Carlos A. Ricketts, in a 1948 pamphlet, calls coca chewers feeble, mentally deficient, lazy, submissive, and depressed. Carlos Enrique Paz Soldán speaks of "a legion of drug addicts . . . if like fatalists we await with folded arms a divine miracle to free our indigenous population from the deteriorating action of coca, we shall be renouncing our position as men who love civilization." Gutiérrez-Noriega, Ricketts, and Luis N. Sáenz proposed to the Second Indigenist Congress at Cuzco, Peru, in 1948 a resolution describing coca as "the greatest obstacle to the improvement of the Indians' health and social condition." Gutiérrez-Noriega writes further of the "spectacular misery" for which coca is the partial cause and false remedy, and declares its use as a substitute for food a "collective crime." [8]

A sardonic comment by Aníbal Prado suggests one reason for this intensity of feeling. Referring to the South American ruling classes, he writes, "The magnanimous gentlemen want the humble people to remain ignorant of their tragedy; they want them to die without knowing of their misery." In other words, coca is "the opiate of the people" in Peru and Bolivia. It keeps the poor at their assigned tasks and suppresses the urge to rebel. Puga considers the dismissal of Gutiérrez-Noriega from his university position in 1948 a political act, and even seems to imply indirectly that his death in an automobile accident was murder. Ricketts speaks of the "amoral interests" that sustain the production and use of coca. Marcel Granier-Doyeux, quoting Puga in the United Nations *Bulletin on Narcotics*, refers to "an inhuman social system and an ignoble custom of earning profits at the expense of the nation's life and future." Friends of coca may unwittingly confirm these opinions by expressing fear that abolition of coca use would cause "gangsterism" and "revolution." [9] The function of economic integration served by coca is not valued highly by those who have no commitment to preserving the integrity of the existing economy; they are unhappy about evidence like Hanna's (Chapter 1) showing that Indians work an average of one day a week just to keep themselves supplied with the drug. This conviction that coca is used as a pacifier for oppressed Indians to prevent radical social reform—similar to the opinion of some blacks that heroin is a device for racial oppression in the United States—attributes to it a kind of harm that falls into our last category, although the burden it implies is a burden not so much in the present social structure as in the passage toward a better one.

In the face of these strong feelings, defenders of coca have felt obliged to claim for it not only harmlessness but numerous medical and social virtues. Even the United Nations Commission rebuked Gutiérrez-Noriega for calling the coca habit an addiction and asserting that it caused serious psychological disturbances. We have described the enthusiastic reports of nineteenth-century investigators like Mortimer, Tschudi, and Mantegazza. A more recent traveler, testifying to the *coqueros'* respectable motives, approvingly quotes a rather prosy Indian folk song:

> No coqueo por vicio
> Ni tampoco por el juicio
> Sino por el beneficio

This may be translated: "I do not chew coca as a vice or out of calculation but because of the good it does." The author of a book sponsored by coca plantation owners and entitled *El Oro Verde de las Yungas (The Green Gold of the Yungas)*—the Yungas is a region in Bolivia where coca is grown—declares coca to have all the virtues and none of the defects of coffee and tea. He even suggests, quoting a Bolivian delegate's speech to the League of Nations in 1932, that it helps to prevent cancer. Richard T. Martin refers rather contemptuously to those who condemn coca as "officials and doctors who have had little if any experience with Indian life." [10] Richard E. Schultes takes a more moderate position. Although he did not recommend the habitual use of coca any more than the use of tobacco or other drugs, he pointed out to us that in Colombia, where the advice of the United Nations has been taken seriously and the effort to prohibit coca has been partly successful, the cultural void is sometimes filled by poisonous, poorly distilled alcoholic concoctions.

Modern scholars' estimation of the role of coca in the Inca Empire has been closely associated with their attitude toward its contemporary use. Mortimer was so impressed by the romance of the Incas that he devoted a large part of his book to the government and culture of their empire, implying without proof that coca was somehow the basis of it all. But in spite or perhaps because of its great cultural and religious significance, coca use was restricted and carefully controlled under the Incas. Gutiérrez-Noriega believed that the widespread use of the drug produced by colonial domination was first an effect and then a cause of the misery of an Indian population oppressed by European masters. Both Mortimer and Gutiérrez-Noriega picture the time of the Incas as a kind of Golden Age; Mortimer attributes their greatness partly to the supposedly pervasive influence of coca, and Gutiérrez-Noriega attributes it partly to the fact that the coca habit had not begun to spread its ravages among the populace.

The political and professional loyalties of the participants in this debate are interesting. Radicals tend to favor restriction or abolition of coca use; conservatives tend to be favorable or indifferent. Physicians today—as they did not in the days of Mantegazza and Mortimer—usually condemn coca; anthropologists, archaeologists, ethnologists, and botanists, even when they are not political conservatives, usually approve of it. Students of small Amazonian and Colombian mountain tribes are more likely to find coca beneficial than students of the Indians of highland Peru and Bolivia. The debate about coca expresses in symbolic form the issue of how much is worth preserving in the traditional indigenous cultures of South America. Like the debate about opium and cannabis in Asia, it is partly a conflict between advocates and opponents of modernization and westernization. Gutiérrez-Noriega associates an attitude of "negativism" and passive resistance to Western culture with the coca habit; he approves of the greater cultural assimilation of the coastal Indians, which includes a low consumption of coca. Martin, in effect agreeing with Gutiérrez-Noriega, considers measures to restrict or abolish the use of coca part of the white man's "attempt to exterminate the Indian's way of life." He declares that to deny the Indians coca would be "as much a disregard for human rights as would be an attempt to outlaw beer in Germany, coffee in the near East, or betel chewing in India." [11]

The irony of this situation is that both sides see themselves as defenders of the Indians against hostile or oppressive white men. Yet Martin finds himself in the company of economic interests with little concern for Indian culture. He also finds himself defending as an essential part of native American culture a practice that became widespread in Peru and Bolivia, and even among some of the more primitive tribes of Colombia and the Amazon, only *after* the European conquest. As for Gutiérrez-Noriega, he is not only in the morally uncomfortable position of trying to save the Indians from themselves, but also under the obligation to help them by making them less Indian.

Is coca an appropriate focal point for this debate about the moral and social condition of the people of Peru and Bolivia? It would be important to know what they themselves think, but no one has gone to very much trouble to find out. Coca at the doses ordinarily taken is probably a more powerful stimulant and euphoriant than tobacco, betel, khat, kola nut, coffee, or tea, so there would be some reason to treat it less casually. But the symbolic importance of the issue for students of Indian life seems to be greater than its actual importance for the Indians themselves. For over 400 years they have been replying, when asked why they use coca, that it suppresses hunger, thirst, and fatigue. Some regard it as an esthetically unpleasing mild vice and others simply as an aid to work. Being deprived of it, especially if they are given a good diet, does not make them unusually wretched or rebellious. According to the United Nations Commission report, the leaders of the Bolivian miners' union, who could be said to represent some part of Indian public opinion, disapprove of the coca habit but would oppose suppressing it without providing more food for the workers.[12] On the whole, it seems that if they had a better diet the Peruvian and Bolivian Indians would use less coca and restrict their use of it more to gaining strength for difficult mountain journeys and relieving minor respiratory and digestive troubles. Meanwhile, the observation by James H. Woods and David A. Downs that coca chewing is no more serious a drug abuse problem than tobacco chewing or coffee drinking in the United States seems to express the opinion of the Indians themselves.[13]

A curious fact about the coca debate is that the social policies advocated by enemies of the drug do not differ greatly from those advocated by its friends. The ruling mood is expressed in Carlos Monge's remark: "Aristotle has warned that governments should be careful not to impose measures that the people are not prepared to accept." [14] The United Nations drug control machinery has not been particularly active, and the United Nations has emphasized social reform rather than suppression as a means to eliminate the coca habit. Everyone opposed to coca seems to agree that better food, better education, and alternative economic opportunities are the best ways to get rid of it, and friends of coca can hardly be opposed to any of those things. Although it is inconsistent to believe both that chewing coca causes malnutrition, ignorance, and poverty, and that one can eliminate it by eliminating malnutrition, ignorance, and poverty, most enemies of coca seem able to live with this inconsistency. In fact, the best reason of all for advocating social reform instead of repression, mentioned by neither side in the debate, is that it would give the people who use coca the capacity to decide what to do about it.

Abuse Potential of Cocaine: Crime and Violence

We have already provided most of the necessary evidence about the *possible* harmful psychological and physiological effects of cocaine. But no list of possible effects, harmful or beneficial, can tell us how often each of them is likely to occur at the doses people are likely to use; and that is what really matters for a rational legal and social policy. The cocaine laws we have now are not based on any careful examination of this question— none has ever been made—but on a 60- or 70-year-old mixture of reasonable medical concern, rumor, and prejudice. Assuming that many more people would use cocaine if they could afford it and obtain it legally, it is hard to say what harmful social consequences this might have and, more important, how they would compare with the harmful effects of drugs that are now much more freely available. The information from the period 1880 to 1930 may not be inaccurate, but it is fragmentary and anecdotal and, naturally, since physicians treating cocaine abusers provide most of it, emphasizes dangers rather than benefits. Nevertheless, we shall try to make a reasonable estimate, using this and later evidence.

Of our seven categories of possible harm, the one which, whether or not it is the most important consequence of drug abuse, provides the best reason for *legal* restrictions on access to a drug, is the fifth, crime and violence. We have reserved the discussion of this subject for this chapter on abuse potential and social policy precisely because it is so much more obviously a matter for government intervention than some of the vaguely defined effects commonly used as justifications for punitive action against users and purveyors of drugs. It clearly involves harm to other people rather than to the drug user himself, and harm of a relatively specific and strictly defined kind. If we suppress a drug because its use is likely to cause crime—that is, of course, crime apart from the possession and consumption of the drug itself—we are not prohibiting anything to anyone for his own good. And by confining our justification for suppression to limited and obvious kinds of damage, as opposed to tyrannically vague categories, we make it easier to protect due process of law from overbearing encroachments by authority.

Consider the firearms analogy again. We have suggested that the most rational justification for restricting access to a drug is that it is a dangerous instrument, like a gun. The kind of danger from drugs with which social policy should be most concerned, then, is the same kind that concerns us in the case of firearms. But the nations and states that restrict or prohibit the private ownership of firearms do so because of the danger that someone will be hurt or killed by a bullet from someone else's gun—not because gun dependence is a bad habit, or because grown men who like to play with guns are immature and socially maladjusted, or because the gun habit produces a dangerous personality type (presumably impulsive and aggressive), or because a man may neglect his wife and children while he oils and polishes his gun collection or keep the neighbors awake by indulging in target practice in the backyard, not even because one can use a gun to kill oneself. It is difficult to impose any legal restrictions at all on firearms in many parts of this country. One can imagine the outrage if they were suppressed for reasons like these. And yet it is considered quite acceptable to suppress psychoactive drugs for analogous reasons.

In other words, it might seem to be little more than common sense to insist that the potential for crime or accident is the best reason for controls on any dangerous instrument; yet this is today far from the most common justification for opposing legal access to psychoactive drugs. In fact, liberals and permissivists suspect sensationalism and appeals to prejudice in any connection of drugs with crime, and conservatives too have largely abandoned arguments referring to crime or accidents that hurt bystanders in favor of emphasis on less clearly defined kinds of medical and social harm. It was not always so. Fear of the violent (often sexually violent) or criminal "dope fiend" once provided drug prohibitionists with their best argument. This line of attack has fallen into partial disrepute for two major reasons. One is that claims about crime are easier to refute than medical and psychiatric justifications for coercive controls. The other is that a grotesquely misplaced fear of violence and crime was used in the early propaganda campaigns against certain drugs, especially marihuana. When it became obvious even to the most stubborn opponents of cannabis that the contention that it caused violence and crime was becoming ineffective, they had to change their approach. By the time the hallucinogenic or psychedelic drugs arrived on the scene, psychiatric justifications for suppression were in the ascendant, and references to crime were few and halfhearted. So today it is a political and moral argument particularly worth emphasizing that in a free society the best reason for limiting access to certain substances by enforcing criminal penalties is that they are likely to be used in a way that will cause criminal acts.

Considering only its psychopharmacological effects, we can say that cocaine certainly has some potential for producing crime and violence. As Bejerot observes, "morphine and heroin. . .have an essentially subduing effect. The effects of central stimulants are decidedly more criminogenic, since on high doses self-confidence increases and inhibitions and judgment are reduced." [15] Some think the original dope fiend—a caricature once used to justify prohibition of marihuana and opiates as well as cocaine—was in fact a hyperactive, paranoid cocaine abuser. Certainly the features of this infamous figure suggest a stimulant rather than a depressant drug, although he might also be a product of alcohol or barbiturates or possibly of the early phase of the opiate abstinence syndrome. Originally the drug fiend may not have been a human being at all but a metaphor for the spirit or essence of the drug, a diabolical analogue of the Peruvian goddess Mama Coca, related to the genie-in-the-bottle of fable and to Demon Rum himself. The phrase "drug fiend" is an example of what Szasz calls objectifying the temptation to use or abuse a drug in the drug itself: the drug "possesses" its user and turns him into a demonic Mr. Hyde. In any case, whatever the source of this fearful image may have been, around the turn of the century cocaine did gain a reputation for causing crime and violence that was a powerful force for its suppression.

Very little is known about the likely as opposed to possible danger of this kind from cocaine. The effects of amphetamines have been studied much more carefully, and since the two kinds of drugs are in many ways psychopharmacologically similar, we can learn something about the potential effects of the one more rarely used and studied from the observed effects of the other. Amphetamine abuse is clearly related to aggressive behavior. The psychomotor stimulation, paranoid ideation, and emotional lability these drugs produce may actually induce aggression directly rather than merely by lowering impulse control, like alcohol and barbiturates. General irritability or sudden changes of mood from congeniality to fierce hostility may cause unprovoked and apparently motiveless assaults. The amphetamine abuser's orientation toward the immediate present and disregard for long-range consequences make him a poor candidate for premeditated criminal acts. But the tendency to react strongly to immediate sensory stimuli, the inclination to refer everything in the environment to oneself that often degenerates into paranoia, and the need to *do* something induced by intense psychomotor stimulation compose a classic portrait of the trigger-happy character.[16]

It is not surprising to find some of the same effects occasionally observed in cocaine users. This may take the relatively mild form described by a man we interviewed: "It makes me feel aggressive—wanting to bite people—assertive." Another interview subject put it this way: "I don't think that it's a source of violence, but I think that if someone does have a violent disposition or a tendency to lose his temper from time to time, it can bring it out or aggravate it, and you just have to learn about that. . . . I'm sure that a drug which is action-oriented, rather than inaction-oriented, is much more likely to spur violence." Maier says that intelligent criminals know that cocaine can give them energy and courage; one of his patients, a burglar, confirmed this by confessing that if he carried a weapon during a burglary he was more likely to use it unnecessarily when under the influence of cocaine. Narcotics agents like to assert that "coke makes a man more dangerous than any other drug," and that it is especially risky to arrest cocaine dealers. An official of the Drug Enforcement Administration epigrammatically observes, "Heroin addicts do the crime to get the drugs. Coke heads take the drug in order to do the crime." Former heroin addicts reported to Irving Soloway that once they were off heroin and "into coke" (or methamphetamine) they began to commit armed robberies and muggings.[17]

In one recent case cocaine abuse, combined with the institutionalized paranoia of illicit drug traffic, was associated with the most extreme forms of violence. The group involved was known as the Company, described in the *Miami Herald* as "the most vicious underworld gang to ever cast a shadow of brutality and lawlessness across South Florida." It had more than 150 members and smuggled hashish, marihuana, and probably heroin as well as cocaine. It was said to be responsible for 37 murders and many beatings and tortures—mostly of members of the gang suspected of treachery, but also of outsiders who knew too much to be allowed to live. Most of the ringleaders are now in jail for the murder of one of their associates; their arrest was based on the testimony of another member who decided he had had enough after this killing and the subsequent fear-induced suicide of his own wife.

The excesses and irrationality of the Company's directors suggest that they were influenced by something more that the usual greed-inspired violence or bizarre notions of honor and revenge associated with organized crime. For example, one man had his legs blown off by a bomb placed in his car because of an unpaid debt of $600. A homicide detective commented, "It wasn't so much the money as the principle of the thing . . . that's what makes these guys so scary." The gang leaders also became openly aggressive toward the police in a self-defeating way, insulting them to their faces and threatening their families. According to a physician who interviewed two members of the gang in custody, they used cocaine together for days at a time and at first regarded the ensuing acute paranoia as a source of amusement rather than an inconvenience. But after several months they became increasingly suspicious of one another and began to kill to resolve their delusions. Our source refers to this ironically as "an underworld mental health program." One of the criminals placed his murders in two categories: those that were "necessary" and those that were "unnecessary" but seemed reasonable at the time he was taking cocaine.[18] Cocaine may not have been the original inspiration for the brutality of these men, who had previous histories of criminal violence, but its psychopharmacological effect enhanced existing tendencies and made their consequences more horrible.

Two kinds of potential connection between cocaine and crime are implied in these reports. First, a central nervous system stimulant may supply the will to go through with any act that requires self-confidence above all—whether it be a public speech, a stage show, or a robbery. It is notorious that alcohol provides similar "Dutch courage" with less psychomotor and intellectual control. Cocaine does not deaden the user to pain and desire, like opiates, but strengthens his will and nerves him to attempt acts that are at and beyond the limits of his capacity. If the opiate addict thinks of himself as a kind of Buddhist attaining Nirvana, the cocaine abuser is more likely to see himself as a Nietzschean superman realizing his will-to-power. This superficially rational egoistic self-confidence can easily pass over into an obviously irrational emotional lability and paranoia that are more dangerous because their effects are less predictable.

Although the suggestion that cocaine use causes crime is obviously something more than police and criminal hyperbole, personality and setting as usual make all the difference. Although many people we interviewed said that someone intoxicated by cocaine might be verbally abusive or otherwise unpleasantly aggressive, no one mentioned any physical violence under its influence. A woman whose husband had gone through a period of abusing various drugs said that cocaine caused mainly verbal unpleasantness, while physical violence was associated with barbiturates. She herself thought well of cocaine in spite of her husband's abuse. In a 1971 study by W. C. Eckerman and several colleagues for the Bureau of Narcotics and Dangerous Drugs (now the Drug Enforcement Administration) on *Drug Usage and Arrest Charges,* 19 percent of those arrested for crimes unrelated to drugs said they had used cocaine, and 9.5 percent said they had used it in the preceding month. Cocaine users were arrested more often for burglary and larceny and less often for crimes against the person than other drug users. Of course, there is no necessary implication that any of the crimes in this study were committed under the influence of any drug. Jared Tinklenberg, commenting on this study and in general on the relation between cocaine and violence, expresses some surprise that it seems to produce "amphetamine-like paranoid assaultiveness" so seldom and concludes that at present it is not a serious crime problem.[19]

The case of the Company suggests that cocaine is seldom associated with crime partly because so little is known about where and how it is used, and also because of high cost, restricted availability, and dilute preparations. But it is wrong to infer in general that if criminals use and sell cocaine, cocaine has made them criminals. The drug can obviously exacerbate tendencies toward paranoia and violence that are already present in its users or encouraged by a criminal milieu. But most violence in the illicit cocaine trade, like the violence in the illicit heroin traffic today and in the alcohol business during Prohibition, is of course not necessarily related to the psychopharmacological properties of the drug. Al Capone did not order murders because he was drunk, and the cocaine dealer "Jimmy" does not threaten his debtors or fear the police because of cocaine-induced paranoia.

We have to rely mainly on an estimate of cocaine's psychopharmacological properties to determine whether violence and crime connected with it would be greater or less if it were legal. It does not cause loss of psychomotor control of the same kind or to the same degree as alcohol or barbiturates, and it is a less powerful and more transient physical and psychological stimulant than the amphetamines, with a much less impressive record of instigating crimes against the person. So it is probably less dangerous in this respect than any of these three drugs. Opiates, on the other hand, cause neither violence nor loss of psychomotor control, and by pacifying the user they make him less interested in crime or any other activity requiring initiative. If they were legalized, practically no crime would be directly associated with their use. The illegality of cocaine produces violence and crime not only by supporting organizations of outlaws but by increasing and in part justifying any paranoid tendencies that the drug itself creates in these men. So it is possible that there would be less violence and crime associated with cocaine if it were legal.

Abuse Potential of Cocaine: Other Factors

We can discuss the other kinds of harm that cocaine may cause more briefly, because we have already described them in analyzing its physiological and psychological effects. We repeat that these other kinds of harm, however important they are to the drug users themselves, should be of less immediate interest to social policy and less subject to coercive controls than potential crime and violence. To run down our list:

(1) Acute psychological effects: psychosis seems to be rare in recreational use, at least among sniffers; the crash is apparently not so hard a landing as the amphetamine crash; the most common problems are insomnia, irritability, and anxiety;

(2) Acute physical effects: overdose death is apparently rare, although this judgment must be qualified by reference to drug mixtures and inadequate reporting; deaths in surgery were at one time more common than deaths on the street, or perhaps only better reported; severe nonfatal acute poisoning from sniffing is probably rare, from injection not so rare; alcohollike or barbituratelike loss of motor control does not occur:

(3) Chronic psychological effects: cases of demoralization and general deterioration with periodic psychoses, like those described by literary men and physicians in the early years of the twentieth century are apparently less common today but would presumably be more so if cocaine were more freely available;

(4) Chronic physical effects: the most common ones today are rhinitis and weight loss; if the drug were more freely available there would probably

be some abuse causing serious malnutrition and debilitation; evidence of brain damage in coca chewers is inconclusive and not based on any observed organic pathology;

(5) Crime and violence: we have discussed this;

(6) Loss of psychomotor control leading to accidents that hurt others: as we have implied, cocaine in acute doses would rarely cause this, although the overreaction of a paranoid abuser to a fancied persecution might have a similar result;

(7) Economic and social burden: insofar as it is possible to estimate this at all, we are inclined to say that cocaine would not become as great a social problem as many other less restricted potentially dangerous instruments, pharmacological and other.

Comparison with Amphetamines

Because of historical and pharmacological resemblances, the relative abuse potential of amphetamines and cocaine is an interesting question. Bejerot writes, "In my opinion the addiction established by the natural central stimulant cocaine is essentially the same disease as addiction based on synthetic central stimulants . . . in all essentials they seem to belong together in the same way as morphine and morphine substitutes." He provides an impressive list of documented similarities between the effects of the two kinds of drugs.[20] Some of the old cases of cocaine abuse might have been less shocking to physicians if they had had amphetamines for comparison. Historically, there are striking resemblances between the career of cocaine up to 1914 and the career of amphetamines in the last 40 years. It is hard to resist mentioning the saying that we are condemned to repeat the history we forget, for that is exactly what has happened in the case of amphetamines. Although the continued use of amphetamines for weight reduction, depression, narcolepsy, and hyperactivity in children will probably prevent the nearly total outlawry to which cocaine has been subjected, recent legal restrictions have actually put them in a position very much like the one cocaine reached a few years before the Harrison Act.

But we are more interested in differences than similarities. To say that cocaine is to amphetamines as morphine is to morphine substitutes is to ignore the fact that in its chemical structure, and to a lesser extent in its pharmacological activity, cocaine is quite different from the synthetic central stimulants. There is little agreement about what this means as far as its potential dangers are concerned. Bejerot and many others state that the difference between the dose that produces the desired stimulant effect and the dose that produces undesirable toxic effects is less for cocaine than for amphetamines, but we have found no clear evidence of this. In some laboratory experiments cocaine administered intravenously by animals produced more physiological and psychological toxic effects than dextroamphetamine or methamphetamine, apparently because it was taken in much larger quantities; in other experiments it did not seem to be more toxic than the amphetamines. Jerome Jaffe states that there is no evidence that the abuse potential of cocaine is greater than that of amphetamines. Another authority, J. Robert Russo, contends that "Cocaine is a much more dangerous agent, and quantitative comparison would not be valid. In contrast to the amphetamines, cocaine is capable of inducing severe cytotoxic effects in nearly all tissues, including the brain." [21] But the evidence of cocaine's cytotoxicity at the dosage levels normally used by human beings, except for reversible liver damage, is inconclusive, and there is some indication that amphetamines may cause brain cell degeneration.[22] Any organic pathology attributable to cocaine is probably similar to that attributable to amphetamines and not so severe as the pathology produced by alcohol and barbiturates.

In the absence of systematic comparative studies on human beings, the best or at least most interesting evidence we have suggesting the relative dangers of cocaine and amphetamine intoxication is the comments of people who have used both drugs. The main theme of remarks made in our interviews is that amphetamine effects are longer lasting and less subtle, more "physical"; for example: "Most people move from cocaine and go to speed, or come from speed and go to cocaine, for they're both basically the same. I'll only say speed is a step above the other. . . . You can stay awake three or four days on speed. Cocaine will keep you up, but it won't hold you as long, not as long." "Speed took me out of bounds. It took me farther than I wanted to go." "I took speed for a while, and that was just a total disaster. . . . I used to get really paranoid, and I used to feel so terrible after shooting up some speed, or snorting some speed. . .my nervous system is just too delicate to handle amphetamines, although I really like the rush. . .but the price was just far too great." (This man still uses cocaine.) "Cocaine. . .gives you energy without making you excessively nervous, where on speed your body tenses. . . . Pure coke, if used properly, will not give you the jitters and so forth." "Speed immediately sets my body shaking, which is something I just cannot tolerate. . . . It's more a physical thing than mental; coke is more mental. . . . When my husband couldn't afford coke he started taking speed. . . . I tried it once and that was it."

Here are some further remarks: "It's not a teeth-gritting, only-in-one-direction amphetamine energy. . . . Amphetamine was like going down a highway, you don't really feel high, you feel forceful, whereas with cocaine you may not even feel forceful." "Speed has demonstrably dynamic physical effects. . .that rush you get from speed you don't get from coke." "Cocaine is sort of like speed. . .it doesn't last as long, though, it's not as physical, it just gets your head cranked up instead of your body." "It's a tiny bit like speed. Speed seems to be more physical, and cocaine is more psychological, more subtle." "Coke is probably most similar to speed. . .they both produce energy. . . . But I don't get physical problems from coke, like tensing of the jaw, or gnashing the teeth, or anything like that, that you get with speed, and that, right away, is a reason why I don't do it. Plus, speed just seems to stay with you *so* long, and you do really get strung out. I do believe speed is physically addicting; now I may be wrong, but I think that's true. And I don't think coke is. . . . I would consider it infinitely superior to speed." "I think that eventually cocaine will warp your perceptions, and your mind, and your emotions, and make you incoherent. That's what speed did to me." "The difference between speed and coke is, if we take the levels of physical, mental or intellectual, and psychic or spiritual, coke's accent is on the last two, speed's accent is on the first two."

All the people we interviewed who had used both drugs preferred cocaine to amphetamines. They confirm the opinion of an amphetamine abuser interviewed in the Haight-Ashbury district in the late 1960s: "Coke is too good for the amateurs around here. . . . But let me tell you, it's worth the hassle. I'll take candy any time I can get it. Man, that shit's the speed freak's caviar." [23] Perhaps, then, cocaine is more dangerous because more attractive. More likely, it is less dangerous for the very reasons that it is more attractive: it does not take the user "up" too far or for too long a time or produce as many subsidiary toxic effects along with the central stimulation, and for similar reasons it may produce fewer psychoses. Since the drugs partially substitute for each other, less amphetamine on the market may mean more cocaine and vice versa. That is how it apparently was in the 1930s and 1940s, with cocaine in decline and amphetamines in the ascendant; it may come to be so again today if amphetamine availability declines. In any case, if cocaine can be harmful to the user to make him dangerous to the people around him, amphetamines are potentially more harmful and more dangerous in exactly the same ways.

Policy Alternatives

There are several ways of posing the question of drug policy. One is to ask how the benefits of using a drug can be made to outweigh its detriments and try to impose a form of regulation that will insure this. Any attempt to do so for cocaine will have to take into account that it is far from innocuous. Both human beings and animals will sometimes continue to use it despite its doing considerable damage. Bejerot overstates the case greatly in calling cocaine abuse "the most severe of all forms of toxicomania and the most difficult to cure," but the animal experiments we have described show that both its attractive power and its potential toxic physical effects and disruptive effect on behavior are high.[24] A woman we interviewed said, "If you can administer it well, it does have a beneficial effect. But the thing I've seen about the drug culture, where they had access to it, it was automatically like pushing a button for excess. Very rarely have I seen anyone use it well, when there was a quantity around." If cocaine were more easily available, or if more people took to using it intravenously, there might be considerable serious abuse. Bejerot's angry report on the 1966–1968 Swedish experiment in liberal prescription of amphetamines suggests the nature (although it exaggerates the severity) of the problems cocaine might pose. On the other hand, the danger of abuse is less in the case of sniffing, which will certainly remain the most common method of administration. Some of our interview subjects who admitted the potential dangers of cocaine nevertheless regarded it as "a life-force not comparable to anything else" or a "utility drug" that "builds with the stones you've accumulated" from other drugs or other experiences. For them the benefits clearly outweigh the detriments. Others would agree instead with the man who told us, "It's a sad place to have your happiness." But supposing that is true of cocaine, it is true of other drugs as well. Without good reason we have permitted people to find their happiness in some of these "sad places" and not others. The dangers of cocaine are not of the nature or degree that the law now implies and the public now assumes. There is little evidence that it is likely to become as serious a social problem as alcohol (or firearms) or as serious a health problem as tobacco. This is the kind of consideration to be taken into account in deciding how to prevent a drug from doing more harm than good.

But there is also the issue raised so insistently by Szasz: Is it any business of the rest of us to use coercion to prevent someone from harming himself with a drug, especially when the danger is not probable but only possible, or even remote? As we observed before, to answer this question fully would be to develop a complete social philosophy. But anyone at all concerned with individual liberty should be suspicious of present policies on cocaine. Even if the drug could be considered to do more harm than good from some unattainable "objective" point of view, it is wrong to allow the law to define harming oneself in a way rejected by the person ostensibly doing the harm. That is what the disease-crime model of drug abuse does in implicitly justifying the present treatment of cocaine and other drugs. It creates not only victimless crimes but victimless diseases. There is no reason why drugs should be treated differently from other useful though possibly dangerous instruments.

The most humane and sensible way to deal with these substances is to create a social situation in which they can be used in a controlled fashion and with moderation. This has been called "domestication." Norman Zinberg and others have elaborated the idea in a series of papers on the social basis of drug abuse prevention. Caffeine is handled this way in most societies. It is taken in such forms and quantities and incorporated into household and workplace ritual in such a way that it does relatively little damage. Some cultures manage alcohol and opium in the same way; occasions for their use are assigned by custom and taught by older to younger generations in a way that preserves society from their potential deleterious effects. Even in our society 95 percent of alcohol users have learned how to control their intake; in spite of the almost insuperable obstacles placed in their way by the law and accepted attitudes, many opiate users are able to do it as well.[25] This compromise solution would make all drug habits virtues in the Aristotelian sense of a prudent mean. Unfortunately, domestication cannot be instituted by decree. Certain characteristics of the drugs themselves and the way they are manufactured may make it difficult; for example, alcoholism has probably become a more serious problem since alcohol was first concentrated as distilled liquor, and opium is more dangerous in the form of morphine and heroin. Besides, if a society does not have the habit of moderation in using a drug, legal action and persuasion are not likely to introduce it against the force of history and cultural tradition.

It is not easy to estimate either the prospects for domestication of cocaine or the relative importance of the chemical nature of the drug itself and the attitudes and traditions surrounding it in the answer to this problem. The distinction between coca and cocaine is essential here. Chemically and pharmacologically, they are as different as weak beer and vodka. The coca leaf and its extract seem to have been fairly well tamed both in South America and in this country at the turn of the century, and it is not obvious that they are more medically dangerous or socially damaging than coffee. Cocaine is another matter. Although abuse of this drug does not have cultural roots as deep as abuse of alcohol, any powerful stimulant may be dangerously attractive in our society. It is sometimes said that people take less interest in a drug that no longer has the allure of the forbidden. No doubt legalizing cocaine would take away a little of its special charm for its users and horror for its enemies, and the drug would lose some of its factitious glamour; but that does not mean that less would be used or that fewer people would abuse it. Nevertheless, it might be easier to domesticate than alcohol or amphetamines.

The opposite policy of prohibition and repression has been chosen in the case of cocaine. There is no doubt that it often works; that is, it can cut down the supply of a drug severely, raise the price, and therefore palliate the harmful (while also reducing the beneficial) consequences of using it. The strict amphetamine laws passed by Japan in the mid-1950s and Sweden in the late 1960s after a period of free abuse are an example. A more famous but less understood example of repressive drug legislation is the prohibition of alcohol in the United States from 1920 to 1933. Everyone knows about the unfortunate side effects of Prohibition; they were exactly the same as the side effects of our present legislation on opiates, cocaine, marihuana, and other drugs. It is not so well known that in those years the total consumption of alcohol was greatly reduced, and with it the amount of alcoholism and its attendant horrors, including serious diseases like cirrhosis of the liver. Repressive legislation has also undoubtedly cut down consumption of cocaine and the opium derivatives, with effects that are equally hard to evaluate socially.*

---

* Richard Ashley, in *Cocaine: Its History, Uses, and Effects* (New York: St. Martin's Press, 1975), p. 50, contests this proposition, but his reasoning is not persuasive. Relying on a record of the amount of coca leaves imported in 1906 and an estimate of the illicit cocaine traffic supplied by a Drug Enforcement Administration official in 1974, he concludes (after doubling the latter estimate on the ground that police figures are always too low, although one might as reasonably expect them to be too high) that Americans used as much cocaine per capita in 1973 as they did in 1906. Even if this is true—and the admitted unreliability of such figures makes it hard to tell—there is reason to believe that without the laws against it consumption of cocaine, like consumption of most other commodities in our more affluent society, would be much *greater* per capita today than it was at the turn of the century. In any case, doubts that prohibition cuts down consumption of a drug should be dispelled by the evidence of an enormous drop in the death rate from cirrhosis of the liver in the 1920s provided in the table that H. Kalant and O. J. Kalant have printed in *Drugs, Society, and Personal Choice* (Don Mills, Ontario: Paperjacks, 1971), p. 106.

The most interesting conclusion suggested by Ashley's figures is that cocaine consumption, as opposed to cocaine talk, is restricted to relatively few people. Accepting his estimate of 40,000 pounds a year coming into the United States illegally, we can calculate that this will supply only 720,000 "moderate" cocaine users (on Ashley's definition, those who take about a gram in an evening every two weeks), considerably less than 1 percent of the adult population. In fact, however, the distribution of cocaine, like that of any luxury commodity, is extremely inegalitarian. A few people use a great deal and a large number have tried it once or twice but have no access to more or cannot afford it. As we pointed out, the main reason our interview subjects gave for not using more was the price. Ashley's own sample of 81

The principle behind prohibition laws is that it is unwise or impossible to count on moderation in the use of certain drugs. But the decision is rarely made on any rational basis. It might be reasonably contended that prohibition is justified if the price we pay in the loss of individual freedom and relatively harmless pleasure for the majority in order to limit the abuse of a drug by a minority is not too high. But that is not the way we justify our drug laws. Instead, certain substances are transformed in the public mind from chemicals that can be used or misused into "bad drugs" with a Circean power to enslave and bestialize. The opiates, for example, are less toxic and less potentially debilitating and socially disruptive than alcohol, so the case for opiate prohibition, that "artificial tragedy with real victims" described by Marie Nyswander, may actually be less convincing than the case for alcohol prohibition. It has often been plausibly contended that the small amount of illegal opiate use today causes far more harm than the large amount of legal use 70 years ago. That is partly because of the pharmacological peculiarity that the absence of opiates (for an addict) is more disruptive and dangerous than their presence; there is no parallel in the case of cocaine. Nevertheless, it is hard to see how the argument for repressive legislation against cocaine is better than the argument for repressive legislation against alcohol. Most nations have decided, in effect, to tolerate the dangers of acute alcohol intoxication and the miseries of alcoholism for the sake of the pleasure most of their citizens take in a few drinks. The United States made that decision explicitly in 1933. There is no obvious reason why the pleasures and dangers of cocaine should be regarded in an entirely different way.

An argument in favor of prohibition suggested by the decline of alcoholism in the 1920s is that, whatever its express justification, it has the effect of imposing moderation for most people by keeping the availability of a drug down. It is hardly easier to develop a damaging cocaine habit today than to become an alcoholic solely by drinking expensive brandy. Even our interview subject who called cocaine a "utility drug" was not in favor of free over-the-counter sales, although he considered existing laws too harsh; he believed that ignorance caused by past abuse and subsequent suppression would only create more abuse. Cocaine right now is

---

contained 4 heavy chronic users who took two to four grams a day. A mere 10,000 such people, averaging three grams daily, would be consuming almost two-thirds of the total illicit supply. Everyone who is anyone may be using cocaine in substantial quantities, but the proportion of the population qualifying as anyone in this sense is minuscule. Ashley's sample included 15 cocaine dealers and 22 musicians—not exactly a cross-section of the populace or even of the fashionable upper and upper-middle class. Even if Ashley's estimate of the total supply is several times too low, it is implausible to contend that if the drug were legal and cheap it would not be used habitually by many more people and heavily by some who now use it occasionally. To avoid misinterpretation, we must insist again that this is not by itself an argument against legalization.

not a serious social problem and, like coca in South America, may be doing as much good as harm. Criminal penalties in a situation like this can have the effect of highway speed limits: the laws are often disobeyed and only sometimes enforced, but they make people wary and serve as moderating guidelines.

But even if we ignore the questions of individual liberty raised by Szasz, the criminal law remains a very expensive and clumsy instrument for handling the problems raised by a drug like cocaine. That would be true even if the law were used in a subtle and flexible way, but it is not. It is wielded aggressively and stupidly, assuming powers it is permitted nowhere else in our lives. Use of prohibited drugs is restricted not to those who can take best advantage of them but to those who know best how to circumvent the law (for example, serious research on the therapeutic applications of psychedelic drugs has been almost completely cut off, but street use continues at about the same level as ten years ago). Because the law is unevenly enforced and makes no distinction between use and misuse, it produces anomalies in apprehension and sentencing that are not only outrageous but pointless. And there are also the familiar incidental consequences of severe repressive legislation—poisonous adulterants, infection, overdoses caused by ignorance, organized criminal violence, costly and often oppressive police apparatus.

The financial and human cost of arrest and imprisonment under drug prohibition laws is enormous and rarely reckoned. It is impossible to determine the cost of the cocaine laws in losses through arrest, imprisonment, and expenditures by police and penal systems, because the Federal Bureau of Investigation does not list cocaine separately in its Uniform Crime Reports. But if we look at the data for marihuana, another drug culturally defined as "bad" whose harmfulness has been exaggerated, we find that the price we have been paying is high. In 1973, according to the FBI, 420,000 Americans were arrested for violations of the cannabis laws; in 1974 this figure rose to 445,000. Even though the majority did not go to jail, the cost must also be reckoned by what being arrested on a drug charge does to a person—to his career, his relationships with others, his self-image, his or his family's savings. And enough do go to prison to make this an important factor in the analysis; for example, in Texas just before the marihuana laws were liberalized in 1973, 800 people (mostly young) were in jail on marihuana charges, serving an average sentence of 9.7 years.[26] No psychopharmacological property of marihuana could possibly be as damaging to those young people or as costly to society as putting them in prison for ten years. The administrative cost of arrests alone is a great burden. In California, for example, the cost of each marihuana arrest in 1974 was nearly $1,500 and rising; by extrapolation, the total cost of marihuana arrests and prosecutions in the United States was about $600,000,000.[27] All this money comes out of the limited resources that also maintain law-enforcement programs protecting persons and property from threats that are more genuine.

But theoretical and practical arguments of the kind we have been outlining have been made for years without much effect. It is easy to say that it would be rational to examine whether the benefits of legalized cocaine would be so much less in proportion to the potential harm than the benefits of legalized alcohol, refined sugar, or handguns. But serious public debate about this question is unlikely, because what a society regards as rational is inseparable from its traditions and self-image; our traditions and self-image are incompatible with any way of classifying social problems that would make a debate like that possible. Raising the subject of illicit drugs sometimes seems to produce an anxiety and abdication of intelligence in those who do not use them that is more disturbing than any effect of the drugs themselves on those who do. We believe that people would be more alert to the consequences of the way they think about drugs if they were more conscious of the social background of their ideas. A narrative history like the one provided in the first part of this book is of some help. In the last chapter we would like to add some more explicitly theoretical reflections on the historical conditions that have brought us to our present pass with regard to psychoactive drugs in general and cocaine in particular.

10

DRUGS AND CULTURE: COCAINE AS A HISTORICAL EXAMPLE

WE HAVE WRITTEN of drug abuse in general and cocaine abuse in particular as though they were at least potentially clearly conceived problems with commonsense answers. Even so, we have been unable to avoid repeated hints that the drug problem may be a misconception, that to center one's interest on drugs, drug users, and the prevention of drug use is to mislocate the issues, and that even those who worry most about drugs are often really concerned about something else. Common sense may be no more than the shared prejudices of a whole society or culture. Our own common sense, for example, makes it necessary for authorities to debate gravely whether possession of marihuana should be decriminalized, even though countless obviously more harmful and dangerous commodities are sold freely. The real issue is often not the chemical substances but blacks, or hippies, or the Establishment, or musical preferences, or sexual habits, or attitudes toward work. If advocates of marihuana prohibition now display a certain defensive irritability instead of their former self-confident aggressiveness, it is not simply because more has been learned about cannabis as a drug but because the reigning cultural mood has begun to change. Still, an extraterrestrial being who believed our declarations that drugs were a major problem and then contrasted our attitudes toward cannabis with our attitudes toward alcohol and tobacco would have to conclude that we were insane. Instead, any extraterrestrial intelligence with an anthropological bent would no doubt decide that the drug problem was not a genuine social issue at all but the center of a murky cloud of symbolically projected passions.

By overemphasizing the drugs themselves as a source of misery we sometimes carve up reality conceptually in an ineffectual and misleading way. Users of the drugs often insist on this. Bejerot complains that Preludin (phenmetrazine) users in Stockholm could rarely be made to admit that the habit itself was bad; they worried only about not being able to obtain the drug. (He attributes this to a drug-induced dependence that prevents them from recognizing that they have a bad habit—a concept we rejected in Chapter 8.) Duster points out that the heroin addicts he studied do not regard their consumption of opiates with moral disapproval and self-reproach, although they feel degraded by the need for thievery to support the habit that is forced on them by the law. The addict "can certainly feel that it is wrong to lure a square into addiction under the present circumstances of the illegality of the drugs, yet he can approve of the idea of drug use in an abstract or ideal world." [1] Marihuana smokers proclaim even more uniformly and vociferously (and more plausibly) that they have no drug problem except the hostility of those who do not use cannabis.

Such opinions are not always merely self-serving declarations by immature and irresponsible drug-dependent persons; many impartial students of drug use agree. Thomas Szasz, for example, combines polemical vigor in analyzing social issues with an Olympian, if not extraterrestrial, detachment about the merits of the drugs themselves. We have noted that he considers the terms *drug addiction* and *drug dependence* as they are ordinarily used to be misleading because they overemphasize the importance of the chemicals themselves as opposed to the way society regards the habit of using them. In one of his polemical analogies he compares the chapter on drug abuse in Goodman and Gilman's standard reference work on pharmacology to a chapter on prostitution in a textbook on gynecology. That is, drugs cause drug abuse only in the sense that the female sexual organs cause prostitution, and it is absurd to confound physiological and social questions under a single heading in this way. Chein, in his study of heroin addicts, points out that the main goal (in fact, we would say, consuming passion) of opiate policy, to get the addict off the drug, "confounds a relatively minor symptom with the disease." [2] (We need not agree with him that there is in fact a disease and that it is properly described as the addict's personality disorder.) With other drugs, as with cannabis, the problem is often the attitude of those who do not use the drug toward those who do.

As we observed before, probably the main evidence that the drug problem is not what it seems is the apparently irrational way we choose which drugs to regard as a problem and the discrepancies in our definitions of the *kind* of problem each drug is said to present. The hypothetical space creature could not clarify this matter by experimentally observing the effects of the drugs; he would have to be a considerable anthropologist and terrestrial historian to understand it. Consider, for example, our contrasting attitudes toward barbiturates and opiates, both used as sedatives and anxiety relievers. Duster sensibly remarks, "If a drug dulls the senses and relieves anxiety, then a moral stand on whether that should be done under certain circumstances should be taken. Instead, in ignorance we have categorically defined heroin use as morally different from barbiturate use due to its social-legal base, not its physiological base." Barbiturate use (or tranquilizer use) has not until recently been a moral issue for respectable society any more than opiate use was in 1900. As Peter Laurie points out, although there were 8,000 cases of barbiturate poisoning in 1969 in Great Britain alone, no publicly recognized barbiturate problem existed and barbiturates were not a drug menace evoking moral fervor.[3] Of course, barbiturate (and amphetamine) use is now being conceived more moralistically, and as more legal restrictions are imposed public attitudes toward these drugs may go the same way public attitudes toward opiates—and cocaine—went in the first two decades of the twentieth century. The point is that such attitudes have never been based on anything as naively obvious as a dispassionate examination of the physiological and psychological effects of the various substances.

In spite of all these indications that the notion of a drug problem *as it is now defined* misplaces the issues, the emotional (and financial) investment of our society in the existence of this phantom remains enormous. The slick magazine of the Drug Enforcement Administration, with its four-color photographs paid for out of tax money, is an embodiment of the government's permanent commitment to defining drug abuse (i.e., use of certain selected illicit psychoactive substances) as a major issue and advertising it to the public as one. Zinberg and Robertson point out that in the early 1970s 90 percent of the public, in a poll, associated drugs with corruption and moral decay; they also note the tendency of nonusers to see all illicit drugs as alike even after the differences have been explained to them carefully and with great publicity.[4] Even rebels against the prevailing attitudes are inclined to see drugs, or at least some drugs, as a problem in themselves, while blaming the problem, of course, on social conditions or conspiracies. The John Birch Society may believe that Communists run the heroin traffic, and the New Left may believe that the CIA runs it, but they agree on the fundamental idea that heroin as such is a menace, and an important one. Of course, misuse of drugs is dangerous, but the emphasis should be on "misuse" and not on "drugs." When an adult beats a child we do not talk about a physical strength abuse problem, although this is obviously a misuse of physical strength. We might refer to a violence problem, but that properly removes the emphasis from the mere existence of superior physical strength. Nor do we refer to the danger of drowning as the deep water problem. Why the obsessional emphasis on certain drugs?

The answer often given, correct as far as it goes but incomplete, is that drugs are symbols charged with cultural tensions. The determination of an Asian government to eliminate the use of opium or of a South American government to stop coca chewing represents symbolically an aspiration to modernize and westernize. Conventional people in our own society displace their repressed and alienated anxieties onto illicit drugs and their users in a form of scapegoating: rebels define their difference by using exotic, often illegal drugs and scorning the ones most commonplace in their societies; both have something to be righteous about. Users of disapproved drugs become dope fiends because they are possessed by demons, especially sexual demons, that the rest of us have cast out. Fear and envy of outsider groups like adolescents and racial minorities are focused on the chemicals their members ingest or are believed to ingest.

All this is familiar by now. But it does not tell us what we are most interested in: why the focus is *drugs*, and why some drugs rather than others. Most responses to this question are polemical and rather thoughtless. Official authorities and some drug educators tell us that illicit drugs will enslave us and make us criminal or insane. Advocates of illicit drug use tell us that the substances they ingest have unique pleasure-giving or mentally liberating properties desperately hated and feared by the oppressive authorities. Nowhere is the traditional overemphasis on the pharmacological properties of the substances themselves more misleading. Conventional opinion about illicit drugs is irrational, but the unconventional response is no less hortatory, no less a call to arms rather than an analysis, and just as much beside the point. The psychedelic (mind-manifesting) properties of alcohol, nicotine, and caffeine are just as genuine as those of LSD; their powerful consciousness-expanding or consciousness-changing (perhaps we should say consciousness-shaping or consciousness-distorting) effects have been recognized for millennia and used by many cultures in rituals and religious observances. South American shamans induce ritual trances by means of tobacco; the divine frenzy of Dionysus is inseparable from its chemical medium, wine; Arab mystics used coffee in their devotions as the Kogi use coca. It is true that each psychoactive drug has a unique set of pharmacological properties, but that is not the reason why social or cultural groups so often insist on the great difference—usually the *moral* difference—between their own and their neighbors' drugs. It is the unique historical circumstances in which each drug comes into use in a given society that makes the difference; it is the fact that psychoactive drugs have proved to be particularly appropriate vehicles for some of our deepest passions that makes the difference moral. We must now discuss the reasons why they are so appropriate.

For the sake of clarity in exposition, we have discussed drug dependence and drug abuse as though we knew exactly what a drug is. Unfortunately, there is little more certainty about what substances to classify as drugs than about what activities should be defined as drug abuse or what habits should be called drug dependence. The WHO, which can apparently be counted on for symptomatic ineptitude and unhelpfulness in these matters, defines a drug as "any substance that, when taken in the living organism, may modify one or more of its functions." [5] In other words, practically anything. Perhaps we should exclude foods, i.e., anything oxidized in the body to produce energy. But then salt should be called a drug and alcohol should not. But salt is, after all, necessary to life. Then should pepper be called a drug? It is neither food nor necessary to life. Are an ink stain used to detect fungus and a gauze bandage of a certain kind drugs? The Federal Drug Administration says that they are, because they are in the *U.S. Pharmacopoeia*. More interesting from the point of view of social policy, are food preservatives and no-calorie sweeteners drugs? Does a vitamin become a drug if it is taken in large quantities or in pills rather than in food? This has become an issue between the FDA and vitamin lovers. Refined sugar is more dangerous than many substances classified as drugs; why is it not subject to similar restrictions? Why are aspirin on the one hand, and marihuana on the other, considered to be drugs while coffee is only a drink and tobacco only a smoke? When is alcohol a drug and when is it not one?

Obviously it is hopeless to seek a consistent principle here. The popular usage is not the medical one, and in both popular and medical usage the term is vague. The confusion is compounded by the fact that to the public "drug" means both "medicine" (as in phrases like "wonder drug") and "psychoactive substance used for pleasure," usually illicit but sometimes including alcohol; and further by the fact that some of the so-called prob-

lem drugs are also used as medicines. If anything is clear in all this confusion, it is that labeling something a drug is a social process that involves
imposing certain kinds of formal and informal restriction and regulation
on its consumption. The issue is determined by the authority under
which the substance is used and the received justification for its use. For
example, there is the case of Hostetter's Bitters, a 32 percent alcohol concoction sold in the nineteenth century as a patent medicine. In 1885 the
Internal Revenue Commissioner declared that it could not be taxed as an
alcoholic beverage if it was dispensed in the bottle as medicine; in 1905
the decision was reversed and Bitters was pronounced to be an alcoholic
drink.[6] The pharmacological effects of the alcohol in Bitters had not
changed, but now they were no longer being defined as the effects of a
drug. To understand where we stand now with respect to substances
usually called drugs, and especially those with the most complex and
profound effects—the psychoactive ones—we have to know how and why
these social definitions are imposed.

Even after it has been decided what substances are drugs, their use
may be assigned to different conceptual categories in different cultures.
The spectrum includes magic, religion, medicine, recreation, madness,
disease, vice, and crime. In twentieth-century Western society we are
anxious to keep these categories separated, and that is one of the reasons
why psychoactive drugs are such a disturbing element for us. In the process of assigning meanings to the experiences they engender, we are
confused by too many possibilities and the need to make too many distinctions. It is a matter of intellectual principle or ideological conviction
with us that medicine or therapy is one thing, fun another, religious ritual
and madness still others. This attitude is reflected in separate formal and
informal institutions regulating illness, recreation, religion, and so on; a
great deal of the controversy over psychoactive drugs involves the conflicting claims of representatives of these institutions, officially sanctioned or outlaw, to be the rightful judges of the true meaning of drug
use. The disease-crime model of drug use, applied capriciously by the law
to some drugs and not others, represents an ineffectual attempt by organized medicine and police to impose their own definitions on society. It has
only increased the acrimony and confusion.

It is not a new observation that one man's religion can be another
man's psychopathology or crime, one man's pleasure another's disease or
vice, and so on. And the effects of psychoactive drugs have always been
ambiguous and variable, because they are realized by way of the idiosyncratic complexities of the central nervous system and affected greatly by
set and setting. But the loss of moral authority that established institu-

tions have recently undergone in industrially advanced countries, together with an efflorescence of psychoactive drug technology, has made the potential ambiguity greater and the insistence of conflicting claims to the authority to transform this ambiguity into socially accepted fixed meanings more intense. Fifty years ago Maier gave special thanks to the Zürich police in his foreword and classified the occasional use of cocaine as a syndrome he called "periodic endogenous cocaine addiction." He was announcing that to use cocaine without medical authority is a disease and should be a crime; this attitude is as common today as it was in his time. On the other side, there is the recently established publication *High Times*, which advertises itself as "the magazine for the multiple drug abuser." By mockingly adopting the terminology of the medical and police establishments, the editors are saying, "You call it a crime or a disease; we call it fun, and we have as much right to define it as you." *High Times* is the precise cultural counterpart of the Drug Enforcement Administration's magazine *Drug Enforcement;* it even has the same slick paper and color photography. These two publications are competing not for the same audience, but for the same subject matter; they are advocating different methods of control not only on the use of psychoactive drugs but also on the interpretation of that use. This interpretation is now so freely contested that practically anyone can put in a claim.

In opposition to the officially sanctioned medicine-disease-crime model, drugs may be thought of as recreational or ritual instruments. The recreational use of (some) psychoactive drugs, stigmatized by official policy as an irresponsible addiction to kicks or thrills, is praised by the opposition as delightful and liberating highs or trips. Fun may also be more solemnly rebaptized pleasure and associated with a supposed philosophy of hedonism manifested in the use of illicit drugs that contradicts a presumed Judaeo-Christian ethic of self-denial. Another line of reinterpretation is to place drug use in the category of religion or in that of consciousness-expansion, which might be said to lie on the border between religion and fun. Flight of ideas becomes creativity, regression becomes mystical experience, hallucinations become visions, psychosis becomes prophecy or some other extraordinary manifestation of the power of mind, disease becomes health in the sense of true wholeness, crime becomes social rebellion.

Although these interpretations have the virtue of being less rigid than the official ones they attempt to combat, their claim to truth is also dubious. For example, there is no single Judaeo-Christian ethic behind the various conflicting ethical principles that have been adhered to at different times by Jews and Christians, and if there were any such ethic it is by no means clear what its position on psychoactive drugs would be. The people whose story is told in the Bible, after all, certainly used alcohol and probably cannabis and opium as well. In any case, a formal philosophy of hedonism has nothing whatsoever to do with a devotion to pleasures of the moment or of the simpler sensual kind. Neither Epicurus nor Jeremy Bentham was recommending even that we should "turn on," much less that we should "drop out." However doubtful the intellectual interest or truth of the conflicting interpretations of psychoactive drug use now being promoted by various social groups may be, they testify to the power of generating meanings that substances affecting the central nervous system possess, and also to the fact of modern history that practically nothing is any longer accepted as unquestionable.

The problem of greatest historical interest, partly because of the prima facie irrationality involved, is the way public opinion, both official and rebellious, has chosen to place the use of some drugs in one cultural category and of other pharmacologically similar ones in an entirely different category. Most explanations of this are either true, but partial and historically unspecific—such as those that refer to cultural and emotional, especially sexual, tensions, and to scapegoating—or they are thoughtless and designed mainly to annoy the established authorities—such as those that rely on the qualities of the chemicals themselves. From the official or respectable point of view, the use of morphine and codeine in certain restricted contexts is medicine; at other times it is disease or crime (like the use of heroin in any situation), and anyone who classifies it as fun or recreation has a personality disorder or character defect. The use of cannabis, medicinal in the nineteenth century, is now disease (perhaps only a mild one) and crime (usually a misdemeanor); again there is something wrong with anyone who classifies it as fun. To use cocaine except as a local anesthetic is also disease and crime, although it too was once a medicine. Amphetamines prescribed by a physician for depression are, of course, medicine; amphetamines prescribed by a layman for himself because he wants to feel better are disease and crime. The use of alcohol, self-prescribed, is fun, unless you use too much, and then it is disease—alcoholism; it is no longer medicine, as it was in the nineteenth century. Use of psychedelic or hallucinogenic drugs is disease, somehow allied to madness, and crime.

The attitudes of the self-designated cultural opposition are subtler and more ambiguous and varied (and often more sensible), but one pattern might be as follows. The use of opiates is probably disease but should not be a crime; use of cannabis and cocaine is primarily fun; use of psychedelics lies somewhere between fun and religion; the prescription sedatives and stimulants are mainly medicine but occasionally fun or disease. The status of alcohol is indeterminate; for example, when the advertisements for *High Times* flatter its readers as "multiple drug abusers," you can be sure that alcohol is not included in the honorific category "drug," yet alcohol does seem to be fun sometimes. Certain drugs are commonly placed by both culture and counterculture in a category on the border between medicine and fun that might be called therapeutic in a loose sense: they start the user going in the morning, or keep him going during the day, or ward off insomnia at night. Coffee, many prescription drugs, and to some extent tobacco are in this category.

We have some interesting evidence on the place of cocaine in this scheme of social definitions. Two of the men we interviewed, both of them black, had used a considerable amount of cocaine but had never tried amphetamines. Since amphetamines were unlikely to have been more inaccessible than cocaine even in the black street culture, we were curious about the reasons. One of them told us that "Most blacks don't like pills" and added that he had never used amphetamines or barbiturates because he was "scared of it. Didn't like to take medicine." He had accepted the official definitions of amphetamine as medicine and cocaine as forbidden fun but modified their intention. For him cocaine was "safe" and "clean" because it was not synthetic and did not come in pill form, hedged about with a medical ritual that implied danger. That cocaine was an illicit drug was a minor matter; he respected and feared the medical profession, but he knew better than to respect the law (although he may have feared it). Some related evidence on the social status of cocaine comes from a 1974 study of drug users at a navy drug rehabilitation facility. The motives they gave for using cocaine and marihuana were heightened sexual pleasure and having a closer relationship with someone. They used amphetamines, on the contrary, mostly to relieve depression, to improve the quality of their speech, or to clarify their thoughts. Barbiturates and opiates were mainly for "getting through the day" or relieving anxiety.[7] In other words, cocaine and marihuana were (loosely) fun, while the other drugs were (again, loosely) therapy. The part played in this distinction by established social definitions and, even more important, available sources, should be as obvious as the part played by these factors in the distinction made by our interview subject between cocaine and "pills." If one obtains amphetamines or barbiturates for the first time in a therapeutic context, e.g., on prescription from a physician, one may continue to associate them with therapy even when using them illicitly. As for cocaine, it is not only no longer officially a therapeutic drug but also very expensive. At $80 a gram, few people can afford to use this short-acting chemical just to keep them going, and it has to be saved for special situations.

Another important cultural category we have not considered is vice. The use of this word in high-level discourse about drugs is rare today; quasi-medical and pseudopsychiatric terminology is more popular. Only a few marginal social groups like Mormons, Black Muslims, and fundamentalist Protestants insist that psychoactive drug use is always a vice. Szasz has suggested that the concept be reintroduced into the debate about drugs in an intellectually purified form, free of its confused popular association with questions of so-called decadence and sexual activity that are of no moral moment. He believes that we should avoid psychiatric terminology and instead examine with the help of broad ethical principles whether a particular person's drug habit is in fact good (a virtue) or bad (a vice). Whatever the merits of this as an ethical theory, it is irrelevant to a sociological consideration of vice in the popular sense and its connection with drugs. For even moderately sophisticated people, the word has a somewhat ludicrous Gay Nineties sound and is associated with police vice squads and a prurient interest in other people's sexual behavior that is felt to be undignified. Nevertheless, the idea that some use of some drugs or any use of other drugs is a vice lies behind most popular attitudes on the matter and is also a secret motive of much of the more sophisticated official discourse. Often the words are about drug abuse or drug dependence and the unexpressed thoughts are about vice in this socially and morally trivial sense.

There are two varieties of drug vice: the exotic, fascinating, tempting, and debasing kind that goes with illicit drugs and has something to do with desired but disapproved sexual activity; and the homey, domesticated kind connected with legal drugs, which is regarded with less fascinated interest and more tolerance. The sexual effects of alcohol, for example, are an object of amusement or annoyance; the sexual effects of marihuana as imagined by people who do not use it are an object of secret envy and expressed horror. In the same way, the hallucinations of alcoholism ("pink elephants and snakes") are almost a joke, while "reefer madness" is (or was) a cause for terror. Even within the subdivision of mild or domesticated vice, some revealing distinctions are made. Coffee, for example, although it is a therapeutic substance used to get through the day, is also a vice (in this case we have not made a virtue of necessity); therefore it is not permitted to children. (The situation is very different in Latin America.) Coca-Cola, on the other hand, even though it contains the same drug—caffeine—and also an enormous amount of the refined sugar that is possibly even worse for children's health than most psychoactive drugs they are likely to use, is not subject to this restriction. Obviously the health justifications given for reserving coffee for adults are spurious. Similar spurious justifications were once given for prohibiting tobacco to women while permitting it to men. In effect, there is an intimate relationship between vice in the popular sense and privilege. A vice is something that one "gets away with" because of a certain social status, high, low, or merely eccentric. Recreational use of cocaine, in particular, is a privilege of this kind, a vice practiced only by those who have a certain amount of money or keep a certain kind of company. In this respect it is similar to sexual license. Vice as a cultural category may be said to lie on the border between fun and crime. The same uneasy passages between amused tolerance (for mild vice or fun) and envious hostility (toward exotic vice or crime) that occur in attitudes toward sexual activity also occur in attitudes toward psychoactive drugs. The peculiarity of drug attitudes is that the difference between mild and severe vice depends less on any observable effects of the substance ingested than on its label and legal status.

The complexity of the relationships between sexuality and the condition of the central nervous system is mirrored, as we have said, in the connection between psychoactive drugs and sexual passion. Both the intense feeling and the interpretive ambiguities in attitudes toward drug use are like those of attitudes toward sexual love. The interpenetration of psychoactive drug use and sexuality goes beyond banal notions about aphrodisiacs and even the popular equation: illicit drug = illicit sexual pleasure. It is reflected in psychoanalytical commentary like vom Scheidt's on Freud's use of cocaine. It appears in symbolic interpretations of love and passion that involve much more than a concern with the sexual act as such. Sexual passion, like drugs, may be conceived as therapy, fun, vice, crime, disease, or, for a few, religion. Passion is often compared with intoxication, and many cultures have the myth of a love philtre, probably based on experiences with alcohol and other drugs. Intoxication with love, like intoxication with drugs, is both desired and feared, both a cure for all that ails the addict and a disease that can utterly destroy him, both panacea and panapathogen (to use Szasz's terms). All this is old and even archetypal. What is unique with us is the shifting uncertainty and controversy that place in doubt all accepted interpretations of both sexuality and psychoactive drugs.

Our social categories for psychoactive drug use and our dilemmas and internal conflicts about how to distribute various drugs among them are not universal characteristics of the human mind or of certain chemicals but products of a specific historical situation. To understand this situation, some anthropological and historical background is necessary. The conceptual partitions our society has carefully erected do not exist in preindustrial cultures. In particular, the distinctions between magic, religion, and medicine have not always been so clear as we make them, or at least profess to make them. The words *health* and *holiness* have a common Germanic root meaning "whole." The process of medical diagnosis and prognosis has always had something occult resembling divination about it, and disease has usually been considered an instrument of gods or of evil spirits, sometimes independent and sometimes called forth by the victim's enemies or his own moral delinquency or ritual transgressions. Thaumaturgy or wonder working includes alchemy, demoniacal tricks, sympathetic magic used in healing, and the complementary application of drugs and incantations. Shaman, witch doctor, sorcerer, and medicine man are terms for social roles that often overlap. The religious-medical or magical-medical ceremony has the function of restoring the harmony of the soul and the spirit world, repairing the broken whole defined as holiness or health.

An important feature of this primitive way of thinking in terms of spirit powers and their harmony is that it emphasizes the moral ambiguity of all the powers involved in illness and its treatment. Like other magical and religious properties, the power of drugs can be used for both good and evil, to restore the balance or upset it. No room is left for the idea of drugs with complex psychological effects that are either intrinsically good or intrinsically bad. This ambiguity between medicine and poison is preserved in the modern Greek word *pharmaki* and in the old-fashioned English "potion," as well as in our bewilderingly various uses of the term *drug*. The genie in the bottle, the dope fiend, Demon Rum, and the concoction in the smoking retort that turns Dr. Jekyll into Mr. Hyde are residual variations of the magical or religious conception of psychoactive drugs. These may now seem quaint or comical ideas, but more important unacknowledged connections between drugs and magic also exist in the public mind. Szasz likes to point out the resemblance between what we call drug abuse and the forms of illicit healing once denounced as witchcraft. Psychoactive drugs, like gods and demons, have powers that are both terrible and wonderful; like the witches' brews that sometimes contained them, they can be used to heal or destroy; to society they may be panaceas or panapathogens, their users either saviors or scapegoats. Witches were said to manipulate natural forces just as psychoactive drugs are now considered to exert their magical powers through "natural" neurophysiological mechanisms. For Szasz, what we define as drug abuse is essentially a rejected religious, magical, or healing rite. The spirit of the *Malleus Maleficarum* survives in narcotics laws. We will examine this suggestion that old magical and religious attitudes toward drugs continue in disguised form to influence twentieth-century practices when we discuss further the modern medical profession and its rivals for control over the social interpretation of psychoactive drug use.

But first we must consider how this interpretation changed in the nineteenth and early twentieth centuries. Magic and religion were no longer an issue, even in the early nineteenth century, in Europe and the United States. Healers no longer attributed illness to spirits or consciously identified the power of drugs as magical. The scientific revolution of the seventeenth century and the enormous prestige of Newtonian physics in the Enlightenment had convinced physicians that most diseases had physical and chemical causes. But the idea of a medical science created in the image of physics remained only a hope. Even a century after Voltaire's time, his description of physicians as men who poured drugs of which they knew nothing into patients of whom they knew less remained largely accurate. This uncertain situation, together with the growth of manufacturing, capitalist entrepreneurship, and the spirit of liberal individualism, made the nineteenth century a great age of self-medication and competing medical authorities. The proprietary drug industry that in some ways best expresses the state of medical science and the conditions of medical practice at this stage of history had its greatest flowering in the late-nineteenth-century United States.

The story of the patent medicine men, which is so intimately bound up with the history of drugs like alcohol, opium, and cocaine, has often been told but rarely examined with a seriousness appropriate to the subject. We are usually presented with a kind of comedy-melodrama in which ridiculous or villainous quacks are routed at the end by the forces of honesty, truth, organized medicine, progress, and the federal criminal law. Even the most intelligent histories of the patent medicine era, like James Harvey Young's *The Toadstool Millionaires*, which takes some account of the complexity of the actual social situation, tend to rely for their theme and general framework on an alternation of amusing or horrifying anecdotes with praise for the triumph of modern legal and medical regulation of drugs. If we consider the evidence in a book like Young's in the context of historical variations in attitudes toward drugs and the present conflict over their social definition and social control, its moral seems less obvious and less simple.

An analogy commonly used by physicians at the time to discourage the public's interest in entrepreneurial medicine was: Would you trust the repair of your watch to a blacksmith? [8] This is revealing in two ways. It expresses the physician's conception of himself as a repairman of the human body working on mechanistic principles, and it embarrassingly emphasizes his inadequacy in that role. The public knew that the most intelligent and honest physician did not understand remotely as much about the human body as a watchmaker understood about a watch or Newton and Lagrange about the solar system. For them it was a choice between "blacksmiths" and worse. The New Hampshire farmer Samuel Thomson, with his theory of cold as the cause of all disease and his steambaths, enemas, and sweat-producing herbs, was as plausible and probably as clinically effective as the physician Benjamin Rush, with his theory of hypertension as the cause of all disease. If physicians were offering bleeding, emetics, and purges as the therapeutic approach to most illnesses, it is hardly surprising that the public turned to unlicensed healers and untried drugs instead. Eventually physicians like Oliver Wendell Holmes, Sr. (who called the patent medicine men "toadstool millionaires") adopted the French doctrine of therapeutic nihilism, based on the plausible assumption that most current medical practices and materials were useless and it was best to stand back and let nature effect the cure. Holmes may have been right, but the suffering public wanted something more. The powerful personalities and publicity campaigns of the proprietary drug men generated a placebo response; creating conviction, they often gave relief.

The relationship between proprietary drugs and orthodox medicine was complex. Many of the preparations described in pharmacopoeias also appeared—often unlabeled or mislabeled—in patent medicines. There was borrowing in both directions, and both sides searched the same sources in botany and folk medicine. In 1901 it was estimated that 90 percent of the doctors in the United States were prescribing proprietary remedies; Mariani was not the only patent medicine man who won the interest and respect of physicians.[9] Pharmacists like Pemberton, the inventor of Coca-Cola, also concocted proprietary drugs. In other words, the line between legitimate or respectable medicine and the eccentric, profiteering, or lunatic fringe was not at all clear. In spite of the great advances in some fields in the late nineteenth century, neither the state of medical science nor the organization of the medical profession was yet such as to produce a great deal of confidence in the difference between reputable and disreputable or legitimate and illegitimate medicine.

The position of drugs that act on the central nervous system was a peculiar one. They were not specific cures for specific diseases, no one knew the mechanism of their action in the body, and yet they provided relief from suffering in the most varied situations. As we have pointed out, they were the classic panaceas. Using them, one could practice the same kind of medicine as the proprietary entrepreneurs without any need of a gift for publicity, persuasion, or business organization. These drugs, by affecting the mind, created their own persuasion and conviction. Opium, alcohol, or cocaine, like faith in some pharmacologically inactive nostrum, actually did make the pain matter less while nature took its course, often toward a restoration of health. As the cure-alls that really worked, psychoactive drugs were an important part of the armamentarium of both orthodox and proprietary medicine. Even Holmes with his therapeutic nihilism made an exception for opium (and anesthetics). In 1915, the year the Harrison Act went into effect, an article in the *Journal of the American Medical Association* declared opium to be *the* indispensable drug in the pharmacopoeia. A list of the most widely used drugs in 1885 shows opium and its alkaloid morphine (for pain) fourth and alcohol (as a sedative and anticonvulsant) sixth (first was iron chloride for anemia, second, quinine for malaria, third, ether for anesthesia, and fifth, sodium bicarbonate for indigestion). By 1910 the list had hardly changed: morphine was fourth and alcohol fifth. In 1820 the first *U.S. Pharmacopoeia* listed nine varieties of wine, including wine of opium (laudanum) and wine of tobacco; by 1905 it contained twelve varieties, including wine of opium and wine of coca. Hospital pharmacies stocked large amounts of wine, and it was prescribed as an appetite stimulant, diuretic, and sedative, as well as a treatment for psychosomatic illnesses like neurasthenia. Sir William Osler approved its use in "enteric and pneumonic fevers." [10] Conditions had not changed so much since the time when the Greek physician Galen recommended laudanum as a universal remedy. Since so little was known about the causes of specific diseases, and since any disease would produce certain kinds of discomfort that these substances could relieve, they remained indispensable.

The psychoactive drugs, like most strong medicine, could also be powerfully poisonous. This had been recognized for a long time. What was new in the late nineteenth century, and fundamental to the campaign against proprietary medicines, was an increasing inability to tolerate this ambiguity, a demand for conceptual (and legal) restrictions, distinctions, and compartmentalizations. Even in 1915 the consciousness that opium, the indispensable medicine, and opium, the horrible poison, creator of dope fiends, were the same substance was beginning to disappear. Today relatively few people outside the health professions know that the feared and hated opium derivatives morphine and heroin are still the most effective treatment for severe pain available to physicians (although heroin is not used in the United States because of the hysterical reaction against it). The isolation of the opium alkaloids and the invention of the hypodermic syringe greatly increased both the powers and the dangers of the drug and enormously magnified the problem of addiction. But it was not merely technological change and the recognition of a genuine disease problem that caused this transformation of attitudes. Nor was racial prejudice against the Chinese decisive, any more than racial prejudice against blacks was the main reason for the decline of cocaine. There was a new hostility to all psychoactive drugs, including alcohol, because of their indeterminate and apparently uncontrollable powers—powers more reliable than those of the quack nostrums but no less mysterious and potentially monstrous.

What happened to alcohol is especially interesting. It was not a matter of new technology or the discovery of new consequences of chronic abuse. Few people took to injecting alcohol with a hypodermic syringe, and alcoholism was familiar enough. The brigade assigned to alcohol in the war against psychoactive drugs was known as the Prohibition movement, and it gained strength throughout the late nineteenth and early twentieth centuries until, a few years after the Harrison Act, it achieved passage of the Eighteenth Amendment, followed by the Volstead Act. One of the lasting effects of this movement was that alcohol lost its status as a medicine. The Pharmacopoeial Convention of 1916, under prohibitionist pressure, deleted all wines, whiskeys, and brandies from the *U.S. Pharmacopoeia*.[11] Since then alcohol preparations have occasionally been listed, but hospital pharmacies no longer stock wine or whiskey and physicians no longer ordinarily consider them to be therapeutic drugs. It is true that the prescription of alcohol as a medicine was permitted by the Volstead Act, and during the 1920s physicians dispensed it in quantities never equaled before or since. But by that time both physicians and patients knew that they were evading the law. The old ambiguity, perfectly genuine, in the status of proprietary drugs like Hostetter's Bitters no longer existed. Alcohol was firmly established as fun, vice, crime, disease—anything but medicine.

The first federal legislation against proprietary medicines, the Pure Food and Drug Act, represented a compromise between nineteenth-century liberal attitudes and the growing reaction against psychoactive drugs. One of the inspirations for this legislation was a series of articles by Samuel Hopkins Adams entitled "The Great American Fraud," published in *Collier's* magazine in 1905. After examining some of the more eccentric nostrums and the products that contained alcohol as their main active ingredient, he devoted an article to what he called "The Subtle Poisons." He believed these to be "the most dangerous of all quack medicines," because they could deceive even highly intelligent people. They caused narcotic addiction; they made young men into criminals and young women into harlots. The most dangerous of these poisons were catarrh powders that contained cocaine and soothing syrups that contained opium.[12] Influenced by writings like this, the Pure Food and Drug Act forbade interstate shipment of food and soda water containing opium or cocaine. It also required that manufacturers state the presence and amount of certain substances—alcohol, opiates, chloral hydrate, acetanilide, cocaine, and others—on the labels of their proprietary remedies. If they denied that their product contained opium—already a favorite promotional device at the time—the denial must be truthful. This was a truth-in-packaging law, a product of the liberal era; it did not aim to eliminate free self-medication but only to make it safer. The Proprietary Manufacturers' Association did not even regard it as a serious defeat. As Young somewhat tendentiously puts it, "Time was to reveal serious shortcomings in the 1906 law and to challenge optimism about the common man's capacity" to decide how and when to drug himself. This is hardly surprising. Adams was suggesting not that cocaine and opiates be labeled properly but that they should not be sold over the counter at all. It was not long before far more drastic methods for organized repression of opiates and cocaine, and then of alcohol, came to seem necessary.

The Pure Food and Drug Act was directed mainly against fraud; the Harrison and Volstead Acts were directed against particular substances. But the difference is not so great as it seems. The progressive repudiation of the common man's right to make choices about these substances was associated with a half-conscious conviction that *any* curative use of them, outside a few restricted contexts entirely dominated by medical professionals, was ipso facto fraudulent. Correct descriptive labels were not enough, because even intelligent people became "possessed" by these drugs, which had intrinsic powers of deception. The ordinary use of alcohol, opiates, and cocaine was no longer to be considered medicine; it was fun, vice, crime, and disease, in the mixtures and proportions we have discussed. No longer was taking opium to relax or cocaine to feel vigorous a cure. Although drugstores still had soda fountains, they were no longer thought of as medicinal. Pleasure and health were not to be conflated in the manner of the Coca-Cola advertisements of the 1890s—at least, where drugs were concerned, not so openly. (The restorative power of recreation is still recognized and health spas are still patronized.) The nineteenth-century ambiguity between health and pleasure began to seem dangerous, as the primitive ambiguity between health and holiness had long seemed absurd. Today "Dr. Feelgood" has become a term of opprobrium. Physicians dispense amphetamines, barbiturates, and tranquilizers for the same reasons they once prescribed cocaine, opiates, and alcohol; but laymen are not permitted to prescribe these drugs for themselves, and the medical ritual that surrounds their legitimate use is an attempt to discourage any confusion of their effects with the slightly disreputable "pleasure."

Brill, using cocaine as an example, describes the process by which the abuse potential of a psychoactive drug comes to be recognized: a new agent is produced by scientific advance; it is tested in medical settings that do not reveal its capacity to cause dependence; then it appears on the black market as a pleasure drug, and physicians recognize its abuse potential. He points out that all objections raised against restrictions on later drugs were first raised with respect to cocaine: only abuse, not normal use, is harmful; normal people do not become addicted; there is no physical craving; it is better than alcohol (no hangover) and may be a cure for alcoholism; individual variations in reaction are great, so unfavorable experience is not typical; unfavorable acute reactions are caused only by toxic overdoses. (It is not clear whether Brill regards all these objections, some of which seem sensible, as nugatory.) Szasz sardonically describes the same process as the social transformation of panaceas into panapathogens. His account is as follows: first the authorities, discovering that the public wants the substance and will pay for it, tax it for revenue; then it is defined as a medicine, legitimate only for treatment of illness, and it is legally dispensed only on prescription by physicians; this creates a black market and "overprescription" abuses, and there is a demand for more stringent controls; finally, research is said to reveal that the drug has no legitimate therapeutic uses at all and it is banned. With cocaine and heroin this process has gone to completion; with other drugs it has been only partially consummated.[13]

Despite differences of tone and emphasis, Brill and Szasz are telling the same story. Government and organized medicine take control over the manufacture and distribution of a dangerous substance that the public enjoys by defining it as medicinal as opposed to pleasurable and enforcing the distinction legally. If a large enough part of the public refuses to countenance the distinction, this confusion between medicine and pleasure is defined as drug abuse and the substance may be banned entirely to prevent it. For Brill, speaking in the name of protection of the innocent, this development is a good one; for Szasz, speaking in the name of individual liberty, it is bad. We are not concerned here with the difficult question of who is right and to what extent, but with clarifying the nature of the process involved: the creation and enforcement of certain conceptual distinctions.

What the accounts by Brill and Szasz lack is historical specificity. Self-medication with opiates and cocaine was shunted into the categories of vice, disease, and crime, and self-medication with alcohol into the categories of fun, vice, disease, and crime, at a particular stage in the development of the medical profession and of (capitalist) industrial society. Although similar events occurred in Europe, we will concentrate on the United States. The first two decades of the twentieth century are generally known as the Progressive era in American politics. Revisionist historians like Gabriel Kolko (in *The Triumph of Conservatism*) prefer to depict it as the age when large-scale capital consolidated its power against labor and populist threats by means of rationalizing reforms and greater integration with government. Whether one calls the era progressive or conservative—it was probably both, if one retains a certain irony about these words—its main political achievement was to create some order in the chaos of late-nineteenth-century liberal capitalism, which might otherwise have permitted mass misery or possibly revolution. The greatest legislative monument of the age was the Federal Reserve Act, which reorganized the banking system; the Pure Food and Drug, Harrison, and Volstead Acts were also characteristic Progressive legislation.

The attack on proprietary nostrums and indiscriminate use of psychoactive drugs, while it put a stop to many dangerous and fraudulent practices, also served to consolidate the power of the organized medical and pharmaceutical professions and the larger drug companies, in cooperation with the federal government. As Young puts it, there was a "conspiracy," including physicians, pharmacists, chemists, muckraking journalists, and government officials, against the manufacturers of proprietaries.[14] Narcotics laws aided the institutional development of the modern professions of medicine and pharmacy by helping to define their areas of competence and exclude surplus and unlicensed practitioners. The impulse to clean up society and reduce disorder worked against such practices as free self-medication and chaotic small-scale entrepreneurial competition. New standards for medical and pharmaceutical practice were professional hygiene, and insistence on clear and legally enforced categories for psychoactive drugs was regarded as intellectual hygiene.

But the situation involved more than a tendency toward consolidation, large-scale organization, and formal regulation in business, government, and the professions. The great advances in the art and science of medicine were at least as important. Florence Nightingale, whose work in military hospitals during the Crimean War is a well-known milestone in the history of preventive medicine, once wrote, "There are no specific diseases; there are specific disease conditions." [15] The condition of medical science being what it was, she could not easily be contradicted. But then came the rise of synthetic chemistry, experimental physiology, and, above all, bacteriology. The promise of a materialist medicine based on the recognition of specific disease agents for specific diseases seemed about to be fulfilled. The work of men like Pasteur and Koch became a model. Physicians gained a genuine self-respect and a new esprit de corps that went with their growing social power and professional organization. At the same time, the cleanliness necessary to prevent the spread of infectious disease became identified metaphorically with purification to eliminate loose practices and unlicensed practitioners. (Bacteriology is probably the source of the epidemiological model of psychoactive drug abuse, which by an inappropriate metaphor confuses persuasion with infection.) Psychoactive drugs were especially suspect. It required no specialized training or diagnostic ability to dose a patient with alcohol, opium, or cocaine for whatever he complained of and produce a satisfied customer who would come back for more. This was a challenge to their professional standing that physicians would no longer tolerate. It was not that they always had something better to offer; in fact, for many complaints physicians still have no better recourse than psychoactive drugs and emotional support—the medical equivalent of the Rolling Stones' "coke and sympathy" being tranquilizers or antidepressants and sympathy. But physicians now felt justified in restricting use of the more powerful drugs and placing them under their own control, and the observed consequences of misuse of these drugs gave them apparently good reasons for doing so. One does not have to be opposed to this development, like Szasz, to recognize that it could occur only under certain historical conditions, and that something more than a new dawn of probity and prudence was involved.

To use a psychoanalytical metaphor, public and official attitudes toward opiates, cocaine, and, in part, alcohol are "fixated" at this historical stage. The repressive legislation imposed during the Progressive era and, at least for opiates and cocaine, fortified since then, has formed or deformed their image in the popular mind. These drugs were being used indiscriminately as medicine and for pleasure at a time when changes in social structure and medical knowledge decreed that medicine and pleasure were to be divorced and that much of what had been medicine and pleasure was now to be regarded as disease, serious rather than venial vice, and crime. Later psychoactive drugs, often wholly synthetic and developed under the aegis of a much more highly disciplined, organized, and respected medical profession and drug industry, have not suffered even at worst quite the same fate. Opiates and cocaine were treated more harshly than alcohol (cannabis, when it began to come into recreational use in this country, was given the same treatment) mainly because they had been introduced comparatively recently and had acquired no fixed popular image. For all practical purposes they were products of the late-nineteenth-century transitional period. Their identification with threatening lower-class outsider groups was a by-product of this indeterminacy and the associated fear of the unknown. Alcohol, on the other hand, was too familiar to be branded with any equivalent of the narcotic stigma and too closely associated with their own kind of innocent fun in too many respectable people's minds to become a permanently plausible drug menace. Nevertheless, as we pointed out, the Prohibition movement and associated developments in medicine did succeed in effacing one part of alcohol's image—its status as a therapeutic drug.

The alliance of the organized medical and pharmaceutical professions with government and police established during the Progressive era produced a rationalized system of control over psychoactive drugs that has endured until the present. What we have called the disease-crime model of drug abuse corresponds to this institutional arrangement. In spite of its intellectual and moral inadequacies, it serves a protective and conservative function required by the organizations that support it. Arguments about when psychoactive drug use is a crime and when it is a disease are essentially jurisdictional disputes between the medical and police professions, and they are usually settled without any need of principled conflict. The Drug Enforcement Administration and its bureaucratic predecessors have always been able to adjust their views to those of the American Medical Association in a process of mutual accommodation. "Modern medical science, of course, is an indispensable bulwark to effective government regulation," writes Young.[16] Szasz, as usual less than enthusiastic about recent progress, believes that in dealing with psychoactive drugs the medical profession has adopted inappropriate political and penological practices. In pointing out the social origins and inadequacy of the disease-crime model, we do not mean to imply (as Szasz sometimes seems to do) that organized medicine and the police are always wrong in particular cases, but only to emphasize the correspondence between fixed conceptual categories and well-organized established institutional divisions. Sociologically, it does not matter whether the use of cocaine has been made illegal because it is a vice and a disease or is considered to be a vice and a disease because it has been made illegal. The point is that, for better or for worse, the system of concepts and the institutional system provide mutual support.

So, both increased numbers and formality of institutional distinctions and changes in the organization and self-conception of occupational groups have determined contemporary attitudes toward psychoactive drugs. The aspects of this process most relevant to the drug problem are professionalization and specialization. In twentieth-century societies, more and more occupations have come to claim the title of profession. Harold Rosenberg, in a brilliant essay entitled "Everyman a Professional," describes this as "the steady transformation of the whole populace into professionals and semi-professionals." As he puts it, "The professional mass keeps expanding, and as it expands it divides. Old professions break up and each fragment becomes the center of a new constellation. . . . At the same time the trades keep propelling themselves upwards into the professions. . .the kitchen manager becomes a dietician and stockbreeders and policemen conduct 'prestige' campaigns to convince society of the learned nature of their pursuits." "A form of work establishes itself as a profession not only through the complication of its technique," he writes "but through *self-consciousness with regard to this technique*" (our italics).[17]

Medicine, law, and theology, it may be said, are the oldest professions, with the most highly developed technique-consciousness. Although they once overlapped both conceptually and institutionally, the functions of these professions are separated in modern industrial societies as perhaps never before. This situation corresponds nicely to the distinction between drug use as medicine or disease, drug use as vice or crime, and drug use as divine or demoniacal. However, religion, at least organized religion, aside from a few marginal sects, has abandoned the field of drug controls. It may attempt to justify medical and police practices or to serve as a loyal opposition from the left (Kant's term for the function of a university faculty of philosophy), but it no longer imposes its own categories of the divine and the demonic. Of the remaining two, law, expanded to include government administration and police, is probably less interesting, important, and highly respected (although not less powerful) than medicine. As the medical and allied professions expand, divide, and send out new branches they tend to incorporate more and more social functions. This has been called "moral entrepreneurship." Szasz calls it "medical imperialism" or "missionary medicine." In Rosenberg's words, "A profession becomes really top-rank when it can offer its system of technical redefinition as the key to the human situation, that is, as philosophy." [18] Medicine may have come closer than any other profession to achieving this. More kinds of phenomena than ever before are being defined as ill-ness. But Szasz's fears of a medical technocracy are exaggerated: medi-cine has made its way by alliances as much as by conquest. Pharmacy and psychiatry, the divisions assigned to psychoactive drugs, operate in relation to the bordering armies of the criminal law by a delicate alterna-tion of skirmishes and treaty conferences. The license of medical practi-tioners to define illness as a social role remains limited; their right to in-terpret the meaning of psychoactive drug use is being contested from without and, perhaps more important, questioned from within the profession.

Partly because of a general loss of respect for established institutions and partly as a penalty for professional overreaching, organized medical control over psychoactive drugs is now being openly and articulately challenged as never before in the twentieth century. This reaction has taken several forms. One, best represented by Szasz, is a nostalgia for a principled nineteenth-century atomic individualism in which each per-son is free to fulfill his desires as he sees fit until they impede the fulfill-ment of others' desires. This view, although intellectually interesting, is socially ineffectual because of its purely negative content. It defines no positive social function or role for drugs and therefore will probably gain no political constituency. A much more popular attitude is advocated by practitioners of fun like the editors of *High Times*. On the principle that what makes you feel good is good for you, they would like to blur the laboriously drawn distinction between drugs as pleasure and drugs as medicine. They might agree with the nineteenth-century medical journal that said of cocaine, "A harmless remedy for the blues is imperial." Glee-fully seizing on sexual implications of drug use that make anxious con-ventional folk fearful or disapproving, they promote drugs as aphrodis-iacs. This system of attitudes is intellectually amorphous. It indiscriminately attributes medicinal virtues, moral improvement, re-ligious insights, and artistic creativity to use of the chosen illicit drug, and somehow makes them all seem to be varieties of pleasure or fun.

In the search for a new shared attitude to oppose that of the medical profession, drug users may also have recourse to religion. It has often been observed that drug sharing, whether passing the peace pipe or pass-ing the joint, coffee break or cocktail party, is a ritual expression of com-munion and solidarity. When the users of a drug are persecuted or threat-ened, they may want to fortify their solidarity with an ideology. The definition of drug use as disease is an attempt to destroy this solidarity, since the sick do not form a community; each is an individual coping with his own misery. (One of the reasons it makes more sense to call schizophrenia a disease than to define illicit drug use that way is that schizophrenics do not collectively deny this interpretation and offer their own—although nonschizophrenics have been willing to do so in their name.) Religion is a shared system of justification that is stronger than mere fun and can overcome the implication of disease. The drug field abandoned by orthodox religion is appropriated by unorthodox sects, eclectic or syncretic, which challenge the hegemony of established medical practices. Aleister Crowley, a cult leader of former times, spoke of using cocaine and heroin with a purposefully religious attitude. The "LSD priest" (Timothy Leary's term) who leads a "drug cult" is in effect a rival of the physician in imparting and applying expertise on drugs. The old profession of priesthood invades the territory claimed by modern medical professionals; to describe the associated attitudes as antiprofessional is not quite accurate.

This tendency, limited as it has been, toward an openly religious conception of psychoactive drug use implies a return to the practices of primitive cultures that do not separate medicine from religion. The diffuseness of the religious notions of what is often called the drug culture corresponds to the primitive conflation of categories; in fact, these ideas are often accompanied by talk about the virtues of primitive culture and community. Since a self-conscious and ideological primitivism is almost a contradiction in terms, this talk has not amounted to much in practice. But by suggesting that we apply the standards of preindustrial cultures to psychoactive drug use, the opposition to established practices has raised two interesting issues: the dangers of technological advance and the protective function of ritual.

Technical advances, both scientific and organizational, have clearly increased the dangers of drug abuse. The isolation of cocaine was one of those ambiguous achievements of nineteenth-century science that now sometimes appear to have been Pyrrhic victories over nature. The opiate and cocaine problems in their present acute form arise from advances in chemistry and medicine like the production of pure drugs and the invention of the hypodermic syringe. Even without intravenous injection of pure nicotine, large-scale manufacturing has made a drug like tobacco more freely available and therefore more dangerous. This kind of problem is certainly not confined to drugs. Andrew Weil has compared the relation between heroin and opium to the relation between refined white sugar and ordinary vegetable carbohydrate sources. We ourselves, in discussing Brill's conception of psychoactive drugs as a shortcut, made a comparison with the automobile. We can add here, following Weil, that some people, taking a long view, now believe *all* automobile use to be "reckless." To refine the analogy, we might say that coca is like a donkey and cocaine like a supersonic jet plane, with all the dangers to the human and terrestrial ecology that such technological miracles imply. René Dubos remarks that drug-caused disease is a kind of parody of illness produced by technology. Alexander Mitscherlich writes: "Bureaucratic-institutional predominance is the jungle of [a] second nature of technical nihilism out of which pestilence and destructive frenzy can break as formerly hunger and plague came forth from magically experienced first nature." [19] Psychoactive drugs in a concentrated form that can be taken in pills or injected may be a cause as well as a partial remedy for this kind of technological illness.

But the choice between the techniques of Neolithic culture and those of modern industrial society is not simple. As a contrast with Weil's preference for opium over its concentrated derivatives, consider this remark about the early nineteenth-century chemist Sertuerner: "One of the greatest benefactors to humanity, the man who turned treacherous opium into pure and reliable morphine, who started doctors on the way to use pure drugs." [20] This opinion obviously has considerable merit. In any case, it is not more natural to smoke opium than to inject morphine, any more than it is more natural to ride a donkey than to drive a car. The most we can say is that the products and techniques of modern industry have dangers commensurate with their powers. The story of Dr. Jekyll and Mr. Hyde has remained part of popular culture because it is a mythical embodiment of the moral ambiguity of modern technology, creator of monsters and miracles. Dr. Jekyll, the modest and upright practitioner of applied science, benefactor of humanity, turns into the monstrous, overbearing, intolerable product of his own technology. It is fitting that the medium of transformation is a drug, because drugs incorporate both the old religious terrors and hopes and the new scientific ones.

The second issue raised by the new opposition to established forms of drug control is the use of quasi-religious and religious ritual, as opposed to medical and criminal restrictions, in preventing drug abuse. In fact, modern medicine is sometimes likened, always sarcastically and with polemical intent—by Szasz, for example—to a state religion, with imposing organizational strength, an intellectually powerful ideology, an ability to create and sustain faith, and the all-important support of civil authority. Doctors in their white vestments are like Inca priests in their golden robes, and exercise a similar monopoly control over the use of cocaine and other drugs. The Drug Enforcement Administration and the American Medical Association are like the Spanish Crown and its intellectual arm, the Inquisition. The occult practices of diagnosis and prognosis have succeeded the occult practices of haruspication and divination. Medicine has divine powers (medical miracles and wonder drugs) and can also call forth and combat demons ("addictive" drugs). Its images of purity are borrowed from religion. It regards unorthodox healing practices as heresy, blasphemy, or pagan superstition to be eliminated by a mixture of official coercion and missionary activity. Its "self-consciousness with regard to technique" justifies sustaining archaic rituals for outmoded or oppressive purposes; for example, the *Journal of the American Medical Association* publishes only those "experimental" studies on marihuana that reach the preordained ritual conclusion that it is evil—"demoniacal," as the Council of Lima described the effects of coca in the sixteenth century. Medicine thus misleadingly presents a conflict of faiths or moral intrepretations of human experience as a series of objective scientific issues. The Council of Lima was more honest.

Although these complaints have some bite, they also have implications that might be turned against the objectors. It could be argued that in modern industrial societies the medical profession has to provide the same kind of ritual or religious context that makes drug technology relatively safe in primitive cultures. Religious control of drug use may be democratic, as with the Kogi, among whom all adult males participate in the ritual; or it may be elitist, like the Inca god-king's monopoly on coca. The open invitation issued in the United States in the 1960s to "turn on, tune in, and drop out" implies a "democratic" conception of quasi-religious drug control; the practices of physicians, on the other hand, are clearly elitist. But this elitism, like that of the Incas, has virtues as well as vices. The medical ritual that restricts the use of most psychoactive drugs serves a protective function even when it cannot be justified by any "objective" scientific principle. Where this ritual is unavailable, as in the case of alcohol, the effect can be disastrous. The priestly role of physicians in prescribing proper occasions for the use of drugs that affect the mind and warning against possession by the demons in these drugs may sometimes seem pretentious and arrogant, and the powers it implies are obviously subject to abuse. But it is doubtful whether anarchic nineteenth-century individualism, a return to Neolithic cultural and technological forms, or the institution of some kind of openly religious restrictive controls would be better, even if they were possible under present social conditions.

Physicians will continue, *faute de mieux*, to exercise a kind of quasi-religious and quasi-political control over some psychoactive drugs, and they are in a better position than ever to examine rationally how and when to use their authority. The medical profession should now be confident enough of its achievements and social status so that it need not fear a loss of respectability because of runaway drugs. And it does not have to treat the complex and indeterminate powers of psychoactive drugs as a challenge to its own domination of territory it has legitimately staked out; the situation is not the same as it was at the turn of the century. Partly because of outside criticism of their former attitudes and practices, many physicians are beginning to recognize this now. The liberalization of marihuana laws, for example, is occurring not because state legislatures have become convinced that cannabis is fun, and not even primarily because of an abstract concern for civil liberties of the kind promoted by Szasz, but because physicians have begun to convince politicians that cannabis is relatively benign and that the laws against it do more damage than the drug itself. Their power of persuasion in this area has possibly never been greater.

To use this power benevolently and undespotically, they must become more reflective about its forms and extent and more willing to challenge established prejudices. This may imply new attitudes about medicine as a profession and some modifications in institutionally established conceptual categories of the kind we have been discussing. As they question the disease-crime model of drug abuse, physicians will have to reconsider the police alliance or accommodation that, as they have now begun to recognize in the case of cannabis, has often been corrupting. To mention one tragicomic irony: the Pure Food and Drug Act required proper labeling of patent remedies containing psychoactive drugs so that the public would know what it was getting; but now the police, often with tacit medical acquiescence, work to *prevent* users of opiates, cocaine, and other illicit drugs from knowing what is in the substances they ingest, by restricting the work of organizations like the PharmChem Laboratory that analyze the contents of street drugs. The law, instead of trying to make self-medication with these drugs as safe as possible, now seems to want to make it as unsafe as possible: self-medication itself (with certain substances) has become the abuse, vice, or crime of overriding importance. Can this be what physicians intended when they first demanded stricter drug controls, or what they intend now? Another question the medical profession will have to face more openly, if it is not to be charged with hypocrisy or neglect, is that of alcohol. Although alcohol is now sometimes rhetorically (and correctly) proclaimed to be the biggest drug problem of all, practically no one acts on this assumption. The alcohol problem has proved uncontainable by the combined powers of law and medicine, but physicians should at least consider what it implies about the disproportionate attention and emotional fervor devoted to controlling other drugs.

Moderate use of alcohol, unlike moderate use of cocaine and other drugs, has fallen into the category of fun rather than vice or crime; the acquiescence of physicians in this invidious distinction can be extenuated by historical circumstances but not justified by any principle internal to medicine.

The somewhat anarchic insurrection that has broken out in the last two decades against existing intellectual and political controls on psychoactive drug use may help lead physicians to new ideas about the limitations and purposes of medicine. Although there is much talk, both favorable and unfavorable, about the "medical model" of various phenomena, no such thing exists in the same sense that there is a quantum-mechanical model of subatomic particle interactions or a Darwinian model of biological evolution. Medicine derives its explanatory methods from miscellaneous sources, and they are often vague. It has to rely on traditions, skills, and individual clinical genius. René Dubos has identified the "doctrine of specific etiology," based on infectious and dietary deficiency diseases, as the source of most of the great triumphs of modern medicine. This doctrine, as we pointed out, was one reason for the expulsion of the old psychoactive panaceas from medicine. As Dubos suggests, it is now recognized to be insufficient. The body usually produces a few general reactions to many different forms of physical and psychological stress or intrusion—the general adaptation syndrome—rather than unique manifestations of the effects of specific etiological agents.[21] Physicians have tacitly recognized this by bringing psychoactive panaceas back into the medical armamentarium in force since the 1930s. For the vaguely defined vegetative and functional problems that still account for a large proportion of visits to physicians—gastric disturbances, back pains, headache, anxiety, depression, fatigue—they often have no clear causal explanations and no better remedies than these drugs.

Partly because of its incompatibility with physicians' conception of themselves as technicians who know exactly what they are doing, this situation causes a great deal of unease. Physicians are accused by laymen and accuse one another of using pills to resolve problems of living that require more complex and difficult adjustments; they fear that people will come to regard a drug habit of one kind or another as the answer to all their miseries (with something like the "soma" of Huxley's *Brave New World* in the back of their minds). These fears are reasonable. On the other hand, some regard drugs as compensations for the imbalances of technological society, neither more nor less legitimate than other ecological technology; and they have a point, too. (Huxley himself, in his last novel, *Island*, introduced a virtuous or utopian counterpart of "soma.") It is not even obvious what restrictions on psychoactive drug use would be desirable in ideal circumstances, much less what restrictions are possible under present social conditions.

Difficult as these problems are, physicians make them even harder by implicitly basing their assumptions about what they are doing on a conception of drug use with its roots in the doctrine of specific etiology that was so successful in establishing the modern profession. The impersonality of an orthodox medicine that treats the human organism as a collection of machine parts may turn people toward unorthodox healing practices if they want health in the old sense of harmony and wholeness. When medicine in this sense is resurrected, it often takes the form of recreation or religion—for example, illicit drugs and faith healing. But often the difference is illusory. What a physician does when he prescribes a tranquilizer like diazepam (Valium) is not fundamentally different from what a layman does when he prescribes a beer, a marihuana cigarette, or a sniff of cocaine for himself. There remains a large area in which medicine and recreation, diseases and problems of living, overlap. The social category that we have loosely called therapy covers some of this area: it includes, for example, most use of coffee in the United States and coca in Peru. Illness is never entirely definable in biochemical terms; the social aspect obscurely recognized by primitives in their attribution of disease to witchcraft remains important. Physicians can be prudent and sparing in their use of drugs without relying on an unworkable conceptual puritanism as a justification. It is arbitrary to pronounce medicinal use of a given drug good and recreational use bad if the *only* basis for the distinction between medicine and pleasure is the institutional authority of the physician himself. If physicians were more conscious of this continuum between recreation and medicine, they might become more tolerant of laymen's self-medication practices, including the use of drugs that are now illegal. They would also be more often respected and heeded when they warn about the dangers of some of these practices.

In making this argument for greater conceptual flexibility, we are not proposing a return to the undifferentiated primitive categories or suggesting that physicians entirely resign their professional authority to practitioners of fun, leaders of drug cults, or individual choice that would be only abstractly free. But they must be more self-questioning about their own practices and more tolerant of others'. The self-consciousness with regard to technique that makes medicine one of the modern professions, and even the genuine triumphs of its techniques, should not cause it to "offer its system of technical redefinition as the key to the human situation." It may be going too far to agree with the assertion of the physician and statesman Rudolf Virchow at the turn of the century that "Medicine is a social science and politics is nothing more than medicine writ large." [22] But some of the political virtues are also medical virtues, at least where psychoactive drugs are concerned: respect for the opinions of those with whom one disagrees and reliance on persuasion rather than coercion. It is impossible to demonstrate the invalidity of the use of a drug that affects the mind in the same way that one demonstrates the falseness of some nineteenth-century theory of infectious disease. The freezing or fixation of public and professional attitudes toward a drug like cocaine that occurred at the turn of the century served a certain historical purpose, but it has also caused dubious coercive practices—at worst, scapegoating and what amounts to cultural oppression, and even at best an overextension of benevolently despotic authority. Cocaine provides one of the best examples on a small scale of how the morally ambiguous properties of a psychoactive drug, under the weight of the institutional and conceptual requirements of a society at a given historical stage, are molded into a public "moral image" that has the typically ideological characteristic of revealing and concealing at the same time. A task of the medical profession, indifferently performed until now for cocaine as well as other drugs, is to make such images reveal more than they conceal so that we can develop public policies without avoidable illusions.

NOTES

1. *The Coca Leaf*

1. Carlos Gutiérrez-Noriega, "El cocaísmo y la alimentación en el Perú," *Anales de la Facultad de Medicina* 31 (1947): 3.

2. Quoted in Carlos Gutiérrez-Noriega and Victor Wolfgang von Hagen, "Coca: the Mainstay of an Arduous Native Life in the Andes," *Economic Botany* 5 (1951): 146.

3. For information on early use of coca, see Victor Manuel Patiño, *Plantas Cultivadas y Animales Domesticadas en América Equinoccial* (Cali: Imprenta Departamental, 1967), pp. 201–233; María Rostowski de Diez Canseco, "Plantaciones prehispánicas de coca en la vertiente del pacífico," *Revista del Museo Nacional* (Lima) 39 (1973): 193–224; Karl-Dietrich Frombach, "Beitrag zur Verbreitung und Auswirkung des Koka-Kauens in Peru," *Zeitschrift für Tropenmedizin und Parasitologie* 18 (1967): 387–396; Néstor Uscátegui Mendoza, "Contribución al estudio de la masticación de las hojas de coca," *Revista Colombiana de Antropología* 3 (1954): 207–289.

4. On coca use and culture under the Incas, see W. Golden Mortimer, *Peru: History of Coca* (New York: J. H. Vail, 1901), passim; Gutiérrez-Noriega, "El cocaísmo y la alimentación," pp. 11–23; Remedio de la Peña Begué, "El uso de la coca entre los Incas," *Revista Española de Antropología Americana* 79 (1972): 277–304.

5. Rostowski, "Plantaciones prehispánicas," p. 203; Manual A. Fuentes, Mémoire sur la Coca du Pérou (Paris, 1886), p. 12; Mortimer, *Peru: History of Coca*, p. 159; José Augustín Morales, *El Oro Verde de las Yungas: Libro de Propaganda Industrial* (La Paz: Imprenta del Instituto Nacional de Readaptación de Inválidos, 1938), p. 28.

6. Gutiérrez-Noriega, "El cocaísmo y la alimentación," pp. 23–90.

7. H. A. Weddell, quoted in Hans W. Maier, *Der Kokainismus* (Leipzig: Georg Thieme, 1926), p. 3; Carlos Gutiérrez-Noriega and Vicente Zapata Ortiz, *Estudios sobre la Coca y la Cocaína en el Perú* (Lima: Ministerio de Educación Pública, 1947), p. 25; Gutiérrez-Noriega, "El cocaísmo y la alimentación," p. 53.

8. Carlos A. Ricketts, *Ensayos de Legislación Pro-indígena* (Arequipa, 1936); Luis Sáenz, *La Coca: Estudio Medico-social de la Gran Toxicomanía Peruana* (Lima, 1938); Gutiérrez-Noriega and Zapata Ortiz, *Estudios*, p. 127; Carlos Monge, "The Need for Studying the Problem of Coca-leaf Chewing," *Bulletin on Narcotics* 4, no. 4 (1952): 13–15.

9. Mario A. Puga, "El Indio y la Coca," *Cuadernos Americanos* 4 (1951): 39–51; "Ten Years of the Coca Monopoly in Peru," *Bulletin on Narcotics* 14, no. 1 (1962): 10; Juan de Onis, "Cocaine a Way of Life for Many in Bolivia," *New York Times*, February 22, 1972; Joel M. Hanna, "Coca Leaf Use in Southern Peru: Some Biosocial Aspects," *American Anthropologist* 76 (1974): 293.

10. United Nations Economic and Social Council, *Report of the Commission of Enquiry on the Coca Leaf*, May 1950, p. 59; Lúis Vásquez Lapeyre, "Labor de la Comisión Peruana para el Estudio del Problema de la Coca" and "Documentados relacionados con la labor efectuada por la Comisión Peruana para el Estudio del Problema de la Coca," *Perú Indígena* 3 (December 1952): 24–130.

11. "Inter-American Consultative Group on Coca-leaf Problems," *Bulletin on Narcotics*, 7, no. 4 (1965): 37–41; Carlos Avalos Jibaja, "Consultative Group on Coca-leaf Problems," *Bulletin on Narcotics* 16, no. 3 (1964): 25, 33, 35; "Twenty Years of Narcotics Control Under the U.N.," *Bulletin on Narcotics* 18, no. 1 (1966): 32–33.

12. "Twenty Years," p. 32; "Report of International Narcotics Control Board on Its Work in 1972," *Bulletin on Narcotics* 25, no. 2 (1973): 51–56.

13. UNESCO *Report*, p. 55; Balthazar Caravedo Carranza and Manuel Almeida Vargas, *Alcoolismo y Toxicomanías*, pt. III (Lima: Ministerio de Salud, 1972), pp. 5–15; Estadística Agraria (Lima: Oficina de Estadística, 1972).

14. UNESCO *Report*, p. 21; Emilio N. Ciuffardi, "Contribución a la química del cocaísmo," *Revista de Farmacología y Medicina Experimental* 2 (1949): 46; "Ten Years," p. 10.

15. P. O. Wolff, "General Considerations on the Problem of Coca-leaf Chewing," *Bulletin on Narcotics* 4, no. 2 (1952): 4; Hanna, "Coca Leaf Use," p. 291.

16. Richard Evans Schultes, "A New Method of Coca Preparation in the Colombian Amazon," Harvard University Botanical Museum Leaflets 9, no. 17, 1957.

17. Sergio Quijada Jara, *La Coca en las Costumbres Indígenas* (Huangayo-Peru, 1950), pp. 10, 19, 23; UNESCO *Report*, p. 55; A. Bühler, "Die Koka bei den Indianern Südamerikas," *Ciba Zeitschrift* 8 Jahrgang, no. 94 (June 1944): 3348.

18. For this and other folk songs, see Quijada Jara, *La Coca*, pp. 54–57.

19. Uscátegui Mendoza, "Contribución al estudio de la masticación de las hojas de coca," pp. 276–279.

20. Ibid., pp. 267–271; see also Nestor Uscátegui Mendoza, "The Present Distribution of Narcotics and Stimulants Amongst the Indian Tribes of Colombia," Harvard University Botanical Museum Leaflets 18, no. 6 (1959): 280–282.

21. Personal communication.

2. *Early History of Cocaine*

1. Quoted in W. Golden Mortimer, *Peru: History of Coca* (New York: J. H. Vail, 1901), p. 151.

2. Ibid., p. 26.

3. Ibid., pp. 167, 168.

4. Ibid., p. 230.

5. For bibliographical references, see ibid. and Hans W. Maier, *Der Kokainismus* (Leipzig: Georg Thieme, 1926).

6. Quoted in Mortimer, *Peru: History of Coca*, p. 172.

7. Paolo Mantegazza, *Sulle virtú ingieniche e medicinali della coca* (Milan: Autosservazione, 1859).

8. Quoted in Maier, *Der Kokainismus*, p. 7.

9. William H. Prescott, *History of the Conquest of Peru* (Philadelphia: J. B. Lippincott, 1847), 1:143.

10. Quoted in Angelo Mariani, *Coca and Its Therapeutic Applications* (New York: J. N. Jaros, 1890), p. 30.

11. Freud, "Über Coca," *Wiener Zentralblatt für die gesamte Therapie* 2 (1884): 289–314, reprinted in *Cocaine Papers*, edited by Robert Byck (New York: Stonehill, 1974), pp. 53, 65, 52.

12. Charles Fauvel, letter to *New York Medical Journal*, quoted in *The Efficacy of Coca Erythroxylon: Notes and Comments by Prominent Physicians*, 2d ed. (New York and Paris: Mariani and Company, 1888), pages unnumbered; Freud, "Über Coca," in Byck, *Cocaine Papers*, p. 73.

13. Quoted in Ernst Joël and Fritz Fränkel, *Der Cocainismus* (Berlin: Springer Verlag,

1924), p. 10; Charles Gazeau, "Recherches expérimentales sur la propriété alimentaire de la Coca," *Comptes Rendus des Séances de l'Académie des Sciences* 72 (1870): 779–801; Sir Robert Christison, "Observations of Cuca, or Coca, the Leaves of Erythroxylon Coca," *British Medical Journal*, April 29, 1876, pp. 527–531; editorial, *British Medical Journal*, March 25, 1876, p. 387; editorial, *British Medical Journal*, April 22, 1876, p. 518; A. Bordier, *Dictionnaire encyclopédique des sciences médicales* (Paris: Masson, 1876), 18: 161–170.

14. *Mariani Wine of Erythroxylon Coca* (New York and Paris: Mariani and Company, 1880); advertisements in the *Therapeutic Gazette* quoted in Anne E. Caldwell, *Origins of Psychopharmacology from CPZ to LSD* (Springfield, Ill.: Charles C Thomas, 1970), p. 15; W. H. Bentley, *Therapeutic Gazette* 1 (1880): 253–254; Bentley quoted in Caldwell, p. 16; editorial, *Therapeutic Gazette* 1 (1880): 172.

15. See Freud, "Über Coca," in Byck, *Cocaine Papers*, pp. 49–73.

16. Theodor Aschenbrandt, "Die physiologische Wirkung und Bedeutung des Cocainum muriaticum auf den menschlichen Organismus," *Deutsche medizinische Wochenschrift* no. 50, December 12, 1883, pp. 730–732; George Ward, "Use of Coca in South America," *Medical Record* 17 (1880): 497.

17. Ernest Jones, *The Life and Work of Sigmund Freud*, 3 vols. (New York: Basic Books, 1953), 1:81.

18. Freud, "Über Coca," in Byck, *Cocaine Papers*, pp. 49–73. Quotations from p. 60.

19. *Coca Erythroxylon and Its Derivatives* (Detroit and New York: Parke Davis and Company, 1885), in Byck, *Cocaine Papers*, p. 144.

20. Aleš Hrdlicka, "Trepanation Among Prehistoric People, Especially in America," *Ciba Symposia* 1 (1933): 170–177; letter in *Medical Research*, November 15, 1884; Julius Claus, review of "Vorläufige Mitteilungen über Cocain," by Dr. Schroff, *Schmidts Jahrbücher der Medizin* 16 (1862): 297–299; Maier, *Der Kokainismus*, p. 47.

21. H. Knapp, "On Cocaine and Its Use in Ophthalmic and General Surgery," *Archives of Ophthalmology* 13 (1884): 402–408. For the story of Koller's discovery, see Hortense Koller Becker, "Carl Koller and Cocaine," *Psychoanalytic Quarterly* 32 (1963).

22. Quoted in Becker, "Carl Koller and Cocaine," p. 322.

23. William A. Hammond, "Coca: Its Preparations and Their Therapeutical Qualities, with Some Remarks on the So-called 'Cocaine Habit,' " *Transactions of the Medical Society of Virginia*, November 1887, p. 225.

24. Ibid., pp. 213, 214, 226.

25. Mortimer, *Peru: History of Coca*, pp. 491–509 and passim.

26. Ibid., p. 383.

27. See Mariani, *Coca and Its Therapeutic Applications*, p. 56 ff.

28. *Mariani's Coca Leaf* 3 (May 1905): 70–71; quoted in Mariani, *The Efficacy of Coca Erythroxylon*, supplement to 5th ed. (New York and Paris: Mariani and Company, 1888); Mariani, *Coca and Its Therapeutic Applications*, p. 68; *Mariani's Coca Leaf* 3 (March 1905): 52–53.

29. Mortimer, *Peru: History of Coca*, pp. 180, 179.

30. Mariani, *Coca and Its Therapeutic Applications*, p. 62.

31. Charles Perry. "The Star-Spangled Powder: Or, Through History with Coke Spoon and Nasal Spray," *Rolling Stone*, no. 117 (August 17, 1972), p. 26.

32. *Current Literature* 48 (1910): 633.

33. For speculation on this, see Myron G. Schultz, "The 'Strange Case' of Robert Louis Stevenson," *Journal of the American Medical Association* 216 (1971): 90–94.

34. See *Coca Erythroxylon and Its Derivatives*, in Byck, *Cocaine Papers*, p. 129.

35. E. J. Kahn, "The Universal Drink," *The New Yorker*, February 28, 1959, p. 41.

36. Robert C. Wilson, *Drugs and Pharmacy in the Life of Georgia: 1733–1959* (Atlanta: Foote and Davis, 1959), p. 212.

37. Kahn, "The Universal Drink," *The New Yorker*, February 21, 1959, p. 50; February 28, 1959, p. 36.

38. Quoted in David F. Musto, *The American Disease: Origins of Narcotic Control* (New Haven: Yale University Press), p. 254.

39. "Coca-Cola Free from Cocaine," *American Food Journal* 6 (October 15, 1911): 3.

40. *Medical Record*, May 29, 1886, quoted in Maier, *Der Kokainismus*, p. 52; Joël and Fränkel, *Der Cocainismus*, p. 11.

41. Louis Lewin, "Referat: Pharmakologie and Toxicologie," *Berliner Klinische Wo-*

chenschrift, May 18, 1885, pp. 321–322; *New York Medical Journal* 42 (1885): 412; *Medical Record*, May 29, 1886, quoted in Maier, *Der Kokainismus*, p. 52.

42. Albrecht Erlenmeyer, "Über Cocainsucht," *Deutsche Medizinal-Zeitung* 7 (1886): 383–384.

43. Mariani, *Coca and Its Therapeutic Applications*, p. 11; *Mariani's Coca Leaf* 4 (March 1906): 49; Mortimer, *Peru: History of Coca*, p. 17.

44. G. W. Norris, "A Case of Cocaine Habit of Three Months' Duration, Treated by Complete and Immediate Withdrawal of the Drug," *Philadelphia Medical Journal* 1 (1901): 304.

45. Wilder Penfield, "Halsted of Johns Hopkins," *Journal of the American Medical Association* 210 (1969): 2214–2218.

46. J. I. Ingle, "William Halsted, Pioneer in Oral Nerve Block Injection and Victim of Drug Experimentation," *Journal of the American Dental Association* 82 (1971): 46–47.

47. Penfield, "Halsted of Johns Hopkins," p. 2217; Ingle, "William Halsted," p. 47.

48. *New York Medical Journal* 42 (1895): 294.

49. The quotations from Freud's letters can be found in Byck, *Cocaine Papers*, pp. 39, 10, 165. Freud's opinion on Parke's cocaine is quoted on p. 122. The dynamometer test is described in "Beitrag zur Kentniss der Kokainwirkung," *Wiener Medizinische Wochenschrift*, January 31, 1885, pp. 130–133, reprinted in Byck, *Cocaine Papers*, pp. 97–104, as "Contribution to the Knowledge of the Effect of Cocaine."

50. Koller quoted in Becker, "Carl Koller and Cocaine," p. 326; Freud reference in Jones, *Sigmund Freud*, 1:91.

51. Freud, "Bemerkungen über Kokainsucht und Kokainfurcht," *Wiener Medizinische Wochenschrift* 28 (1887): 929–932, reprinted in Byck, *Cocaine Papers*, pp. 171–176, as "Craving for and Fear of Cocaine."

52. Jones, *Sigmund Freud*, 2:189.

53. Ibid., 1:79, 83, 84, 85.

54. Jürgen vom Scheidt, "Sigmund Freud und das Kokain," *Psyche* 27 (1973).

55. Jones, *Sigmund Freud*, 1:308–311.

56. The fullest reference on the subject of Freud's work on cocaine and cocaine's effect on Freud is Byck, *Cocaine Papers*. It includes his letters, his papers, dreams and dream-analyses, the relevant passages from Jones, other historical information, and even a section from a recent book on the cocaine traffic. It is the most accessible and useful source of Freud's letters and papers in English translation, and we have used it for all citations.

57. Sir Arthur Conan Doyle, *The Complete Sherlock Holmes* (New York: Doubleday, 1960), 1:89, 351; 2:622.

58. See also David Musto, "Sherlock Holmes and Sigmund Freud," in Byck, *Cocaine Papers*, pp. 357–370.

59. Maier, *Der Kokainismus*, p. 61; Joël and Fränkel, *Der Cocainismus*, p. 15; "Report of Committee on the Acquirement of Drug Habits," *American Journal of Pharmacy* 75 (1903): 474–487.

60. Thomas G. Simonton, "The Increase in the Use of Cocaine Among the Laity in Pittsburgh," *Philadelphia Medical Journal* 11 (1903): 556–561; W. B. Meister, "Cocainism in the Army," *Military Surgeon* 34 (1914): 344–351; quotation on p. 350.

61. "The Growing Menace of Cocaine," *New York Times Magazine*, August 2, 1908, pp. 1–2; Edward Marshall, "Uncle Sam Is the Worst Drug Fiend in the World," *New York Times*, March 12, 1911; Musto, *The American Disease*, p. 254.

62. Musto, *The American Disease*, p. 18; "Report of Committee on the Acquirement of Drug Habits," pp. 476, 479, 485.

63. "The Cocaine Habit Among Negroes," *British Medical Journal*, November 29, 1902, p. 1729.

64. Musto, *The American Disease*, p. 8.

65. Most of the legal information is from McLaughlin, "Cocaine: The History and Regulation of a Dangerous Drug," *Cornell Law Review* 5 (1973). The reference to drug education is in Maier, *Der Kokainismus*, p. 78. The material on Dr. Tucker and on the New York law of 1907 is in Musto, *The American Disease*, pp. 252 and 103.

66. S. Dana Hubbard, "The New York City Narcotic Clinic and Differing Points of View on Narcotic Addiction," *Monthly Bulletin of the New York City Department of Health* 10 (1920): 40; Musto, *The American Disease*, p. 179.

67. McLaughlin, "Cocaine," gives the legal history up to 1973. Richard Ashley, *Cocaine: Its History, Uses, and Effects* (New York: St. Martin's Press, 1975), pp. 175–186, provides a table of current state and federal laws.

68. McLaughlin, "Cocaine," p. 558; Jean-Marie Pelt, *Drogues et plantes magiques* (Paris: Horizons de France, 1971), p. 122; Maier, *Der Kokainismus*, p. 73; *New York Times*, April 15, 1926, p. 20.

3. *Cocaine in America Today*

1. *Statistisch Jaaroverzicht van Neederlandsch-Indie*, 1920–1938 (Departement van Landbouw, Nijverheid en Handel); George H. Gaffney, "Narcotic Drugs—Their Origin and Routes of Traffic," in *Drugs and Youth: Proceedings of the Rutgers Symposium on Drug Abuse*, edited by J. R. Wittenborn et al. (Springfield, Ill.: Charles C Thomas, 1969), p. 60.

2. Carlos Gutiérrez-Noriega, "El cocaismo y la alimentación en el Perú, *Anales de la Facultad de Medicina* 31 (1948), p. 90.

3. Richard Ashley, *Cocaine: Its History, Uses, and Effects* (New York: St. Martin's Press, 1975), p. 90.

4. *Dawn II Analysis: Drug Abuse Warning Network Phase II Report* (Washington, D.C.: Drug Enforcement Administration, 1974), p. 78.

5. International Narcotics Control Board Statistics Report, 1973.

6. "Highwitness News," *High Times* 7 (December 1975): 110.

7. Nicholas Gage, "Drug-smuggling Logistics Bizarre and Often Fatal," *New York Times*, April 22, 1975, pp. 1, 24; Marc Olden, *Cocaine* (New York: Lancer Books, 1973), p. 65; Ashley, *Cocaine*, p. 200.

8. Gerald T. McLaughlin, "Cocaine: The History and Regulation of a Dangerous Drug," *Cornell Law Review* 57 (1973): 537–538; Richard Woodley, *Dealer: Portrait of a Cocaine Merchant* (New York: Holt, Rinehart and Winston, 1971), p. 47.

9. Anne Crittenden and Michael Ruby, "Cocaine: The Champagne of Drugs," *New York Times Magazine*, September 1, 1974, p. 14; Peter Kihss, "150 Charged in Smuggling of $35 Million in Cocaine," *New York Times*, October 7, 1974, p. 39; Ashley, *Cocaine*, p. 125.

10. "The VOS Caper," *Newsweek*, December 17, 1973, p. 38; Joseph Novitski, "U.S. Drug Watch Shifts Its Focus," *New York Times*, March 12, 1972, p. 17.

11. Irving H. Soloway, "Methadone and the Culture of Addiction," *Journal of Psychedelic Drugs* 6 (1974): 95–96.

12. Woodley, *Dealer*, p. 181; Juan de Onis, "Cocaine a Way of Life for Many in Bolivia," *New York Times*, February 22, 1972, p. 2; Kihss, "150 Charged," p. 39; "Narcotics Officer Is Indicted," *New York Times*, November 22, 1974, p. 41.

13. Woodley, *Dealer*, pp. 1, 10; George R. Gay et al., "An Old Girl: Flyin' Low, Dyin' Slow, Blinded by Snow: Cocaine in Perspective," *International Journal of the Addictions* 8 (1973): 1038; *PharmChem Newsletter* 4 (9), 1975; personal communication, Merck and Co.

14. Irving Spiegel, "17 Indicted in Bronx as Members of Cocaine Ring," *New York Times*, November 24, 1974, p. 30; Michael T. Kaufman, "2 at Columbia Indicted in Sale of Cocaine," *New York Times*, December 21, 1973, p. 1; Dave Anderson, "U.S. Agents Link 4 N.F.L. Stars to Drug Traffic," *New York Times*, June 12, 1973, p. 55; "Ex-Chilean Aide Held in Drug Case," *New York Times*, September 10, 1974, p. 45; "A Tragic Trail's End for the Yankee Mules," *Time*, September 12, 1974, p. 36.

15. "Ex-Cuban Government Official in Cocaine Arrest," *New York Times*, August 25, 1973, p. 92; Olden, *Cocaine*, pp. 58, 139.

16. Nicholas Gage, "Lack of Treaties Hinders U.S. Effort to Curb Drugs," *New York Times*, April 24, 1975, pp. 1, 26.

17. "Professor Cleared of Drug Conspiracy in U.S. Court Here," *New York Times*, July 31, 1975, p. 45.

18. Gage, "Lack of Treaties," p. 26.

19. "The White Goddess," *Time*, April 11, 1949, p. 44.

20. James M. Markham, "Florida Is Becoming Drug Traffic Center," *New York Times*, May 1, 1972, p. 65.

21. L. Vervaeck, "Quelques aspects médicaux et psychologiques de la cocaïnomanie,"

Le Scalpel (Brussels) 76 (1923): 777; Woodley, *Dealer*, pp. 39–40, 42; Olden, *Cocaine*, p. 19.

22. Woodley, *Dealer*, p. 57; Bruce Jay Friedman, *About Harry Towns* (New York: Knopf, 1974), p. 79.

23. *PharmChem Newsletter* 3 (2), 1974; *PharmChem Newsletter* 3 (8), 1975; Ashley, *Cocaine*, p. 139.

24. "Cocaine," *Clinical Toxicology* 7 (1974): 541–543; Ashley, *Cocaine*, pp. 143–146, 198–199.

25. Herbert Abelson, Reuben Cohen, Diane Schrayer, and Michael Rappeport, "Drug Experience, Attitudes, and Related Behavior Among Adolescents and Adults," in *Drug Use in America*, edited by National Commission on Marihuana and Drug Abuse, 4 vols. (Washington, D.C.: Government Printing Office, 1973), 1:627–628; Crittenden and Ruby, "Cocaine," p. 14; Leo Levy, "Drug Use on Campus," *Drug Forum* 2 (1973): 147, 161. John T. Gossett, Jerry M. Lewis, and Virginia Austin Phillips, "Extent and Prevalence of Illicit Drug Use as Reported by 56,745 Students," *Journal of the American Medical Association* 216 (1971): 1464–1470.

26. *Dawn II Analysis*, pp. 79, 89, 95, 184–188, 201.

27. Jerry Hopkins, "Cocaine: A Flash in the Pan, a Pain in the Nose," *Rolling Stone*, no. 81 (April 29, 1971), pp. 1, 6, 8; Crittenden and Ruby, "Cocaine," p. 14; "Guitarist for Rock Group Accused of Having Drugs," *New York Times*, March 30, 1973, p. 31; "Witness Details Workings of Recording Industry," *New York Times*, June 8, 1973, p. 24; Nicholas von Hoffman, "The Cocaine Culture: New Wave for the Rich and Hip," *Washington Post*, April 23, 1975, pp. B1, B6.

28. *Drug Enforcement* 1 (Spring 1974): 38–39; *Drug Enforcement* 1 (Fall 1974): 43.

29. Ernst Joël and Fritz Fränkel, *Der Cocainismus* (Berlin: Springer Verlag, 1924), p. 19; von Hoffman, "The Cocaine Culture," p. B6.

30. "High Is How You Live It," *Washington Post*, April 21, 1975, p. B1; Olden, *Cocaine*, p. 14.

31. Ashley, *Cocaine*, pp. 118–120.

32. *High Times* 7 (December 1975), passim.

33. Crittenden and Ruby, "Cocaine."

34. Domestic Council Drug Review Task Force, *White Paper for the President: Drug Abuse* (Washington, D.C.: Government Printing Office, 1975), pp. 37–39.

35. *Boston Globe*, December 27, 1975, p. 1.

4.   *From Plant to Intoxicant*

1. We are grateful to Timothy Plowman of the Harvard Botanical Museum for looking over the sections on botany and cultivation and advising us on the present status of the taxonomy of the genus *Erythroxylum*. W. Golden Mortimer, *Peru: History of Coca* (New York: J. H. Vail, 1901), has detailed information on botany and cultivation, especially on pp. 235–276.

2. Thomas A. Henry, *The Plant Alkaloids*, 4th ed. (Philadelphia: Blakiston, 1949), gives full details on the coca alkaloids.

3. See G. R. Gay, D. S. Inaba, C. W. Sheppard, J. A. Newmeyer, and R. T. Rappolt, "Cocaine: History, Epidemiology, Human Pharmacology, and Treatment. A Perspective on a New Debut for an Old Girl," *Clinical Toxicology* 8 (1975): 161.

4. For information on cocaine extraction and purification, see Mortimer, *Peru: History of Coca*, pp. 296–313, and Richard Ashley, *Cocaine: Its History, Uses, and Effects* (New York: St. Martin's Press, 1975), pp. 190–193.

5. A. T. Shulgin, "Drugs of Abuse in the Future," in *Drug Use in America: Problem in Perspective*, edited by National Commission on Marihuana and Drug Abuse, 4 vols. (Washington, D.C.: Government Printing Office, 1973), 1:221; S. C. Jordan, A. Lasslo, H. L. Livingston, H. Alperin, and A. Gersing, "Comparative Pharmacology of Cocaine and the Diethylamide Derivative of Cocaine," *Archives Internationales de Pharmacodynamie et Thérapie* 115 (1958): 452–473.

6. Sigmund Freud, "Über Coca," in *Cocaine Papers*, edited by Robert Byck (New York: Stonehill, 1974), p. 68; Mortimer, *Peru: History of Coca*, pp. 427, 428.

7. For the generally accepted theory that cocaine acts mainly by blocking reuptake, see, for example, S. Z. Lander and María L. Enero, "The Potentiation of Responses to Adrenergic Nerve Stimulation in the Presence of Cocaine: Its Relationship to the Metabolic Fate of Released Norepinephrine," *Journal of Pharmacology and Experimental Therapeutics* 191 (1974): 431–443. For experiments suggesting that this explanation is inadequate, see S. Kalsner and M. Nickerson, "Mechanism of Cocaine Potentiation of Response to Amines," *British Journal of Pharmacology* 35 (1969): 428–439; Gertrude D. Maengwyn-Davies and Theodore Koppanyi, "Cocaine Tachyphylaxis and Effects on Indirectly-acting Sympathomimetic Drugs in the Rabbit Aortic Strip and in Splenic Tissue," *Journal of Pharmacology and Experimental Therapeutics* 154 (1966): 481–492; K. Nakatsu and R. J. Reiffenstein, "Increased Receptor Utilization: Mechanism of Cocaine Potentiation," *Nature* 217 (1968): 1276–1277; U. Trendelenburg, "Effect of Cocaine on Pacemaker of Isolated Guinea-pig Atria," *Journal of Pharmacology and Experimental Therapeutics* 161 (1968): 222–231; W. J. Davidson and I. R. Innes, "Dissociation of Isoprenaline by Cocaine from Inhibition of Uptake in Cat Spleen," *British Journal of Pharmacology* 39 (1970): 175–181; Jorgen Scheel-Kruger, "Behavioral and Biochemical Comparison of Amphetamine Derivatives, Cocaine, Benzotropine, and Tricyclic Anti-depressant Drugs," *European Journal of Pharmacology* 18 (1972): 63–73.

8. Maengwyn-Davies and Koppany, "Cocaine Tachyphylaxis"; Trendelenburg, "Effect of Cocaine on Pacemaker"; Scheel-Kruger, "Behavioral and Biochemical Comparison"; Arvid Carlsson, "Biochemical Pharmacology of Amphetamines," in *Abuse of Central Stimulants*, edited by Folke Sjoqvist and Malcolm Tottie (New York: Raven Press, 1969), pp. 305–310; Pierre Simon, Zahier Sultan, Raymond Cherinat, and Jacques-R. Boissier, "La cocaïne, une substance amphétaminique?" *Journal de Pharmacologie* 3 (1972): 129–142; R. M. Post, J. K. Kotin, and F. K. Goodwin, "Effects of Cocaine in Depressed Patients," *American Journal of Psychiatry* 131 (1974): 511–517.

9. Hideo Higuchi, Takaaki Matsuo, and Kiro Shimamoto, "Effects of Methamphetamine and Cocaine in the Depletion of Catecholamines of the Brain, Heart, and Adrenal Glands in Rabbit by Reserpine," *Japanese Journal of Pharmacology* 12 (1962): 48–56; G. Schmidt and P. Neisse, "Zentrale Wirkung von Cocainhomologen und ihre Beeinflussigkeit durch Reserpin oder Adrenolytica," *Archiv für Experimentelle Pathologie und Pharmakologie* 243 (1962): 148–161; Kalsner and Nickerson, "Mechanism of Cocaine Potentiation"; I. R. Innes and G. W. Karr, "Protection Against Induction of Supersensitivity to Catecholamines by Cocaine," *British Journal of Pharmacology* 42 (1971): 603–610.

10. Simon et al., "La cocaïne, une substance amphétaminique?"

11. A. Wauquier and C. J. E. Niemeegers, "Intracranial Self-stimulation in Rats as a Function of Various Stimulus Parameters. V. Influence of Cocaine on Medial Forebrain Bundle Stimulation with Monopolar Electrodes," *Psychopharmacologia* 38 (1974): 201–210; Janice E. Christie and T. J. Crow, "Behavioral Studies of the Actions of Cocaine, Monoamine Oxidase Inhibitors, and Iminodibenzyl Compounds on Central Dopamine Neurons," *British Journal of Pharmacology* 47 (1973): 39–47.

12. Janice E. Christie and T. J. Crow, "Behavioral Studies of the Actions of Cocaine, Monoamine Oxidase Inhibitors, and Iminodibenzyl Compounds on Central Dopamine Neurons," *British Journal of Pharmacology* 47 (1973): 39–47.

13. Richard L. Hawks, Irwin J. Kopin, Robert W. Colburn, and Nguyen B. Thoa, "Norcocaine: A Pharmacologically Active Metabolite of Cocaine Found in the Brain," *Life Sciences* 15 (1974): 2189–2195; Cary Eggleston and Robert A. Hatcher, "A Further Contribution to the Pharmacology of the Local Anesthetics," *Journal of Pharmacology and Experimental Therapeutics* 13 (1919): 448; F. Fish and W. D. C. Wilson, "Excretion of Cocaine and Its Metabolites in Man," *Journal of Pharmacy and Pharmacology* 21 (1969): 1355–1385; A. R. McIntyre, "Renal Excretion of Cocaine in a Case of Acute Cocaine Poisoning," *Journal of Pharmacology and Experimental Therapeutics* 57 (1936): 133; Guillermo Cruz Sánchez and Angel Guillén, "Eliminación de la cocaína en sujetos no habituados," *Revista de Farmacología y Medicina Experimental* 2 (1949): 15; Fernando Montesinos, "Metabolism of Cocaine," *Bulletin on Narcotics* 17, no. 2 (1965): 11–18.

5. The Acute Intoxication

1. Ernst Joël and Fritz Fränkel, *Der Cocainismus* (Berlin: Springer Verlag, 1924), p. 36; Carlos Gutiérrez-Noriega and Vicente Zapata Ortiz, *Estudios sobre la coca y la cocaína en el Perú* (Lima: Ministerio de Educación Pública, 1947), p. 45.

2. Arno Offerman, "Über die zentrale Wirkung des Cocains und einiger neuen Ersatzpräparate," *Archiv für Psychiatrie* 76 (1926).

3. The first two quotations are from Lawrence Kolb, *Drug Addiction: A Medical Problem* (Springfield, Ill.: Charles C Thomas, 1962), p. 101. The third is from Bruce Jay Friedman, *About Harry Towns* (New York: Knopf, 1974), p. 86; the fourth is from Malcolm X, *The Autobiography of Malcolm X*, written with the assistance of Alex Haley (New York: Grove Press, 1964), p. 134; the fifth is from Thomas McGuane, *Ninety-Two in the Shade* (New York: Farrar, Strauss, and Giroux, 1972), pp. 18, 21. The middle ones are from our interviews. The last four are from Joël and Fränkel, *Der Cocainismus*, p. 56; Richard A. Woodley, *Dealer: Portrait of a Cocaine Merchant* (New York: Holt, Rinehart and Winston, 1971), p. 59; Aleister Crowley, *Cocaine* (Level Press, 1973), no page number; Sigmund Freud, "Über Coca," in *Cocaine Papers*, edited by Robert Byck (New York: Stonehill, 1974), p. 60.

4. The first quotation is from Offerman, "Über die Zentrale Wirkung," p. 609. The others are from our interviews. The Burroughs passages are from *Naked Lunch* (New York: Grove Press, 1959), p. 24, and "Letter From a Master Addict to Dangerous Drugs," *British Journal of Addiction* 53 (1956): 126–127.

5. United Nations Economic and Social Council, *Report of the Commission of Enquiry on the Coca Leaf*, May 1950, p. 21; Emilio Ciuffardi, "Dosis de alcaloides que ingieren los habituados a la coca," *Revista de Farmacología y Medicina Experimental* 1 (1948): 81–99; Ciuffardi, "Dosis de alcaloides que ingieren los habituados a la coca. Nuevas observaciones," *Revista de Farmacología y Medicina Experimental* 1 (1948): 216–231; Joel M. Hanna, "Coca Leaf Use in Southern Peru: Some Biosocial Aspects," *American Anthropologist* 76 (1974): 283.

6. Otto Nieschulz, "Psychopharmakologische Untersuchungen über Cocain und Ecgonin," *Arzneimittel-Forschung* 21 (1971); R. M. Post et al., "The Effect of Orally Administered Cocaine on Sleep of Depressed Patients," *Psychopharmacologia* 37 (1974); Freud, "Uber Coca," in Byck, *Cocaine Papers*, p. 58; Carlos Gutiérrez-Noriega, "Errores sobre la interpretación del cocaísmo en las grandes alturas," *Revista de Farmacología y Medicina Experimental* 1 (1948): 104; Gutiérrez-Noriega and Zapata Ortiz, *Estudios sobre la Coca y la Cocaína*, pp. 57–58.

7. Guillermo Cruz Sánchez and Angel Guillén, "Estudio químico de las substancias alcalinas auxiliares del cocaísmo," *Revista de Farmacología y Medicina Experimental* 2 (1949): 209–215.

8. Guillermo Cruz Sánchez and Angel Guillén, "Eliminación de la cocaína," *Revista de Farmacología y Medicina Experimental* 2 (1949): 8–17.

9. For all data see Emilio Ciuffardi, "Contribución a la química del cocaísmo," *Revista de Farmacología y Medicina Experimental* 2 (1949): 18–93.

10. W. Golden Mortimer, *Peru: History of Coca* (New York: J. H. Vail, 1901), p. 422; Richard T. Martin, "The Role of Coca in the History, Religion, and Medicine of South American Indians," *Economic Botany* 24 (1970): 436.

11. Ernest Jones, *The Life and Work of Sigmund Freud*, 3 vols. (New York: Basic Books, 1953), 1:88.

12. Nieschulz, "Psychopharmakologische Untersuchungen."

13. Ibid., p. 275.

14. Ciuffardi, "Dosis de alcaloides que ingieren los habituados a la coca," p. 84; Guillermo Cruz Sánchez and Angel Guillén, "Toxicidad de la totacoca," *Revista de Farmacología y Medicina Experimental* 2 (1949): 275.

15. Personal communication.

16. Carlos Gutiérrez-Noriega, "Alteraciones mentales producidas por la coca," *Revista de Neuro-Psiquiatría* 10 (1947); quotation is on p. 154.

17. Carlos Gutiérrez-Noriega, "Acción de la coca sobre la actividad mental de sujetos habituados," *Revista de Medicina Experimental* 3 (1944); quotations are on pp. 4 and 10.

18. Quoted in Hans W. Maier, *Der Kokainismus* (Leipzig: Georg Thieme, 1926), p. 8.

19. Francisco Risemberg Mendizábal, "Acción de la coca y de la cocaína en sujetos habituados," *Revista de Medicina Experimental* 3 (1944): 317–328.

20. Personal communication.

21. Gutiérrez-Noriega, "Acción de la coca," pp. 16 and 8; Offerman, "Über die Zentrale Wirkung," p. 629.

22. Aleister Crowley, *Diary of a Drug Fiend* (London: Sphere Books, 1972), p. 81.

23. Kolb, *Drug Addiction*, p. 101.

24. Experiment discussed in Bernard Barber, *Drugs and Society* (New York: Russell Sage Foundation, 1967), p. 37.

25. Malcolm X, *Autobiography*, p. 147; Daniel Odier, *The Job: Interviews with William Burroughs* (New York: Grove Press, 1970), p. 153.

26. See Lester Grinspoon and Peter Hedblom, *The Speed Culture* (Cambridge, Mass.: Harvard University Press, 1975), pp. 70–92; also Bernard Weiss and Victor G. Laties, "Enhancement of Human Performance by Caffeine and the Amphetamines," *Pharmacological Reviews* 14 (1962): 1–36.

27. Maier, *Der Kokainismus*, pp. 128–133.

28. Vicente Zapata Ortiz, "Modificaciones psicológicas y fisiológicas producidas por la coca y la cocaína," *Revista de Medicina Experimental* 3 (1944): 132–162.

29. Freud, "Über Coca," in Byck, *Cocaine Papers*, p. 60.

30. Grinspoon and Hedblom, *The Speed Culture*, pp. 86–92.

31. Freud, "Über Coca," in Byck, *Cocaine Papers*, pp. 63–64.

32. Quoted in Maier, *Der Kokainismus*, p. 13.

33. Sir Robert Christison, "Observations on Cuca, or Coca," *British Medical Journal*, April 29, 1876, pp. 527–531; Martín Cárdenas, "Psychological Aspects of Coca Addiction," *Bulletin on Narcotics* 4, no. 2 (1952): 8; Carlos Gutiérrez-Noriega, "Observaciones sobre el cocaísmo obtenidas en un viaje al sur de Perú," *Revista de Farmacología y Medicina Experimental* 1 (1948): 245.

34. Sigmund Freud, "Contribution to the Knowledge of the Effect of Cocaine," in Byck, *Cocaine Papers*, pp. 96–104; Carlos Gutiérrez-Noriega, "Acción de la cocaína sobre la resistencia a la fatiga en el perro," *Revista de Medicina Experimental* 3 (1944): 329–340; Dore Thiel and Bertha Essig, "Cocain und Muskelarbeit I," *Arbeitsphysiologie* 3 (1930): 287–297; R. Herbst and P. Schellenberg, "Cocain und Muskelarbeit II," *Arbeitsphysiologie* 4 (1931): 203–216.

35. Freud, "Contribution," in Byck, *Cocaine Papers*, p. 100; C. Jacobj, "Die peripheren Wirkungen des Kokains und ihre Bedeutung für die Erklärung des Kokakauens der Indianer," *Archiv für Experimentelle Pathologie und Pharmakologie* 159 (1931): 495–515, quotation on p. 514; Joel M. Hanna, "Further Studies on the Effects of Coca-chewing on Exercise," *Human Biology* 43 (1971): 200–209.

36. Woodley, *Dealer*, pp. 87, 204.

37. UNESCO *Report*, pp. 23, 58.

38. George R. Gay and Charles W. Sheppard, "Sex-crazed Dope Fiends—Myth or Reality?" *Drug Forum* 2 (1973): 125–140, quotation on p. 132; Richard L. Nail, E. K. Eric Gunderson, and Douglas Kolb, "Motives for Drug Use Among Light and Heavy Users," *Journal of Nervous and Mental Disease* 159 (1974): 131–138; E. H. Ellinwood, Jr., "Amphetamine Psychosis: A Multidimensional Process," *Seminars in Psychiatry* 1 (1969): 211.

39. Fred I. Leavitt, "Drug-induced Modifications in the Sexual Behavior of Male Rats," Ph.D. thesis, University of Michigan, 1968, in *Dissertation Abstracts International* 30, no. 1, 410-B; review of *The Pirate* in *New York Times Book Review*, October 27, 1974, p. 51.

40. Maier, *Der Kokainismus*, p. 55; Joël and Fränkel, *Der Cocainismus*, p. 27.

41. L. Vervaeck, "Quelques aspects médicaux et psychologiques de la cocaïnomanie," *Le Scalpel* (Brussels) 76 (1923): 744.

42. Maier, *Der Kokainismus*, p. 171; Vervaeck, "Cocaïnomanie," p. 779; Richard Rhodes, "A Very Expensive High," *Playboy*, January 1975, p. 268.

43. Gay and Sheppard, "Sex-crazed Dope Fiends," p. 139.

44. Vervaeck, "Cocaïnomanie," p. 744; Gutiérrez-Noriega, "Alteraciones mentales," p. 159; Joël and Fränkel, *Der Cocainismus*, p. 27; Maier, *Der Kokainismus*, pp. 96, 97.

45. Rasmus Fog, "Stereotyped and Non-stereotyped Behavior in Rats Induced by Various Stimulant Drugs," *Psychopharmacologia* 14 (1969): 299–305.

46. J. Chalmers da Costa, "Four Cases of Cocaine Delirium," *Journal of Nervous and Mental Disease* 16 (1889): 188–194.

47. Herbst and Schellenberg, "Cocain und Muskelarbeit II," p. 207; Gutiérrez-Noriega and Zapata Ortiz, *Estudios*, pp. 81–85; Freud, "Über Coca," in Byck, *Cocaine Papers*, p. 58; Joel M. Hanna, "Responses of Quechua Indians to Coca Ingestion During Cold Exposure," *American Journal of Physical Anthropology* 34 (1971): 273–277; Gutiérrez-Noriega and Zapata Ortiz, *Estudios*, pp. 86–87.

48. M. Reese Guttman, "Acute Cocaine Intoxication," *Journal of the American Medical Association* 90 (1928): 753–755.

49. Carlos Gutiérrez-Noriega and Vicente Zapata Ortiz, "Cocaínismo experimental. I. Toxicología general, acostumbramiento, y sensibilización," *Revista de Medicina Experimental* 3 (1944): 303; George R. Gay et al., "An Old Girl: Flyin' Low, Dyin' Slow, Blinded by Snow: Cocaine in Perspective," *International Journal of the Addictions* 8 (1973): 1035.

50. J. B. Mattison, "Cocaine Poisoning," *Medical and Surgical Reporter* 65 (1891): 645–650.

51. Emil Mayer, "Toxic Effects Following the Use of Local Anesthetics," *Journal of the American Medical Association* 82 (1924): 876–885.

52. "$2 Million Settlement for Brain Injury from Anesthetic Overdose," *Anesthesia and Analgesia* 53 (1974): 520; S. Dana Hubbard, "The New York City Narcotic Clinic and Differing Points of View on Narcotic Addiction," *Monthly Bulletin of the New York City Department of Health* 10 (1920): 34; Leslie Maitland, "For Singles, Scene Has Sordid Side," *New York Times*, November 1, 1974, p. 37.

53. R. D. Pickett, "Acute Toxicity of Heroin, Alone and in Combination with Cocaine or Quinine," *British Journal of Pharmacology* 40 (1970): 145P–146P.

54. Maier, *Der Kokainismus*, pp. 200, 202, 203.

55. Vervaeck, "Cocaïnomanie," p. 749; Jean Cocteau, *Le grand écart* (Paris: Librarie Stock, 1923), pp. 169–181.

56. *Dawn II Analysis: Drug Abuse Warning Network Phase II Report* (Washington, D.C.: Drug Enforcement Administration, 1974), pp. 79, 89, 184–188.

57. Gay et al., "An Old Girl," p. 1036; "Cocaine," *Clinical Toxicology* 7 (1974): 541.

58. William A. Hammond, "Coca: Its Preparations and Their Therapeutical Qualities, with Some Remarks on the So-called 'Cocaine Habit,' " *Transactions of the Medical Society of Virginia*, November 1887, pp. 215–219.

6. *Effects of Chronic Use*

1. Carlos Gutiérrez-Noriega, "Alteraciones mentales producidas por la coca," *Revista de Neuro-Psiquiatría* 10 (1947), quotation on p. 173; Gutiérrez-Noriega and Vicente Zapata Ortiz, *Estudios sobre la Coca y la Cocaína en el Perú* (Lima: Ministerio de Educación Pública, 1947), pp. 74, 33, 34.

2. Carlos Gutiérrez-Noriega and Vicente Zapata Ortiz, "La inteligencia y la personalidad en los habituados a la coca," *Revista de Neuro-Psiquiatría* 13 (1950): 22–60.

3. J. C. Negrete and H. B. M. Murphy, "Psychological Deficit in Chewers of Coca Leaf," *Bulletin on Narcotics* 19, no. 4 (1967), quotation on p. 17.

4. H. B. M. Murphy, O. Rios, and J. C. Negrete, "The Effects of Abstinence and Retraining on the Chewer of Coca-leaf," *Bulletin on Narcotics* 21 (1969): 41–47.

5. D. Goddard, S. N. de Goddard, and P. C. Whitehead, "Social Factors Associated with Coca Use in the Andean Region," *International Journal of the Addictions* 4 (1969), quotation on p. 580.

6. Carlos Gutiérrez-Noriega and Vicente Zapata Ortiz, "Observaciones fisiológicas y patológicas en sujetos habituados a la coca," *Revista de Farmacología y Medicina Experimental* 1 (1948): 1–31.

7. Alfred A. Buck, Tom T. Sasaki, Jean J. Hewitt, and Anne A. Macrae, "Coca-chewing and Health: An Epidemiological Study Among Residents of a Peruvian Village," *American Journal of Epidemiology* 88 (1968), quotation on p. 175.

8. James E. Hamner III and Oscar L. Villegas, "The Effect of Coca Leaf Chewing on the

Buccal Mucosa of Aymara and Quechua Indians in Bolivia," *Oral Surgery, Oral Medicine, and Oral Pathology* 28 (1969): 287–295.

9. Gutiérrez-Noriega and Zapata Ortiz, *Estudios*, p. 73; United Nations Economic and Social Council, *Report of the Commission of Enquiry on the Coca Leaf*, May 1950, p. 30.

10. Vicente Zapata Ortiz, "The Chewing of Coca-leaves in Peru," *International Journal of the Addictions* 5 (1970): 292; UNESCO *Report*, p. 27.

11. Carlos Monge, "The Need for Studying the Problem of Coca-leaf Chewing," *Bulletin on Narcotics* 4 (1952): 13–15; Joel M. Hanna, "Coca Leaf Use in Southern Peru: Some Biosocial Aspects," *American Anthropologist* 76 (1974): 289; Carlos Gutiérrez-Noriega, "Errores sobre la interpretación del cocaísmo en las grandes alturas," *Revista de Farmacología y Medicina Experimental* 1 (1948).

12. Hans W. Maier, *Der Kokainismus* (Leipzig: Georg Thieme, 1926), p. 138; Ernst Joël and Fritz Fränkel, *Der Cocainismus* (Berlin: Springer Verlag, 1924), p. 35.

13. Charles Perry, "The Star-spangled Powder: Or, Through History with Coke Spoon and Nasal Spray," *Rolling Stone*, no. 115 (August 17, 1972): 26.

14. Richard A. Woodley, *Dealer: Portrait of a Cocaine Merchant* (New York: Holt, Rinehart and Winston, 1971), p. 59.

15. Aleister Crowley, *Cocaine* (Level Press, 1973), pages unnumbered.

16. John Symonds and Kenneth Grant, eds., *The Confessions of Aleister Crowley* (London: Jonathan Cape, 1969), p. 904; John Symonds, *The Great Beast: The Life and Magick of Aleister Crowley* (London: Macdonald, 1971), p. 276.

17. John Symonds and Kenneth Grant, eds., *The Magickal Record of the Beast 666: Diaries of Aleister Crowley 1914–1920* (London: Duckworth, 1972), pp. 207–208, 209, 210, 253, 128, 280.

18. John Symonds, introduction to Aleister Crowley, *Diary of a Drug Fiend* (London: Sphere Books, 1972), pp. 14, 16.

19. Crowley, *Diary of a Drug Fiend*, pp. 53, 45, 67, 68, 108, 65, 114, 182, 229, 247, 303, 309, 316, 315, 379, 378.

20. Introduction to Crowley, *Cocaine*.

21. Joël and Fränkel, *Der Cocainismus*, p. 33.

22. Léon Natanson, "Au sujet des lésions nasales chez les priseurs de cocaïne," *Revue de Laryngologie* 57 (1936); Joël and Fränkel, *Der Cocainismus*, p. 33.

23. Joël and Fränkel, *Der Cocainismus*, p. 33.

24. Pitigrilli, *Cocaine* (San Francisco: And/Or Press, 1974), pp. 119, 51–52, 20, 50, 208.

25. Vladimir Nabokov, "A Matter of Chance," in *Tyrants Destroyed and Other Stories* (New York: McGraw-Hill, 1975), pp. 144, 148.

26. Maier, *Der Kokainismus*, pp. 135, 140, 141.

27. Ibid., pp. 102–112.

28. Joël and Fränkel, *Der Cocainismus*, p. 44; Lawrence Kolb, *Drug Addiction: A Medical Problem* (Springfield, Ill.: Charles C Thomas, 1962), p. 18; Maier, *Der Kokainismus*, pp. 118–119.

29. Maier, *Der Kokainismus*, pp. 193–194; Joël and Fränkel, *Der Cocainismus*, p. 66.

30. Joël and Fränkel, *Der Cocainismus*, pp. 23, 46; Ardrey W. Downs and Nathan B. Eddy, "Effect of Repeated Doses of Cocaine on the Dog," *Journal of Pharmacology and Experimental Therapeutics* 46 (1932): 195–198; E. H. Ellinwood, Jr., "Amphetamine Psychosis: A Multidimensional Process," *Seminars in Psychiatry* 1 (1969): 218.

31. William Burroughs, *Naked Lunch* (New York: Grove Press, 1959), pp. 18–19.

32. Maier, *Der Kokainismus*, pp. 136, 144–145.

33. Ibid., pp. 145–150.

34. Ibid., pp. 150–161.

35. Ibid., pp. 161–178.

36. Ibid., pp. 178–188.

37. D. R. Brower, "The Effects of Cocaine on the Central Nervous System," *Journal of the American Medical Association* 6 (1886): 59–62.

38. W. Mayer-Gross, "Selbstschilderung eines Cocainisten," *Zeitschrift für die gesamte Neurologie und Psychiatrie* 62 (1920).

39. Nils Bejerot, "A Comparison of the Effects of Cocaine and Synthetic Central Stimulants," *British Journal of Addiction* 65 (1970): 35.

40. See Lester Grinspoon and Peter Hedblom, *The Speed Culture* (Cambridge, Mass.: Harvard University Press, 1975), pp. 112–127.

41. K. Heilbronner, "Cocainpsychose?" *Zeitschrift für die gesamte Neurologie und Psychiatrie* 15 (1913): 415–426; Maier, *Der Kokainismus*, p. 143; Grinspoon and Hedblom, *The Speed Culture*, pp. 131–132.

42. Carlos Gutiérrez-Noriega, "Inhibición del sistema nervioso central producido por intoxicación cocaínica crónica," *Revista de Farmacología y Medicina Experimental* 2 (1949).

43. Larry Stein and G. David Wise, "Mechanism of the Facilitating Effect of Amphetamines on Behavior," in *Psychotomimetic Drugs*, edited by Daniel H. Efron (New York: Raven Press, 1970 , p. 124; Gutiérrez-Noriega, "Inhibición," p. 219.

44. L. Vervaeck, "Quelques aspects médicaux et psychologiques de la cocaïnomanie," *Le Scalpel* (Brussels) 76 (1923): 775; Joël and Fränkel, *Der Cocainismus*, pp. 28, 25.

45. A. W. Downs and N. B. Eddy, "Effect of Repeated Doses of Cocaine on the Dog," *Journal of Pharmacology and Experimental Therapeutics* 46 (1932): 195–198; Carlos Gutiérrez-Noriega and Vicente Zapata Ortiz, "Intoxicación crónica por la cocaína. I. Efectos sobre el crecimiento y reproducción de las ratas," *Revista de Medicina Experimental* 5 (1946): 84, 79, 69; Downs and Eddy, "Effect of Repeated Doses of Cocaine on the Rat," *Journal of Pharmacology and Experimental Therapeutics* 46 (1932): 199–200.

46. William F. Geber and Lee C. Schram, personal communication.

47. Gutiérrez-Noriega, "Inhibición," pp. 224–228; Vincent Marks and P. A. L. Chapple, "Hepatic Dysfunction in Heroin and Cocaine Users," *British Journal of Addiction* 62 (1967): 189–195.

48. Maier, *Der Kokainismus*, pp. 203–206.

49. J. R. Wittenborn et al., eds. *Drugs and Youth: Proceedings of the Rutgers Symposium on Drug Abuse* (Springfield, Ill.: Charles C Thomas, 1969).

50. A. Friedlander, "Über Morphinismus und Cocainismus," *Medizinische Klinik* 9 (1913): 1581; Joël and Fränkel, *Der Cocainismus*, p. 49; Burroughs, "Letter From a Master Addict to Dangerous Drugs," *British Journal of Addiction* 53 (1956): 127; John Langrod, "Secondary Drug Use Among Heroin Users," *International Journal of the Addictions* 5 (1970): 614; Gay et al., "An Old Girl: Flyin' Low, Dyin' Slow, Blinded by Snow: Cocaine in Perspective," *International Journal of the Addictions* 8 (1973): 1038; David E. Smith and George R. Gay, eds., "It's So Good, Don't Even Try It Once": *Heroin in Perspective* (Englewood Cliffs, N.J.: Prentice-Hall, 1972), p. 74; R. C. Stephens and R. S. Weppner, "Patterns of 'Cheating' Among Methadone Maintenance Patients," *Drug Forum* 2 (1973): 310; Carl D. Chambers, W. J. Russell Taylor, and Arthur D. Moffet, "The Incidence of Cocaine Abuse Among Methadone Maintenance Patients," *International Journal of the Addictions* 7 (1972): 427–441; Max Glatt, "Drug Use in Great Britain," *Drug Forum* 1 (1972): 291; Nils Bejerot, *Addiction and Society* (Springfield, Ill.: Charles C Thomas, 1970), p. 152.

51. Daniel Odier, *The Job: Interviews with William Burroughs* (New York: Grove Press, 1969, 1970), p. 129; Perry M. Lichtenstein, "Narcotic Addiction," *New York Medical Journal* 100 (1914): 622–626, in *Narcotic Addiction*, edited by John A. O'Donnell and John C. Ball (New York: Harper & Row, 1966), p. 28; Louis Lewin, *Phantastica: Narcotic and Stimulant Drugs*, trans. P. H. A. Wirth (New York: Dutton, 1931), pp. 81, 83; Gerald Deneau, et al., "Self-administration of Psychoactive Substances by the Monkey," *Psychopharmacologia* 16 (1969): 41; Peter Laurie, *Drugs: Medical, Psychological and Social Facts* (Baltimore: Penguin Books, 1969), p. 38.

52. Walter R. Cuskey, Arnold William Klein, and William Krasner, *Drug-Trip Abroad: American Drug-Refugees in Amsterdam and London* (Philadelphia: University of Pennsylvania Press, 1972), pp. 80–83, quotation on pp. 82–83.

53. Kolb, *Drug Addiction*, p. 101; Burroughs, "Letter From a Master Addict," p. 127.

54. Magnan and Saury, "Trois cas de cocaïnisme chronique," *Comptes rendus des Scéances et Mémoires de la Société de Biologie*, 9ème Série, no. 1 (1889): 60–63.

55. William Burroughs, *Junkie* (New York: Ace Books, 1953), p. 119; M. W. Nott, "Potentiation of Morphine Analgesia by Cocaine in Mice," *European Journal of Pharmacology* 5 (1968): 93–99.

56. Joël and Fränkel, *Der Cocainismus*, p. 81; Mark Mehta, *Intractable Pain* (London: W. B. Saunders Company, 1973), p. 71.

57. Bejerot, *Addiction and Society*, p. 34.

7. Cocaine in Medicine and Psychiatry

1. Horacio Fabrega and Peter K. Manning, "Health Maintenance Among Peruvian Peasants," *Human Organization* 31 (1973).

2. W. Golden Mortimer, *Peru: History of Coca* (New York: J. H. Vail, 1901), pp. 491–509.

3. Ernst Joël and Fritz Fränkel, *Der Cocainismus* (Berlin: Springer Verlag, 1924), p. 81; Irving H. Leopold, "Can a Cocaine Eyedrop Solution Induce Drug Dependence?" *Journal of the American Medical Association* 212 (1970): 1219.

4. J. M. Ritchie and Paul Greengard, "On the Mode of Action of Local Anesthetics," *Annual Review of Pharmacy* 6 (1966): 405–430.

5. Felicity Reynolds, "The Pharmacology of Local Analgesic Drugs," in *A Practice of Anesthesia*, 3rd ed., edited by W. D. Wylie and H. C. Churchill-Davidson (Chicago: Year Book Medical Publishers, 1972), p. 1162.

6. Henry K. Beecher and Donald P. Todd, *A Study of the Deaths Associated with Anesthesia and Surgery* (Springfield, Ill.: Charles C Thomas, 1954), pp. 20–22.

7. Mavis Buck, "A Method of Local Anesthesia for the Correction of Simple Fracture of the Nose," *British Journal of Plastic Surgery* 18 (1965): 363–368.

8. Nicholas L. Schenck, "Local Anesthesia in Otolaryngology: A Re-evaluation," *Annals of Otology, Rhinology, and Laryngology* 84 (1975): 65–72.

9. *The Merck Index: An Encyclopedia of Chemicals and Drugs*, 8th ed. (Rahway, N.J.: Merck, 1968), p. 275.

10. Freud, "On the General Effect of Cocaine," in *Cocaine Papers*, edited by Robert Byck (New York: Stonehill, 1974), pp. 116–117.

11. Hans Berger, "Zur pathogenese des katatonischen Stupors," *Münchner Medizinische Wochenschrift* 68 (1921): 448–450; Ulrich Fleck, "Über Cocainwirkung bei Stuporosen," *Zeitschrift für die gesamte Neurologie und Psychiatrie* 92 (1924): 84–118, quotations on p. 97.

12. Gustav Bychowski, "Zur Wirkung grosser Cocaingaben auf Schizophrene," *Monatsschrift für Psychiatrie und Neurologie* 58 (1925): 329–344.

13. Arno Offerman, "Über die zentrale Wirkung des Cocains und einiger neuen Ersatzpräparate," *Archiv für Psychiatrie* 76 (1926): 626 ff.

14. August Jacobi, "Die psychische Wirkung des Cocains in ihrer Bedeutung für die Psychopathologie," *Archiv für Psychiatrie und Nervenkranken* 79 (1927), quotation on p. 402.

15. Erich Lindemann and William Malamud, "Experimental Analysis of the Psychopathological Effects of Intoxicating Drugs," *American Journal of Psychiatry* 13 (1934), quotation on p. 873; Morton A. Rubin, William Malamud, and Justin M. Hope, "The EEG and Psychopathological Manifestations in Schizophrenics as Influenced by Drugs," *Psychosomatic Medicine* 4 (1942): 355–361.

16. R. M. Post et al., "Effects of Cocaine in Depressed Patients," *American Journal of Psychiatry* 131 (1974); Post et al., "The Effect of Orally Administered Cocaine on Sleep of Depressed Patients," *Psychopharmacologia* 37 (1974).

17. E. H. Ellinwood, Jr., "Behavioral and EEG Changes in the Amphetamine Model of Psychosis," in *Neuropharmacology of Monoamines and Their Regulatory Enzymes*, edited by Earl Usdin (New York: Raven Press, 1974), p. 281; Solomon H. Snyder, "Catecholamines in the Brain as Mediators of Amphetamine Psychosis," *Archives of General Psychiatry* 27 (1972): 171.

18. E. H. Ellinwood, Jr., "Amphetamine Psychosis: I. Description of the Individuals and Process," *Journal of Nervous and Mental Disease* 144 (1967): 274; Ellinwood, "Amphetamine Psychosis: A Multidimensional Process," *Seminars in Psychiatry* 1 (1969): 209–218; Snyder, "Catecholamines," pp. 170, 173.

19. Hans W. Maier, *Der Kokainismus* (Leipzig: Georg Thieme, 1926), p. 143; Joël and Fränkel, *Der Cocainismus*, pp. 61–62.

20. E. H. Ellinwood, Jr., and Abraham Sudilovsky, "Chronic Amphetamine Intoxication: Behavioral Model of Psychosis," in *Psychopathology and Psychopharmacology*, edited by Jonathan O. Cole, Alfred M. Freedman, and Arnold J. Friedhoff (Baltimore and London: Johns Hopkins University Press, 1973), pp. 51–70, quotation on p. 68.

21. Ellinwood, "Amphetamine Psychosis," p. 214; E. H. Ellinwood, Jr., and M. Marlyne Kilbey, "Amphetamine Stereotypy: The Influence of Environmental Factors and Prepotent Behavioral Patterns on Its Topography and Development," *Biological Psychiatry* 10 (1975): 5, 11.

22. Snyder, "Catecholamines," pp. 173–174.

23. Ellinwood and Sudilovsky, "Chronic Amphetamine Intoxication," p. 56; Ellinwood and Kilbey, "Amphetamine Stereotypy," p. 15.

24. Snyder, "Catecholamines," pp. 174–175; M. C. Wilson and C. R. Schuster, "Effects of Stimulants and Depressants on Cocaine Self-administration Behavior in the Rhesus Monkey," *Psychopharmacologia* 31 (1973): 291–304.

25. Snyder, "Catecholamines," pp. 176–177.

26. David L. Garver, R. Francis Schlemmer, James W. Maas, and John M. Davis, "A Schizophreniform Behavioral Psychosis Mediated by Dopamine," *American Journal of Psychiatry* 132 (1975): 33–38.

27. Ellinwood, "Amphetamine Psychosis," p. 223; E. Eidelberg, H. Lesse, and F. P. Gault, "An Experimental Model of Temporal Lobe Epilepsy: Studies of the Convulsant Properties of Cocaine," in *EEG and Behavior*, edited by G. H. Glaser (New York: Basic Books, 1963), pp. 272–283.

28. Ellinwood, "Behavioral and EEG Changes," pp. 282, 291.

29. Snyder, "Catecholamines," p. 175.

30. Post, "Cocaine Psychosis: A Continuum Model," *American Journal of Psychiatry* 132 (1975): 228–230.

31. For a summary of recent work on stimulants and psychosis, see Solomon H. Snyder, Shailesh P. Banerjee, Henry I. Yamamura, and David Greenberg, "Drugs, Neurotransmitters, and Schizophrenia," *Science* 184 (1974): 1243–1253.

8. *Cocaine and Drug Dependence*

1. See Lester Grinspoon and Peter Hedblom, *The Speed Culture* (Cambridge, Mass.: Harvard University Press, 1975), p. 151.

2. William Burroughs, *Naked Lunch* (New York: Grove Press, 1959), p. 65; Burroughs, "Letter From a Master Addict to Dangerous Drugs," *British Journal of Addiction* 53 (1956): 127; Carlos Gutiérrez-Noriega, "Observaciones sobre el cocaísmo obtenidas en un viaje al sur de Perú," *Revista de Farmacología y Medicina Experimental* 1 (1948): 241; Gutiérrez-Noriega and Vicente Zapata Ortiz, "Estudio de habituados a la coca en estado de abstinencia," *Revista de Farmacología y Medicina Experimental* 3 (1950): 71.

3. M. H. Seevers, "Characteristics of Dependence on and Abuse of Psychoactive Substances," in *Chemical and Biological Aspects of Drug Dependence*, edited by S. J. Mulé and Henry Brill (Cleveland: CRC Press, 1972), p. 17.

4. See Grinspoon and Hedblom, *The Speed Culture*, pp. 160–164; R. M. Post et al., "The Effect of Orally Administered Cocaine on Sleep of Depressed Patients," *Psychopharmacologia* 37 (1974): 64.

5. Grinspoon and Hedblom, *The Speed Culture*, p. 164.

6. J. D. P. Graham, "Recent Theories About the Pharmacological Basis of Tolerance and Dependence," *British Journal of Addiction* 67 (1970): 83; Grinspoon and Hedblom, *The Speed Culture*, pp. 154–155; M. Fink, "Electrophysiology of Drugs of Dependence," in Mulé and Brill, *Drug Dependence*, p. 384.

7. Nils Bejerot, *Addiction and Society* (Springfield, Ill.: Charles C Thomas, 1970), p. 81.

8. Ernst Joël and Fritz Fränkel, *Der Cocainismus* (Berlin: Springer Verlag, 1924), p. 72; Hans W. Maier, *Der Kokainismus* (Leipzig: Georg Thieme, 1926), p. 210; A. W. Downs and N. B. Eddy, "Effect of Repeated Doses of Cocaine on the Rat," *Journal of Pharmacology and Experimental Therapeutics* 46 (1932): 200; Bejerot, *Addiction and Society*, p. 66.

9. A. L. Tatum and M. H. Seevers, "Experimental Cocaine Addiction," *Journal of Pharmacology and Experimental Therapeutics* 36 (1929): 401–410; Robert M. Post and Richard T. Kopanda, "Cocaine, Kindling, and Reverse Tolerance," *Lancet*, February 15, 1975, 409–410; Carlos Gutiérrez-Noriega, "Inhibición del sistema nervioso central producida por

intoxicación cocaínica crónica," *Revista de Farmacología y Medicina Experimental* 2 (1949): 224, 227.

10. H. O. J. Collier, "Drug Dependence: A Pharmacological Analysis," *British Journal of Addiction* 67 (1972).

11. *WHO Technical Report Service* 407 (1969): 61.

12. *WHO Technical Report Service* 406 (1970): 41; ibid. 437 (1970): 13.

13. *WHO Technical Report Service* 273 (1964): 14–15.

14. Thomas Szasz, *Ceremonial Chemistry: The Ritual Persecution of Drugs, Addicts, and Pushers* (Garden City, N.Y.: Anchor Press, 1974), p. 3.

15. Norman E. Zinberg and John A. Robertson, *Drugs and the Public* (New York: Simon and Schuster, 1972), p. 61.

16. Maier, *Der Kokainismus*, p. 153; Bruce Jay Friedman, *About Harry Towns* (New York: Knopf, 1974), p. 92.

17. Burroughs, *Naked Lunch*, pp. 19, 24; "Letter From a Master Addict," p. 127; Richard Rhodes, "A Very Expensive High," *Playboy*, January 1975, p. 270.

18. Personal communication.

19. William A. Hammond, "Coca," *Transactions of the Medical Society of Virginia*, November 1887, p. 220; Joël and Fränkel, *Der Cocainismus*, p. 78.

20. Isidor Chein et al., *The Road to H* (New York: Basic Books, 1964), pp. 238, 243, 237.

21. Seevers, "Characteristics of Dependence," pp. 15, 16; Theresa Harwood, "Cocaine," *Drug Enforcement* 1 (Spring 1975): 24.

22. E. H. Ellinwood, Jr., "Amphetamine and Stimulant Drugs," in *Drug Use in America: Problem in Perspective*, edited by National Commission on Marihuana and Drug Abuse, 4 vols. (Washington, D.C.: Government Printing Office, 1973), 1:143–144.

23. Gerald Deneau et al., "Self-administration of Psychoactive Substances by the Monkey," *Psychopharmacologia* 16 (1969); Roy Pickens and Travis Thompson, "Cocaine-reinforced Behavior in Rats: Effects of Reinforcement Magnitude and Fixed-ratio Schedule," *Journal of Pharmacology and Experimental Therapeutics* 161 (1968): 122–129; Roy Pickens, "Self-administration of Stimulants by Rats," *International Journal of the Addictions* 3 (1968): 215–221.

24. Travis Thompson and Roy Pickens, "Stimulant Self-administration by Animals: Some Comparisons with Opiate Self-administration," *Federation Proceedings* 29 (February 1970): 6–12.

25. Tomoji Yanagita, "An Experimental Framework for the Evaluation of Dependence Liability of Various Types of Drugs in Monkeys," *Bulletin on Narcotics* 25, no. 4 (1973).

26. Charles Perry, "The Star-spangled Powder," *Rolling Stone*, no. 117 (August 17, 1972), p. 24.

27. Zinberg and Robertson, *Drugs and the Public*, p. 97.

28. Fink, "Electrophysiology of Drugs of Dependence," pp. 380, 384–385.

29. Collier, "Drug Dependence," p. 277.

30. Brill, "Introduction," Mulé and Brill, *Drug Dependence*, p. 6.

31. Ibid.

32. Bejerot, *Addiction and Society*, p. 23.

33. Szasz, *Ceremonial Chemistry*, pp. 82, 84–85.

34. Burroughs, "Letter From a Master Addict," p. 131.

35. Bejerot, *Addiction and Society*, pp. 165–166.

36. Troy Duster, *The Legislation of Morality: Law, Drugs, and Moral Judgment* (New York: Free Press, 1970), pp. 117–129, quotations on pp. 128 and 156.

37. Szasz, *Ceremonial Chemistry*, p. 52.

38. Duster, *Legislation of Morality*, p. 213.

39. Carl N. Edwards, *Justice Administration and Drug Dependence* (Boston: Justice Resource Institute, 1973), p. 44.

40. Szasz, *Ceremonial Chemistry*, pp. xvii, 111; Thomas Szasz, "Alcoholism: A Socio-ethical Perspective," *Washburn Law Journal* 6 (1967): 264.

41. Szasz, *Ceremonial Chemistry*, pp. 153–174.

9. Abuse Potential of Coca and Cocaine: The Policy Debate

1. WHO Technical Report Service 407 (1969): 6.

2. Nils Bejerot, Addiction and Society (Springfield, Ill.: Charles C Thomas, 1970), p. xvii.

3. Thomas Szasz, Ceremonial Chemistry: The Ritual Persecution of Drugs, Addicts, and Pushers (Garden City, N.Y.: Anchor Press, 1974), p. 176.

4. Quoted in ibid., p. 100.

5. Isidor Chein et al., The Road to H (New York: Basic Books, 1964), pp. 333-334.

6. Troy Duster, The Legislation of Morality: Law, Drugs, and Moral Judgment (New York: Free Press, 1970), p. 104.

7. Carlos Gutiérrez-Noriega, "El cocaísmo y la alimentación en el Perú," Anales de la Facultad de Medicina 31 (1948): 73.

8. Mario A. Puga, "El Indio y la coca," Cuadernos Americanos 4 (1951): 51; Carlos A. Ricketts, "La Coca: Problema de Prevención Social" (Arequipa; 1948), passim; Paz Soldán. "Un Memorandum sobre la Situación Actual de la Coca Peruana" (Lima: Ediciones La Reforma Médica, 1936); Gutiérrez-Noriega, "El cocaísmo y la alimentación," pp. 35, 54.

9. Quoted in Jorge Bejarano, "Further Considerations on the Coca Habit in Colombia," Bulletin on Narcotics 4, no. 3 (1952): 3; Puga, "El Indio y la coca," p. 44; Marcel Granier-Doyeux, "Some Sociological Aspects of the Problem of Cocaism," Bulletin on Narcotics 14, no. 4 (1962): 2; Sergio Quijada Jara, La Coca en las Costumbres Indígenas (Huangayo, Peru, 1950), p. 60.

10. United Nations Economic and Social Council, Report of the Commission of Enquiry on the Coca Leaf, May 1950, p. 38; Von Merzbacher, "Über Koka und Kokakauer: Reiseerinnerungen aus Südamerika," Münchner Medizinische Wochenschrift 76 (1929): 2018; José Augustín Morales, El Oro Verde de Las Yungas: Libro de Propaganda Industrial (La Paz: Imprenta del Instituto Nacional de Inválidos, 1938), pp. 46, 183; Richard T. Martin, "The Role of Coca in the History, Religion, and Medicine of South American Indians," Economic Botany 24 (1970): 424.

11. Gutiérrez-Noriega, "El cocaísmo y la alimentación," pp. 28, 29; Martin, "Role of Coca," p. 436.

12. UNESCO Report, p. 38.

13. James H. Woods and David A. Downs, "The Psychopharmacology of Cocaine," in Drug Use in America: Problem in Perspective, edited by National Commission on Marihuana and Drug Abuse, 4 vols. (Washington, D.C.: Government Printing Office, 1973), 1:130.

14. Carlos Monge, "La necesidad de estudiar el problema de la masticación de las hojas de coca," Perú Indígena 3 (7 and 8): 135.

15. Bejerot, Addiction and Society, p. xv.

16. See Lester Grinspoon and Peter Hedblom, The Speed Culture (Cambridge, Mass.: Harvard University Press, 1975), pp. 182-205.

17. Hans W. Maier, Der Kokainismus (Leipzig: Georg Thieme, 1926), pp. 233, 127, 231; Marc Olden, Cocaine (New York: Lancer Books, 1973), pp. 60, 20; Irving Soloway, "Methadone and the Culture of Addiction," Journal of Psychedelic Drugs 6 (1974): 97.

18. Edna Buchanan and James Camp, "Killing One of Their Own Led to Gang's Fall," Miami Herald, October 13, 1975; personal communication from John D. Griffith, M.D.

19. Jared Tinklenberg, "Drugs and Crime," in National Commission, Drug Use in America, 1:263-264.

20. Bejerot, Addiction and Society, pp. 81, 76-78.

21. Jerome H. Jaffe, "Drug Addiction and Drug Abuse," in The Pharmacological Basis of Therapeutics, 4th ed., edited by Louis S. Goodman and Alfred Gilman (London and Toronto: Macmillan, 1970), p. 293; J. Robert Russo, Amphetamine Abuse (Springfield, Ill.: Charles C Thomas, 1968), p. 12.

22. See Grinspoon and Hedblom, The Speed Culture, p. 143.

23. Quoted in David E. Smith and John Luce, *Love Needs Care* (Boston and Toronto: Little, Brown, 1971), p. 244.

24. Bejerot, *Addiction and Society*, pp. 66–67; Tomoji Yanagita, "An Experimental Framework for the Evaluation of Dependence Liability of Various Types of Drugs in Monkeys," *Bulletin on Narcotics* 25, no. 4 (1973): 64.

25. See Richard Jacobson and Norman E. Zinberg, *The Social Basis of Drug Abuse Prevention* (Washington, D.C.: Drug Abuse Council, 1975); Wayne M. Harding and Norman E. Zinberg, "The Effectiveness of the Subculture in Developing Rituals and Social Sanctions for Controlled Drug Use," in *Drugs, Rituals, and Altered States of Consciousness*, edited by Brian M. du Toit (1976, in press); Norman E. Zinberg and Richard Jacobson, "The Natural History of 'Chipping,' " *American Journal of Psychiatry* 133 (1976): 37–40.

26. *Marijuana in Texas: A Report to the Senate Interim Drug Study Committee* (Austin, 1972), pp. 42, 49.

27. California Legislature Senate Select Committee on Control of Marijuana, *Marijuana: Beyond Misunderstanding* (May 1974), p. 118.

10.   *Drugs and Culture: Cocaine as a Historical Example*

1. Nils Bejerot, *Addiction and Society* (Springfield, Ill.: Charles C Thomas, 1970), p. 247; Troy Duster, *The Legislation of Morality: Law, Drugs, and Moral Judgment* (New York: Free Press, 1970), pp. 210, 193.

2. Thomas Szasz, *Ceremonial Chemistry: The Ritual Persecution of Drugs, Addicts, and Pushers* (Garden City, N.Y.: Anchor Press, 1974), p. 10; Isidor Chein et al., *The Road to H: Narcotics, Delinquency, and Social Policy* (New York: Basic Books, 1964), p. 14.

3. Duster, *Legislation of Morality*, p. 236; Peter Laurie, *Drugs: Medical, Psychological, and Social Facts* (Baltimore: Penguin Books, 1969), p. 70.

4. Norman E. Zinberg and John A. Robertson, *Drugs and the Public* (New York: Simon and Schuster, 1972), pp. 29, 39.

5. *WHO Technical Report Service* 407 (1969): 6.

6. James Harvey Young, *The Toadstool Millionaires* (Princeton, N.J.: Princeton University Press, 1961), p. 119.

7. Richard L. Nail, E. K. Eric Gunderson, and Douglas Kolb, "Motives for Drug Use Among Light and Heavy Users," *Journal of Nervous and Mental Disease* 159 (1974): 131–138.

8. Young, *Toadstool Millionaires*, p. 72.

9. Ibid., p. 158.

10. Szasz, *Ceremonial Chemistry*, p. 77; Mickey C. Smith and David A. Knapp, *Pharmacy, Drugs, and Medical Care* (Baltimore: Williams and Wilkins, 1972), p. 161; Salvatore P. Lucia, *A History of Wine as Therapy* (Philadelphia: Lippincott, 1963), pp. 145, 148, 170.

11. Lucia, *A History of Wine*, p. 148.

12. See Young, *Toadstool Millionaires*, pp. 221, 244.

13. Henry Brill, "Recurrent Patterns in the History of Drugs of Dependence and Some Interpretations," in *Drugs and Youth: Proceedings of the Rutgers Symposium on Drug Abuse*, edited by J. R. Wittenborn et al. (Springfield, Ill.: Charles C Thomas, 1969), pp. 12–14; Szasz, *Ceremonial Chemistry*, pp. 142–143.

14. Young, *Toadstool Millionaires*, p. 225.

15. Quoted in René Dubos, *Man Adapting* (New Haven: Yale University Press, 1965), p. 353.

16. Young, *Toadstool Millionaires*, p. 251.

17. Harold Rosenberg, "Everyman a Professional," in *The Tradition of the New* (New York: Horizon Press, 1959), pp. 62–63.

18. Rosenberg, "Everyman a Professional," p. 67.

19. Dubos, *Man Adapting*, p. 347; Alexander Mitscherlich, "Über Etablierte Unfreihei-

ten im Denken der unbewussten Freiheit," in *Krankheit als Konflikt I* (Frankfurt-am-Main: Suhrkamp, 1966), p. 126.

20. Milton Silverman, "The Conquest of Pain: Sertuerner and Morphine," in *Readings in Pharmacy*, edited by Paul Doyle (New York: Wiley, 1967), p. 235.

21. Dubos, *Man Adapting*, pp. 326, 328.

22. Quoted in Mitscherlich, "Die Krankheit der Medizin," *Krankheit als Konflikt I*, p. 52.

BIBLIOGRAPHICAL ESSAY

A brief commentary on some of the items in the bibliography may be useful at this point. The collection of Freud's cocaine papers (1884–1887) edited by Byck is the most useful source on Freud and cocaine; it includes background material from Freud's own letters and dreams, Ernest Jones' biography, and other writings. Freud's essays themselves are still a reasonably good introduction to coca and cocaine. Mariani's pamphlet (1890) gives an idea of how enthusiastic physicians and drug vendors regarded coca at that time. Mortimer's large volume (1901) is often cited as a classic in the field, possibly because most studies of drugs are so inadequate that it seems monumental by comparison. It lacks the form, coherence, and definitive quality of a classic; much of it has little connection with the title. We think that Mortimer is being given credit more for his intentions and his industry than his achievement. Nevertheless, his accounts of the botany and cultivation of coca and the chemistry of the coca alkaloids (though outdated in some respects) are the fullest available; so is his information on the therapeutic uses of coca in late-nineteenth-century medicine. His great love for coca is evident throughout. The bibliography is extensive; it covers not only early historical, ethnographic, botanical, and experimental work on coca and cocaine but also many topics that are only peripherally related: the history and culture of the Incas, South American geography and travel books, general biological and psychological theories, medical treatises of various kinds, and so on.

The experiments by Fleck (1924), Bychowski (1925), Jacobi (1927), and Lindemann and Malamud (1934) are representative of the tentative early work on the effects of cocaine on mental patients; Post and his colleagues (1974) have continued this work. It would be useful to have more extensive and elaborate controlled experiments like Offerman's (1926), which studies normal as well as depressed or schizophrenic subjects. The papers by Deneau, Yanagita, and Seevers (1969) and Yanagita (1973) are representative of experiments on the voluntary injection of cocaine and other psychoactive drugs by laboratory animals. The results are striking, but their applicability to normal human patterns of cocaine use is dubious.

Joël and Fränkel's book (1924) and Maier's (1926) are thorough studies of cocaine abuse based on clinical experience. Like most other works on cocaine, both for and against, they do not explicitly convey the information that they are telling only one side of the story. The bibliography in Maier's book is useful for anyone who wants to explore further the early social history of cocaine and the psychiatric and medical problems it created at that time.

The studies by Gutiérrez-Noriega and Zapata Ortiz (1944–1949) contain much helpful information and some plausible conclusions but have to be used with caution. Gutiérrez-Noriega had strong moral feelings about coca use and considered it an intrinsic part of an oppressive social system, so he tended to find ill effects everywhere he looked. Unlike many other writers, he identifies the effects of coca with those of cocaine. The United Nations *Report* (1950) is obviously the work of a committee trying to cope with opposing political pressures. It does not go so far as Gutiérrez-Noriega would have liked, but it does condemn coca and advocate measures (rather gentle and gradual ones) to abolish its use. The political position it takes on coca is reformist and was opposed by the conservative Peruvian and Bolivian governments of the time. The controlled studies made in the 1960s by Negrete, Murphy, and Rios, by the Goddards and Whitehead, and by Buck, Sasaki, Hewitt, and Macrae have added considerably to the knowledge of coca use and its social context. It seems to be neither as damaging as its enemies have feared nor as beneficial as its friends have hoped to prove it. But the debate will undoubtedly continue, because retrospective research of this kind, always produces ambiguous results. It is worth noting that even when they conclude that coca may be harmful, these researchers do not consider it so serious a problem that coercive measures are necessary. Other interesting recent work on coca includes the paper by Nieschulz (1971) on possible effects of subsidiary coca alkaloids, which is unfortunately inconclusive; the review essay by Martin (1970), a strong defender of the drug; and Hanna's article (1974), which discusses the local economic context of coca use. The article by Fabrega and Manning (1973) is valuable not only for its information about the uses of coca and other herbal drugs but also for its general approach to the subject of folk medicine.

A recent book on cocaine is Ashley's carefully researched history and connoisseur's handbook (1974). Ashley speaks largely for and to the illicit cocaine user, with appropriate polemical flourishes against the medical and government establishments; he provides advice on such matters as storage, testing for purity, and legal dangers. The book is good on recent social history and the contemporary cocaine culture but more sketchy on uses and effects of the drug. Although he does not discuss coca at any length, Ashley seems to agree with Gutiérrez-Noriega that coca is not much different from cocaine while disagreeing with him about everything else. Despite a tendency to ignore or discount any reports of ill effects from cocaine, the book is useful because it presents a reasoned position unpreoccupied by the concerns of physicians, psychiatrists, and law-enforcement officials that dominate most published studies of illicit drugs.

The anthology edited by Andrews and Solomon (1975) contains a number of articles and excerpts, all except one previously in print, with an introduction by the editors. Their main contention, reflected to a great extent in the choice of articles, is that coca in its natural form is a much better drug than pure cocaine—chemically different, more useful, and less subject to abuse. Many of the pieces are drawn from obscure sources or appear for the first time in English. The translations from Mantegazza's 1859 monograph on the medicinal virtues of coca and Natanson's paper on nasal lesions in cocaine sniffers are useful, and the original essay by Andrew Weil on his coca and cocaine experiences in South America is particularly interesting.

Firsthand reports on cocaine intoxication and the social context of cocaine use are found in many of these scholarly works and also in accounts by Hammond (1887), Mayer-Gross (1920), Crowley (1972, original 1922), Burroughs (1959), Woodley (1971), and Friedman (1974). Hammond's paper, which we have quoted extensively, is reprinted in full in Byck's edition of Freud's cocaine papers. It is a remarkable account from an era when physicians were bolder (or more foolhardy) about self-experimentation than they are now. The story told by "Dr. Schlwa" in Mayer-Gross' article gives an idea of the effects of intravenous abuse of morphine and cocaine in combination. Crowley and Burroughs have written descriptions of cocaine intoxication based on their own experience in essay, memoir, and fictional form. Crowley's are more extensive; they include not only the book cited in the bibliography but others referred to in the footnotes and occasional passages in works not mentioned here. Woodley's study of a Harlem cocaine dealer and Friedman's novel provide a plausible picture of how cocaine works in some contemporary American lives; they create the atmosphere and portray the social background necessary for understanding or even fully defining the effects of a psychoactive drug in actual use.

We have also included a few more general works on drugs and drug abuse that we found helpful in defining our attitudes toward cocaine. Duster's study of heroin addicts in a compulsory treatment program (1970) explains some of the oddities and injustices of the way we treat illicit drug users in the language of the sociology of deviance; it also provides interesting documentation on how members of the most despised and feared class of drug users see themselves and how others see them. Brecher's book (1972) is an unusually sensible, clearly written survey that makes a point of including the "nondrug drugs" along with the medical and illicit ones. The report of the National Commission on Marihuana and Drug Abuse (1973) may be an early sign of changing official attitudes; the articles on cocaine and on drugs and crime are particularly useful. Jacobson and Zinberg (1975) effectively advocate preventive methods for dealing with drug abuse that are more consonant with the standards of rationality we try to apply to other social problems than the present established techniques. Szasz's polemic (1974) reintroduces the themes of his earlier work on mental illness in a field where they are more plausible. His conclusions, expressed with strong conviction and a sense of certainty, are genuinely thought-provoking, even more when they suggest intellectually fruitful reservations and opposition than when it is easy to agree with them. Young (1971) provides the best history of the patent medicine era and an excellent source of material on the social roots of contemporary attitudes toward drugs. Dubos (1965) places the present situation of medicine in a larger historical and scientific context.

SELECTED
BIBLIOGRAPHY

Andrews, George, and Solomon, David, eds. *The Coca Leaf and Cocaine Papers*. New York and London: Harcourt Brace Jovanovich, 1975.

Ashley, Richard. *Cocaine: Its History, Uses, and Effects*. New York: St. Martin's Press, 1975.

Becker, Hortense Koller. "Carl Koller and Cocaine." *Psychoanalytic Quarterly* 32 (1963): 309–373.

Bejerot, Nils. *Addiction and Society*. Springfield, Ill.: Charles C. Thomas, 1970.

———. "A Comparison of the Effects of Cocaine and Synthetic Central Stimulants." *British Journal of Addiction* 65 (1970): 35–37.

Brecher, Edmund M., and the editors of *Consumer Reports*. *Licit and Illicit Drugs: The Consumers Union Report on Narcotics, Stimulants, Depressants, Inhalants, Hallucinogens, and Marijuana—Including Alcohol, Nicotine, and Caffeine*. New York: Little, Brown, 1972.

Buck, Alfred A.; Sasaki, Tom T.; Hewitt, Jean J.; and Macrae, Anne A. "Coca-chewing and Health: An Epidemiological Study Among Residents of a Peruvian Village." *American Journal of Epidemiology* 88 (1968): 159–177.

Burroughs, William. "Letter From a Master Addict to Dangerous Drugs." *British Journal of Addiction* 53 (1956): 119–131.

———. *Naked Lunch*. New York: Grove Press, 1959.

Bychowski, Gustav. "Zur Wirkung grosser Cocaingaben auf Schizophrene." *Monatsschrift für Psychiatrie und Neurologie* 58 (1925): 329–344.

Chein, Isidor; Gerard, Donald L.; Lee, Robert S.; and Rosenfield, Eva. *The Road to H: Narcotics, Delinquency, and Social Policy*. New York: Basic Books, 1964.

Ciuffardi, Emilio N. "Contribución a la química del cocaísmo." *Revista de Farmacología y Medicina Experimental* 2 (1949): 18–93.

Collier, H. O. J. "Drug Dependence: A Pharmacological Analysis." *British Journal of Addiction* 67 (1972): 277–286.

Crowley, Aleister. *Diary of a Drug Fiend*. London: Sphere Books, 1972. Originally published by William Collins Sons and Company, London, 1922.

Cruz Sánchez, Guillermo, and Guillén, Angel. "Eliminación de la cocaína en sujetos no habituados." *Revista de Farmacología y Medicina Experimental* 2 (1949): 8–17.

Deneau, Gerald; Yanagita, Tomoji; and Seevers, M. H. "Self-administration of Psychoactive Substances by the Monkey." *Psychopharmacologia* 16 (1969): 30–48.

Dubos, René. *Man Adapting.* New Haven: Yale University Press, 1965.

Duster, Troy. *The Legislation of Morality: Law, Drugs, and Moral Judgment.* New York: Free Press, 1970.

Ellinwood, E. H., Jr. "Amphetamine Psychosis: A Multidimensional Process." *Seminars in Psychiatry* 1 (1969): 208–226.

———. "Amphetamine Psychosis: I. Description of the Individuals and Process." *Journal of Nervous and Mental Disease* 144 (1967): 273–283.

———. "Behavioral and EEG Changes in the Amphetamine Model of Psychosis." In *Neuropharmacology of Monoamines and Their Regulatory Enzymes,* edited by Earl Usdin. New York: Raven Press, 1974.

Ellinwood, E. H., Jr., and Kilbey, M. Marlyne. "Amphetamine Stereotypy: The Influence of Environmental Factors and Prepotent Behavioral Patterns on Its Topography and Development." *Biological Psychiatry* 10 (1975): 3–15.

Fabrega, Horacio, and Manning, Peter K. "Health Maintenance Among Peruvian Peasants." *Human Organization* 31 (1973): 243–256.

Fleck, Ulrich. "Über Cocainwirkung bei Stuporosen." *Zeitschrift für die gesamte Neurologie und Psychiatrie* 92 (1924): 84–118.

Friedman, Bruce Jay. *About Harry Towns.* New York: Knopf, 1974.

Freud, Sigmund. *Cocaine Papers.* Edited by Robert Byck. New York: Stonehill, 1974.

Gay, George R.; Sheppard, Charles W.; Inaba, Darryl S.; and Newmeyer, John A. "An Old Girl: Flyin' Low, Dyin' Slow, Blinded by Snow: Cocaine in Perspective." *International Journal of the Addictions* 8 (1973): 1027–1042.

Goddard, D.; de Goddard, S. N.; and Whitehead, P. C. "Social Factors Associated with Coca Use in the Andean Region." *International Journal of the Addictions* 4 (1969): 41–47.

Grinspoon, Lester, and Hedblom, Peter. *The Speed Culture: Amphetamine Use and Abuse in America.* Cambridge, Mass.: Harvard University Press, 1975.

Gutiérrez-Noriega, Carlos. "Acción de la coca sobre la actividad mental de sujetos habituados." *Revista de Medicina Experimental* 3 (1944): 1–18.

———. "Alteraciones mentales producidas por la coca." *Revista de Neuro-Psiquiatría* 10 (1947): 145–176.

———. "El cocaísmo y la alimentación en el Perú." *Anales de la Facultad de Medicina* 31 (1948): 1–90.

———. "Errores sobre la interpretación del cocaísmo en las grandes alturas." *Revista de Farmacología y Medicina Experimental* 1 (1948): 100–123.

———. "Inhibición del sistema nervioso central producida por intoxicación cocaínica crónica." *Revista de Farmacología y Medicina Experimental* 2 (1949): 191–235.

———, and Zapata Ortiz, Vicente. *Estudios sobre la Coca y la Cocaína en el Perú.* Lima: Ministerio de Educación Pública, 1947.

Hammond, William A. "Coca: Its Preparations and Their Therapeutical Qualities, with Some Remarks on the So-called 'Cocaine Habit.' " *Transactions of the Medical Society of Virginia,* November 1887, pp. 212–226.

Hanna, Joel M. "Coca Leaf Use in Southern Peru: Some Biosocial Aspects." *American Anthropologist* 76 (1974): 281–296.

Jacobi, August. "Die psychische Wirkung des Cocains in ihrer Bedeutung für die Psychopathologie." *Archiv für Psychiatrie und Nervenkranken* 79 (1927): 383–406.

Jacobson, Richard, and Zinberg, Norman E. *The Social Basis of Drug Abuse Prevention.* Washington, D.C.: Drug Abuse Council, 1975.

Joël, Ernst, and Fränkel, Fritz. *Der Cocainismus.* Berlin: Springer Verlag, 1924.

Jones, Ernest. *The Life and Work of Sigmund Freud.* 3 vols. New York: Basic Books, 1953.

Kolb, Lawrence. *Drug Addiction: A Medical Problem.* Springfield, Ill.: Charles C Thomas, 1962.

Laurie, Peter. *Drugs: Medical, Psychological, and Social Facts.* Baltimore: Penguin Books, 1969.

Lewin, Louis. *Phantastica: Narcotic and Stimulant Drugs.* Translated from 2d German ed. by P. H. A. Wirth. New York: Dutton, 1931.

Lindemann, Erich, and Malamud, William. "Experimental Analysis of the Psychopathological Effects of Intoxicating Drugs." *American Journal of Psychiatry* 13 (1934): 853–881.

McLaughlin, Gerald T. "Cocaine: The History and Regulation of a Dangerous Drug." *Cornell Law Review* 57 (1973): 537–573.

Maier, Hans W. *Der Kokainismus.* Leipzig: Georg Thieme, 1926.

Mariani, Angelo. *Coca and Its Therapeutic Application.* New York: J. N. Jaros, 1890.

Martin, Richard T. "The Role of Coca in the History, Religion, and Medicine of South American Indians." *Economic Botany* 24 (1970): 422–438.

Mayer-Gross, W. "Selbstschilderung eines Cocainisten." *Zeitschrift für die gesamte Neurologie und Psychiatrie* 62 (1920): 222–233.

Mortimer, W. Golden. *Peru: History of Coca: "The Divine Plant" of the Incas with an Introductory Account of the Andean Indians of Today.* New York: J. H. Vail, 1901.

Mulé, S. J., and Brill, Henry, eds. *Chemical and Biological Aspects of Drug Dependence.* Cleveland: CRC Press, 1972.

Murphy, H. B. M.; Rios, O.; and Negrete, J. C. "The Effects of Abstinence and Retraining on the Chewer of Coca-leaf." *Bulletin on Narcotics* 21, no. 2 (1969): 41–47.

Musto, David F. *The American Disease: Origins of Narcotic Control.* New Haven: Yale University Press, 1973.

Natanson, Leon. "Au sujet des lésions nasales chez les priseurs de cocaïne." *Revue de Laryngologie* 57 (1936): 215–233.

National Commission on Marihuana and Drug Abuse. *Drug Use in America: Problem in Perspective.* 4 vols. Washington, D.C.: Government Printing Office, 1973.

Negrete, J. C., and Murphy, H. B. M. "Psychological Deficit in Chewers of Coca Leaf." *Bulletin on Narcotics* 19, no. 4 (1967): 11–17.

Nieschulz, Otto. "Psychopharmacologische Untersuchungen über Cocain und Ecgonin." *Arzneimittel-Forschung* 21 (1971): 275–283.

Offerman, Arno. "Über die Zentrale Wirkung des Cocains und einiger neuen Ersatzpräparate." *Archiv für Psychiatrie* 76 (1926): 600–633.

Post, Robert M. "Cocaine Psychoses: A Continuum Model." *American Journal of Psychiatry* 132 (1975): 225–231.

Post, R. M.; Kotin, J. K.; and Goodwin, F. K. "Effects of Cocaine in Depressed Patients." *American Journal of Psychiatry* 131 (1974): 511–517.

Post, R. M.; Gillin, J. C.; Wyatt, R. J.; and Goodwin, F. K. "The Effect of Orally Administered Cocaine on Sleep of Depressed Patients." *Psychopharmacologia* 37 (1974): 59–66.

Quijada Jara, Sergio. *La Coca en las Costumbres Indígenas.* Huangayo-Peru, 1950.

Snyder, Solomon H. "Catecholamines in the Brain as Mediators of Amphetamine Psychosis." *Archives of General Psychiatry* 27 (1972): 169–179.

Szasz, Thomas. *Ceremonial Chemistry: The Ritual Persecution of Drugs, Addicts, and Pushers.* Garden City, N.Y.: Anchor Press, 1974.

Tatum, A. L., and Seevers, M. H. "Experimental Cocaine Addiction." *Journal of Pharmacology and Experimental Therapeutics* 36 (1929): 401–410.

United Nations Economic and Social Council. *Report of the Commission of Enquiry on the Coca Leaf.* Official Records, 5th year, 12th session, Special Supplement, Vol. 1, May 1950.

Uscátegui Mendoza, Néstor. "Contribución al estudio de la masticación de las hojas de coca." *Revista Colombiana de Antropología* 3 (1954): 207–289.

Vervaeck, L. "Quelques aspects médicaux et psychologiques de la cocaïnomanie." *Le Scalpel* (Brussels) 76 (1923): 741–749, 769–780, 797–806.

Vom Scheidt, Jürgen. "Sigmund Freud und das Kokain." *Psyche* 27 (1973): 385–430.

Wittenborn, J. R.; Brill, Henry; Smith, Jean Paul; and Wittenborn, Sarah A., eds. *Drugs and Youth: Proceedings of the Rutgers Symposium on Drug Abuse.* Springfield, Ill.: Charles C Thomas, 1969.

Woodley, Richard A. *Dealer: Portrait of a Cocaine Merchant.* New York: Holt, Rinehart and Winston, 1971.

Yanagita, Tomoji. "An Experimental Framework for Evaluation of Dependence Liability of Various Types of Drugs in Monkeys." *Bulletin on Narcotics* 25, no. 4 (1973): 57–64.

Young, James Harvey. *The Toadstool Millionaires.* Princeton: Princeton University Press, 1961.

Zinberg, Norman E., and Robertson, John A. *Drugs and the Public.* New York: Simon and Schuster, 1972.

INDEX